The Impact of Networks on Unemployment

The Impact of Networks on Unemployment

J.M. Hurst

J.M. Hurst

ISBN 978-1-137-02537-1   ISBN 978-1-349-66890-8   (eBook)
DOI 10.1057/978-1-349-66890-8

Library of Congress Control Number: 2016950058

© The Editor(s) (if applicable) and The Author(s) 2016
The author(s) has/have asserted their right(s) to be identified as the author(s) of this work in accordance with the Copyright, Designs and Patents Act 1988.
This work is subject to copyright. All rights are solely and exclusively licensed by the Publisher, whether the whole or part of the material is concerned, specifically the rights of translation, reprinting, reuse of illustrations, recitation, broadcasting, reproduction on microfilms or in any other physical way, and transmission or information storage and retrieval, electronic adaptation, computer software, or by similar or dissimilar methodology now known or hereafter developed.
The use of general descriptive names, registered names, trademarks, service marks, etc. in this publication does not imply, even in the absence of a specific statement, that such names are exempt from the relevant protective laws and regulations and therefore free for general use.
The publisher, the authors and the editors are safe to assume that the advice and information in this book are believed to be true and accurate at the date of publication. Neither the publisher nor the authors or the editors give a warranty, express or implied, with respect to the material contained herein or for any errors or omissions that may have been made.

Cover image © Tetra Images / Alamy Stock Photo
Cover design by Samantha Johnson

Printed on acid-free paper

This Palgrave Macmillan imprint is published by Springer Nature
The registered company Macmillan Publishers Ltd, London

The original version of this book was revised: The Author information and some content was printed in error in October 2016 and has been removed from the book and the content updated in December 2016. The erratum to this book is available at DOI 10.1057/978-1-349-66890-8_9

Preface

Networks continuously shape and change the social and economic dynamics of a place. Think of the historical trajectory of governing networks, from those assisting feudal systems to parish vestries organising levies and disbursing parochial incomes and donations, then to parish councils and local government. In recent decades, 'governance' networks have permeated every policy avenue, supposedly for better outcomes, yet hardly anyone considers the network impact. Despite these networks, human potential is frittered away and many local problems sustained. Britain's historic vote in 2016 to exit Europe's grand partnership has deepened economic uncertainty and challenges local networks. But British and European unemployment policy, as cascaded through network policy, had already produced social and geographic division. Neighbourhood unemployment, for example, has blighted millions of lives and drained welfare resources in the absence of effective networks that coordinate a job supply. Britain has a long history of neighbourhood unemployment, but neither the market nor government has eased the problem. Article 23 of the 1948 Universal Declaration of Human Rights advocates the 'human right to work', but not to work itself, and the 1953 European Convention on Human Rights protects some aspects of the right to work, but these are not wide-ranging obligations.[1] Economic growth cannot deliver good local jobs across the country if the chain of supply and demand is broken or retains inferior employment legislation. As the economic system creates jobs, but also limits jobs for efficiency gains and profit margins, and job competition casts surplus labour aside. Instead of a work guarantee, government funds a 'welfare industry' of advisers and trainers to support people to find work in places where jobs are scarce or where there is a skills mismatch.

Networks for reducing unemployment have also assisted multi-sector governance, meaning public, private and voluntary-sector involvement in the activity of governing to steer market liberalisation (i.e., privatisation of public services), and a multitude of regeneration schemes with revival packages for wealth stimulation. Yet network policy has not proved transformative enough in areas of decline, even during periods of economic growth (Lawless et al. 2011). In order to achieve significant change, unemployment policies would need to lever many actors and policy domains to overcome barriers. If network outcomes are weak for unemployed citizens, network thinking might be misconceived. This book shifts the focus from abstract concepts of 'governance' to how networks deal with real-world intractable problems such as unemployment and gain insights about the network impact.

For example, political cycles from the mid-1980s onwards had European Union mandates to tackle uneven regional development and coordinate different types of capitalism, urban regeneration and welfare reforms. Yet European economics and successive governments' local and subregional networks failed to adequately address area based unemployment (Moore et al. 1989; Bassett 1996; Richardson 2000). From the late 1990s, the New Labour government replaced Conservative-led growth and enterprise networks with local networks to coordinate strategies of public policy institutions, quasi-autonomous non-governmental organisations (quangos), local agencies and national programmes for tackling neighbourhood renewal themes: unemployment, education and training, health and crime reduction (Hastings 1996; Geddes 2008; Durose and Rees 2012). However, neighbourhood unemployment increased as national unemployment fell to 4.8% in 2004 (this includes 1.39 million unemployed people, 652,000 vacancies or 2.1 unemployed heads per job vacancy). Then after 2010, the Conservative–Liberal Democrat coalition government shifted to subregional economic networks, based on private sector-led growth and post-recession recovery (Harrison 2011). From 2015, these networks retained under the Conservative government, but English devolution policy may fracture governance arrangements. In 2011, national unemployment rose to 8.1%, and fell to 4.9% the month before the EU referendum in 2016 (this includes 1.65 million unemployed people, 748,000 vacancies or 2.2 unemployed heads per job vacancy) (ONS 2016). However, this unemployment reduction requires a more nuanced reckoning as the hallmark of high employment growth is unstable jobs, low incomes, worse youth employment prospects, and an increase of 'in-work' benefit claimants and sickness benefit claimants in areas with high long-term unemployment.

Universities and colleges have also benefited from a surge in formal education in recent years, but their knowledge networks are not steering local labour conditions. Indeed, neighbourhood unemployment, skills mismatch and the quality and supply of jobs remain sizeable problems and all of the above factors have frustrated local authorities.

The book does not represent unemployed people per se or make detailed theories about the causes and effects of unemployment (see Garraty 1978; Oswald 1997). Nor does it deliberate over welfare measures, expenditure (see Barr 1998: 182–201), welfare systems (Spicker 1988), structural dysfunction in global financial crises, access to capital markets and future growth models (Hay 2011), the quality of employment support or success in placing people in work or training (see Gray 2000: 307). However, it does analyse network change in the unemployment policy field over the decades and link theoretical traditions to original empirical research, and it is the first investigation to enquire why different formal and informal networks in two English local authorities have struggled to overcome neighbourhood unemployment, whilst studying the outcomes achieved. Besides focusing on data collection during Labour's network campaign, a period of economic expansion, it also tracks network development prior to and following the 2008 recession.

Despite modest signs of economic recovery, the Bank of England is puzzled as to which factors are weakening labour productivity, product and process innovation (Barnett et al. 2014). Largely, policymakers uphold wage flexibility to retain labour, while muting the role of factors associated with the new risky economy (including low pay, job churn and insecure employment). The cost of 'flexicurity' (flexibility and security in labour market functioning) have introduced uncertainty for long-term public expenditure. Labour market statistics and the unemployment count are based on Labour Force Survey estimates and require interpretation to avoid misunderstandings. The UK employment count includes people who work one paid hour or more a week, people on government-supported work experience or training programmes, full-time students with a part-time job and people who work unpaid in a family business but benefit from the family business's profits (ONS 2012: 4–5). A third of the rise in total employment since 2010 relates to people registering as self-employed, many of whom may still be looking for work and 'hidden unemployed'. The unemployment count, as defined by the International Labour Organisation (ILO), includes anyone looking and available for work, whether claiming benefits or not. The UK claimant count refers to people on Jobseeker's Allowance (JSA), a welfare benefit introduced in 1996 for people actively seeking work. From 2013, Universal Credit is gradually replacing JSA and merging six benefit streams to support the transition between work and employment. The separate count of Jobseeker's Allowance (JSA) claimants is more reliable but subject to assessment, appeals and sanctions. A recent decrease in JSA claims coincided with a rise in benefit sanctions and it is unclear whether sanctioned claimants left the unemployment system to find work. Consequently, the parliamentary Work and Pensions Committee investigate 'off-benefit' flow outcomes against the performance management practices of 'Jobcentre Plus' (JCP) (the UK's statutory employment service that involves 'sign and go' conditionality for working-age benefit claimants and a range of labour market support, and which manages job centres and provides a job-match function and training programmes) (WPC 2014). Furthermore, the number of Employment Support Allowance (ESA) claimants assigned to the Work-Related Activity Group to help people with health problems return to work, or classified Fit for Work is falling. However, the proportion of people assigned to the Support Group for those unable to work because of health conditions or a severe disability is rising (DWP 2014: 5). Unemployment can devastate self-confidence, impair health, induce self-harm, affect all members of a household and strain local services. Interventions treating unemployment as a psychological attitudinal problem will not set social wrongs right; neither will moving people towards insecure employment or 'work experience' where 70% of participants will return to JCP, or long periods without vocational rehabilitation. Unemployed people, like other targets of discrimination throughout history are susceptible to conceptual prejudices, taboos, isolation and institutionalisation that thwart their economic integration. Government's preoccupation with the 'moral hazard' of welfare hides the embarrassing inadequacy of economic growth and market failure to provide enough work.

Thus, a more integrated explanation of the widest range of factors influencing network variables in the policy field is required. Although the book challenges network theory and practice, it does not intend to diminish the efforts of network participants, some of whom support the neediest people in society with little network influence, some of whom felt excluded, and some officials were socialised into processes they did not agree to, and/or believed networks had achieved very little. Others expected the state-mandated network structure to steer prosperity. Rather, this book represents the widest testimony about the struggle to reduce unemployment and achieve network impact.

The book will be of interest to scholars, graduates and network researchers in public administration and management, political science, local government studies, economics, social policy and urban geography. It should also be informative for practitioners, including government officials, local policymakers, private and voluntary-sector organisations, interest groups, think tanks and housing associations, and readers outside the UK who are dealing with unemployment, particularly in Europe and North America.

The Structure of the Book

The book explores an empirical puzzle, that of why neighbourhood unemployment persists despite the efforts of networks, and answers two questions: 'What kinds of factors affect network outcomes in areas of high unemployment?' and 'What kind of network outcomes result?' Part I links the empirical enquiry to a theoretical and methodological investigative framework and a policy environment. Part II presents case studies. The research applies mixed methods (qualitative and quantitative measures), including social network analysis (SNA) to uncover formal and informal networks and test and analyse the impact of five factors on network outcomes at three levels. Eighty-six semi-structured interviews give a voice to network practitioner experiences. The comparative study demonstrates why networks associated with unemployment delivered socially suboptimal outcomes for four neighbourhoods in two local authorities.

Chapter 1 clarifies concepts and the investigative approach for assessing the influence of five factors on socially optimal or suboptimal network outcomes: central environment, area-based factors, network structure, network processes and the individual's actions. These are understood through outcome-indicators at three impact levels: neighbourhood, organisational/participant and network. Chapter 2 reviews the relevance of network theories, and of structure, agency and processes, namely network governance and network management. Chapter 3 considers the role of unemployment policy and governance structures for overseeing labour market problems. Chapter 4 examines eight network trends in English urban regeneration policy from 1945 that have failed to combat unemployment. The next two chapters profile case sites, including overall socio-economic policies, local network culture, case networks, sociograms visualising network connectivity and factor comparison of case impact. Chapter 5 details five case studies in Tower Hamlets, an inner-city London borough (Labour-controlled). Chapter 6 details five case studies in Great Yarmouth, a seaside town in the east of England (Conservative-controlled). Chapter 7 assesses network performance perceptions, case network outcomes and actors' preferred outcomes. Finally, Chapter 8 models the causal constraints and considers the research limitations.

Anticipating the conclusions discussed in later chapters, the book argues that current network arrangements cannot overcome the problem of neighbourhood unemployment. Without rethinking networks as supporting radical policy reform within a framework of civil rights for full employment, future outcomes will merely resemble the past failures of economists. Network policy and governance theory has grown in stature and raised expectations for handling unemployment issues, but the policy interests, competitive mechanisms and behaviours of actors seeking funding thwart coherent implementation at ground level. The cases studied yielded similar findings: low connectivity; weak issue representation; preoccupation with governance outcomes, institutional and welfare reforms and short-term programmes for alleviating unemployment; unsatisfactory resource distribution; and a lack of socio-economic policy integration. Neither formal nor informal networks led to much action to combat neighbourhood unemployment because central government and participants were overly self-interested or preoccupied with external controls, policy change and network-level outcomes. The analysis concluded that high-reputation networks are associated with low performance; many participants lacked interest in unemployment or had other interests besides tackling unemployment, and lacked local knowledge. Despite networks' preoccupation with governance outcomes, collaboration was weak and joined-up working was more an aspiration than a reality. Moreover, joined-up governance theories emphasise network functioning while overlooking operational difficulties on the ground. Network governance, network management and decision-making processes appear to undermine the conditions for socially optimal outcomes to prevail.

This book developed from doctoral research undertaken at Birkbeck College, London University. I am especially grateful to my former supervisors Professor Peter John and Dr Dionyssis Dimitrakopoulos for their excellent comments over many years. Professor Deborah Mabbett added insights at a critical point in my studies. Professor Jonathan Davies and Dr Nancy Holman from the examiners board offered helpful comments. The book project is my own undertaking and any oversights or errors are my responsibility alone. I would like to thank the publishers at Palgrave Macmillan, especially the assistance of Jemima Warren, anonymous reviewers and copy-editors and Springer Nature production team. The case studies depended on interviews with 86 network associates from the London Borough of Tower Hamlets and the Great Yarmouth Borough Council, all of whom remain anonymous. I am indebted to them for their cooperation. Finally, special thanks are due to those whose friendship, good wishes and shelter helped sustain my research.

J.M. Hurst

Notes

1. O'Connell (2012) explains the case law content and protected aspects of 'the right to work' under the European Court of Human Rights. However, little has been done to develop an international mechanism on the right to work itself. For a history of the 'work' concept in social science, see Karlsson (2004). The Job Guarantee is another idea, and links to the 'social-welfare/right-to-work tradition' and 'Keynesian and post-Keynesian economic theory' (see Mitchell and Muysken 2008: 19, 251–9; Harvey 2013: 39–58).

Bibliography

Barnett, A., Batten, S., Chiu, A., Franklin, J. and Sebastiá-Barriel, M. (2014). The UK Productivity Puzzle. *Bank of England Quarterly Bulletin*, Q2.

Barr, N. (1998). *The economics of the welfare state* (3rd ed.). Stanford, CA: Oxford University Press/Stanford University Press.

Bassett, K. (1996). Partnerships, business elites and urban politics: New forms of governance in an English city? *Urban Studies, 33*(3), 539–555.

Department for Work and Pensions (DWP). (2014). Employment and support allowance: Outcomes of work capacity assessments, Great Britain. *Quarterly Official Statistics Bulletin*, 11 December 2014. London: DWP.

Durose, C. & Rees, J. (2012). The rise and fall of neighbourhood in the New Labour era. *Policy and Politics, 40*(1), 38–54.

Garraty, J. A. (1978). *Unemployment in history: Economic thought and public policy.* New York: Harper and Row.

Geddes, M. (2008). Government and communities in partnerships in England: The empire strikes back? In M. Considine & S. Giguère (Eds.), *The theory and practice of local governance and economic development* (pp. 100–125). Basingstoke: Palgrave Macmillan.

Gray, A. (2000). The comparative effectiveness of different delivery frameworks for training of the unemployed. *Journal of Education and Work*, *13*(3), 307–325.

Harrison, J. (2011). *Local Enterprise Partnerships*. Loughborough: Loughborough University.

Harvey, P. (2013). Wage policies and funding strategies for job guarantee programs. In M. J. Murray & M. Forstater (Eds.), *The job guarantee: Toward true full employment* (pp. 39–58). New York: Palgrave Macmillan.

Hastings, A. (1996). Unravelling the process of 'Partnership' in Urban Regeneration Policy. *Urban Studies*, *33*(2), 253–268.

Hay, C. (2011). Pathology without crisis? The strange demise of the Anglo-Liberal growth model. *Government and Opposition*, *46*(1), 1–31.

Karlsson, J. C. (2004). Ontology of work. In S. Fleetwood & S. Ackroyd (Eds.), *Critical realist applications in organisation and management studies* (pp. 90–112). London: Routledge.

Lawless, P., Overman, H. G., & Tyler, P. (2011). *Strategies for underperforming places*. In SERC Policy Paper 6, London.

Mitchell, W. F., & Muysken, J. (2008). *Full employment abandoned: Shifting sands and policy failures*. Cheltenham: Edward Elgar.

Moore, C., Richardson, J. J., & Moon, J. (1989). *Local partnership and the unemployment crisis in Britain*. London: Unwin Hyman.

O'Connell, R. (2012). The right to work in the European Convention on Human Rights. *European Human Rights Law Review*, (2), 176–190.

Office for National Statistics (ONS). (2012, September). *Interpreting labour market statistics*. Available at https://www.ons.gov.uk/

Office for National Statistics (ONS). (2016). *UK labour market: July 2016. Statistical bulletin*. Available from https://www.ons.gov.uk/

Oswald, A. J. (1997). *The missing piece of the unemployment puzzle*. Inaugural Lecture. Department of Economics, University of Warwick.

Richardson, J. (2000). Government, interest groups and policy change. *Political Studies*, *48*, 1006–1025.

Spicker, P. (1988). *Principles of social welfare*. London: Routledge.

Work and Pensions Committee (WPC). (2014, January). *The role of Jobcentre Plus in the reformed welfare system*. Second Report of Session 2013–14, HC479. Norwich: Stationery Office.

Contents

Preface — vii

List of Acronyms and Abbreviations — xvii

List of Figures — xxi

List of Tables — xxiii

Part I Framework for Investigating Network Impact — 1

1 Why Network Impact? — 3

2 Theoretical Background — 47

3 Unemployment Policy Context — 63

4 Urban Regeneration Policy and Governing Networks — 127

Part II	Investigating and Analysing Network Impact	159
5	Inner-City Network Cases	161
6	Seaside Town Network Cases	211
7	Network Impact: Performance and Outcomes	259
8	Conclusions: Modelling Suboptimal Network Outcomes	281

Erratum to: The Impact of Networks on Unemployment	E1

Appendix 1:	UK Unemployment Rates and Trends	297
Appendix 2:	Research Design and Methodology	301
	Case Selection Criteria	303
	Networks Uncovered	305
	Research Subjects	307
	Data Collection	309
	Data Analysis	311
Appendix 3:	Interviewing Strategy	313
Appendix 4:	Employment-Related Funding Streams and Provision in Tower Hamlets 2004–2008	317
Appendix 5:	Primary Networks in Great Yarmouth (2004–2008)	319
Index		323

List of Acronyms and Abbreviations

ALMP	Active Labour Market Policies
BME	Black, Minority Ethnicity
BL	Business Link
CN	Central and Northgate wards
CAB	Citizens Advice Bureau
CoPA	Committee of Public Accounts
CDP	Community Development Project
CPAG	Community Plan Action Groups
DBEIS	Department for Business, Energy and Industrial Strategy
DBIS	Department for Business, Innovation and Skills
DCLG	Department for Communities and Local Government
DfE	Department for Education
DfES	Department for Education and Skills
DoE	Department of Employment
DIUS	Department for Innovation, Universities and Skills
DfIT	Department for International Trade
DTI	Department of Trade and Industry
DWP	Department for Work and Pensions
EEDA	East of England Development Agency
EIL	East India and Lansbury wards
ESA	Employment and Support Allowance
ESOL	English for speakers of other languages
EZ	Enterprise Zone
EU	European Union
ERDF	European Regional Development Fund
ESF	European Social Fund
ESIF	European Structural and Investment Funds

GOE	Government Office for the East of England
GYBC	Great Yarmouth Borough Council
GYLSP	Great Yarmouth Local Strategic Partnership
GDP	Gross Domestic Product
HO	Home Office
IB	Incapacity Benefit
ILO	International Labour Organisation
IMF	International Monetary Fund
IMD	Index of Multiple Deprivation
JCP	Jobcentre Plus
JSA	Jobseekers Allowance
LSC	Learning and Skills Council for England
LAP	Local Area Partnerships
LEGI	Local Economic Growth Initiative
LEP	Local Enterprise Partnership
LGA	Local Government Association
LSP	Local Strategic Partnership
LBTH	London Borough of Tower Hamlets
LDA	London Development Agency
LSOA	Lower Super Output Area
MSC	Manpower Services Commission
NAIRU	Non-Accelerating Inflation Rate of Unemployment
NAP	National Action Plans
NAO	National Audit Office
NALEP	New Anglia Local Enterprise Partnership
NPM	New Public Management
NRF	Neighbourhood Renewal Funding
NRU	Neighbourhood Renewal Unit
NCC	Norfolk County Council
NEET	Young people aged from 16 to 24 who are not in employment, education or training
ODPM	Office for the Deputy Prime Minister
OECD	Organisation for Economic Cooperation and Development
ONS	Office for National Statistics
Poplar HARCA	Poplar (Housing and Regeneration Community Association)
RDA	Regional Development Agency
RGF	Regional Growth Fund
SB	Spitalfields/Banglatown and Bethnal Green South wards
SC	Southtown and Cobholm wards
SRB	Single Regeneration Budget
SME	Small and medium-sized enterprises
SNA	Social Network Analysis

THP	Tower Hamlets Partnership
TEC	Training and Enterprise Council
UK	United Kingdom
UKCES	UK Commission for Employment and Skills
UP	Urban Aid Programme
URC	Urban Regeneration Company
VCS	Voluntary, community and social enterprise sector
WNF	Working Neighbourhood Fund
WNP	Working Neighbourhood Pilot

List of Figures

Fig. 1.1	A conceptual framework for investigating network impact	22
Fig. 3.1	Stakeholders in the unemployment policy field	109
Fig. 4.1	A timeline of British political influence in central-local governing network structure and central unemployment/employment policy infrastructure from 1960	133
Fig. 5.1	Actual contacts in Case A (CSPG) relating to the unemployment issues in SB	185
Fig. 5.2	Actual contacts in Case B (WNP) relating to the unemployment issues in EIL	188
Fig. 5.3	Actual contacts in Case C (PAN) relating to the unemployment issues in EIL	190
Fig. 5.4	Actual contacts in Case D (EGO)	192
Fig. 5.5	Actual contacts in Case E (SB) relating to the unemployment issues in SB	194
Fig. 6.1	Actual contacts in Case A (EF) relating to the unemployment issues in CN	236
Fig. 6.2	Actual contacts in Case B (WNP) relating to the unemployment issues in CN	238
Fig. 6.3	Actual contacts in Case D (SC) relating to the unemployment issues in SC	241
Fig. 6.4	Actual contacts in Case E (BW) relating to the unemployment issues in SC	242
Fig. 8.1	Two explanatory models of factors and key variables that impede network outcomes	283

Appendix 1	Claimant count historical estimates 1881–2015 (experimental statistics)	297
Appendix 1	Labour Force Survey: ILO unemployed over 12 months in the UK 1992–2015 aged 16–64	298
Appendix 1	Labour Force Survey: ONS: All unemployed in the UK 1992–2015 aged 16–24	299
Appendix 1	Labour Force Survey: ONS part-time workers: Could not find full-time job	299

List of Tables

Table 1.1	Three levels of network outcomes associated with unemployment support	29
Table 3.1	Unemployment-related themes affecting network functioning	69
Table 5.1	Working-age claimants in the LBTH ward clusters—November 2004 to November 2014	170
Table 5.2	Numbers of contacts in the LBTH case networks	182
Table 5.3	Contacts in case networks aware of the unemployment issues in EIL, SB or borough-wide only (%)	199
Table 6.1	Working-age claimants in the GYBC ward clusters—November 2004 to November 2014	220
Table 6.2	Numbers of contacts in the GYBC case networks	232
Table 6.3	Contacts in case networks aware of the unemployment issues in CN, SC or borough-wide only (%)	249
Table 7.1	Networks associated with unemployment in the LBTH—by contacts associated with supporting unemployment issues	260
Table 7.2	Networks associated with unemployment in the GYBC—by contacts associated with supporting unemployment issues	261
Table 7.3	Actual networks perceived to be effective in tackling unemployment—identified by network contacts in the LBTH	263

Table 7.4	Actual networks perceived to be effective in tackling unemployment—identified by network contacts in the GYBC	265
Table 7.5	Six outcome indicators and thirty empirical referents for assessing case network outcomes at three levels of analysis: neighbourhood, network and organisational/participant in two local authorities 2004–5	268
Appendix 1	Unemployment rates in the UK 1973–2015 (percentage of total labour force)	298

PART I

Framework for Investigating Network Impact

CHAPTER 1

Why Network Impact?

Over the past few decades, different types of network aimed at tackling neighbourhood unemployment in UK localities that have fallen behind economically, have failed to alleviate this problem, even when the economy was buoyant. Unemployment devastates neighbourhoods, yet scholars offer little explanation as to why networks struggle to overcome unemployment and networks lose sight of their own impact. Although 'prescriptive' networks appear to be an efficient method for addressing complex policy problems, our understanding about policy failure in the network environment and the networks themselves, as the 'barriers to change', is incomplete (Scharpf 1978: 345–50; Rhodes and Marsh 1992: 15).

Classic studies from the 1970s described some of the inter-organisational limitations in policymaking, coordination, decision-making, and implementation (Tuite 1972; Hanf and Scharpf 1978; Pressman and Wildavsky 1973). Recent literature has raised doubts about network effectiveness, and further understanding is required, but much less is known about network impact and the factors influencing network outcomes over time (Geddes 2000; Davies 2007; Entwistle et al. 2007; Klijn et al. 2010; Turrini et al. 2010; McGuire and Agranoff 2011). Studies of networks and partnerships in various policy fields suggest that impact and outcomes are difficult to research empirically (Dowling et al. 2004; Kenis and Provan 2009; Isett et al. 2011). Consequently, 'network impact' is an under-theorised concept, and realistic accounts of network performance and the variable network outcomes that arise in localities are lacking, particularly in neighbourhoods where unemployment has risen steadily over decades.

There are many ways in which government and non-state sectors would like networks to impact, but well-intentioned networks may still be ineffective. Hence this book aims to examine theoretical sticking points in the literature, and define the challenges and impacts of real-world networks through original research. As Börzel (2011) claims, there is 'more research on successful network governance arrangements than there is on failures' (2011: 57). In short, 'network governance' refers to the principles by which actors from different policy contexts relate to each other to govern and implement policy (John and Cole 2000: 81–2).[1] However, governance theories appear preoccupied with network functions, rather than how networks connect to political, economic and social objectives (Giguère and Considine 2008: 6; Fawcett and Daugbjerg 2012: 201). Theories about 'governance networks' can also be faulted for saying too little about network impacts for citizens in places coordinating local economic and social development (Torfing 2005; Klijn and Kopenjan 2012). Some sceptics claim that '*all* governance structures fail' (Rhodes 2000: 83, italic in the original) because they disconnect from the 'real limitations of the market, state and mixed economy' (Jessop 2000: 11). Others raise expectations for self-governing networks that operate within limits set by external power (government and formal institutions), or the use of network management to steer outcomes (Rhodes 1997; Hajer and Versteeg 2005; Provan and Kenis 2008).

Yet network-centric theories and critiques, from network formation and network structure to network dynamics, virtually ignore an impact perspective. Scholars such as Hay and Richards (2000) relate the British general election cycle to network evolution in higher tiers of government, not impact on the ground. Still looking upwards, the term 'meta-governance' deliberates on 'the governance of governance' and whether the state (competent or incompetent) or indeed any sector has a legitimate functional role to shape rules and bureaucratically control the governance sphere (Jessop 2000; Papadopoulos 2007: 474; Sørensen and Torfing 2009). For O'Toole (2008: 223) 'Changing or at least influencing outcomes is precisely what public authorities have in mind when they engage in efforts at meta-governance.' Swyngedouw

---

[1] 'Actor' is a gender-neutral concept, interchangeable with 'network participant'; both terms represent the policy sector and organisation in which they function to tackle local unemployment.

(2005: 2001) distinguishes three 'orders of governance'. First, institutions in the meta-governance sphere define 'grand principles' of governmentality. These include international organisations promoting neo-liberal policy, from the World Trade Organisation and the Organisation for Economic Cooperation and Development (OECD), to the European Union (EU). Second, governance hierarchies will codify and formalise these principals, while the third order implements the governance at ground level. Network theories, however, say little about governance in the shadow of welfare policy, governing institutions, weak economic growth, or market failure to create work, and finding out what impact networks have gained on the ground becomes an empirical matter.

Institutions engaging business organisations in networks and 'partnerships', expect joint working to have an economic impact. (Moore et al. 1989). Others see networks as empowering organisations with similar values, while disempowering others (Taylor 2003). Sørensen and Torfing (2008a) claim that network theorists are more interested 'in the role of governance networks in enhancing governance efficiency' than 'the impact of governance networks on democracy' (2008a: 233). But networks are not successful modes of democratic accountability (Sullivan and Skelcher 2002: 150–3; Dryzek 2008). The 'chains of delegation' between constitutional authority, governance, managers and actual delivery of services has already 'lengthened beyond the practical limits of answerability to citizens and their representatives' (Lynn 2008: 8). Network meeting minutes or reports are often unpublished, and the representation of unemployed people can be superficial. Then again, 'democratic anchorage of networks' is just one aspect of network impact.

Network instability is another feature, linked to electoral cycles and policy change, department and institutional (re)structure or termination, EU policy and regional and local politics. Moreover, the social and economic regeneration and employment assistance policies that networks handle for reducing unemployment appear flawed, are difficult to coordinate, and neighbourhood barriers to work persist. Networks handling unemployment need to be realistic, as no economic model can accommodate all people who want to work. In Europe, the youth employment market has collapsed and youth talent accumulates 'with nowhere for it to go' (Mourshed et al. 2014: 49). In Britain, the Conservatives pre-election campaign in 2015 announced that unemployed youth must work unpaid for benefits, but withheld the information that youth programmes had already failed to organise real work for them. Perversely, equitable training would further strain job competition, as graduates outnumber graduate jobs and infiltrate low-paid jobs or compete for unpaid internships, and many people feel underemployed (Livingstone 2004). In 2013, 3.06 million 'underemployed' part-time workers wanted a full-time job (see Appendix 1). Taking into account regional variation, the problem affects young people, low-wage sectors and around one in five self-employed people (ONS 2012, 2013; MacInnes et al. 2013: 8–9). Following the 2008 recession, industries reduced workforce hours to avoid mass redundancies, and real consumer wages fell (Bell and Blanchflower 2013: 7). Consequently, poverty increased among working-age adults from 7.5 million in 2007/2008 to 7.9 million in 2011/2012; moreover, 5 million people were paid less than the living wage (see MacInnes et al. 2013: 14). Between 2011 and 2013, 4.8 million different people claimed JSA, and one-half of male claimants and one-third of female claimants made new claims within 6 months of their previous claim (2013: 9, 84, 106–7). This churning effect results from low-paid, insecure jobs, and weak bargaining power and employment protection for employees.

In the post-war years, 'social-regulatory institutions' and trade unions organised mutual constraints, employment stability and beneficial obligations to protect the working classes from capitalism's 'self-destructive dynamic' (Streeck 1989: 90). Subsequently, flexible labour markets fractured trade unions, but low-waged 'flexible' service jobs are still unable to support a family with children (Scharpf 2000: 101). Network governance thinking, however, ignores labour protection issues; rather the focus is on organisations' communication and interaction *beyond* institutional power and legal frameworks; whereby 'Network governance is neither market nor government nor civil society, it is a hybrid organizational form' according to Bogason and Zølner (2007: 4–5). Since networks find it hard to change or reform neo-liberal institutions' rulings, they largely reaffirm the status quo by addressing problems selectively so as to implement government's preferred policies. Network governance delimit policy choices to further field opinion. A good example is that of networks supporting industry to implement the green economy while overlooking the carbon costs of social welfare and wasting livelihoods in places without sufficient jobs.

For Denzau and North (1994: 5) the problems of economic development 'require an understanding of the shared mental models and ideologies that have guided choices.' Therborn (1986: 133–4) indicates that governments choose unemployment to lower inflation, adjust labour structure, reduce wages and break trade union power. Another governmental approach is to blame unemployed people for lacking competitive skills, despite a scarcity of skilled jobs. Similar attitudes permeate certain networks and relate to ideologies from the past which activate emotional judgement, associated with stigmatisation theory and leadership denial (Link and Phelan 2001). In brief, for the last 200 years, following any financial crisis the UK and US governments have recycled a 'perversity thesis' that switches the reality and responsibility for the situation from being 'the fault of the economy to the fault of the poor' (see Somers and Block 2005: 255–66). The 'conversion narrative' claims welfare hurts the job-impoverished, discourages them from working and erodes moral character (2005). Politicians and media frequently overestimate the number of welfare cheats and stigmatise unemployed people as 'undeserving' of welfare benefits and 'work-shy,' while underplaying the problems of corporate cheats and tax evaders, corporate welfare (massive subsidies to firms) and wasteful government expenditure (Baumberg et al. 2012). The OECD's policy advice for growth and equality, contradicts itself, as they note in 'many countries high economic growth has often been accompanied by a rise in inequality' (OECD 2009: 48; 2014). Furthermore, the welfare 'stick': benefit entitlement deterrents, activation strategies, sanctions and work programmes is not working. Against this backdrop, this book argues that networks cannot create a buoyant equitable economy, that neighbourhood unemployment problems have outgrown network governance solutions and that the literature takes an unrealistic view of what networks can achieve. Steering a middle course between naive and nihilistic options, the government must facilitate debate on the purpose of work which looks beyond capitalist limitations, and make economic reorganisation and social production a national priority so that individual citizens who desire to work achieve the right to do so.

Toward Network Realism: A Post-Positivist Approach

This book's research design uses actors' perceptions of network performance, a discerning critical realist epistemology and triangulation to minimise bias. The approach does not claim absolute truth, nor does it rely on the mechanical scientific world of empiricism (the foundation of positivism) to quantify reality. Rather, the emphasis is on identifying the perceived 'external "reality" and the social construction of that "reality"' (Marsh and Furlong 2002: 31). Networks represent the 'deeper organizing principles that ... produce and reproduce the structures that shape action' (Kilduff and Tsai 2003: 112). Thereby, 'underlying structural mechanisms' facilitate and constrain social, political or economic conditions, and explain why some groups have limited access to networks, or lack control over society's structural inequalities (May 2001: 11–12; Marsh and Smith 2001: 530, 536). However, the structure-agency dialectic is constant, and 'concerns the issue of to what extent we as actors have the ability to shape our destiny as against the extent to which our lives are structured in ways out of our control; the degree to which our fate is determined by external forces' (McAnulla 2002: 271). Using criteria and network theory, this study appraises and explains the mechanisms causing network outcomes in particular contexts, and therefore takes a hermeneutic approach, while also blending positivist, interpretive and realist epistemological positions. The forces influencing networks are not always observable; they range from political ideology, interpretation of local customs and history, and dominant discourses, to patterns of control, resource distribution by certain groups within networks or the constraints of capitalist economic relations (Marsh and Smith 2001: 529–30). Hence, causal explanations entail the observable and testable, and unobservable and interpretable dimensions of history, values, structure, processes and concept meanings (Dowding 2001: 97; Marsh and Furlong 2002: 18). Readers may contest the methods used, and judge the quality of the facts and narratives. However, generalisations are limited to a small-N study, the book's qualitative data carries 'health warnings', and technique limitations, missing variables or theories will be future research considerations.

This chapter proceeds as follows. First, it deals with network definitions. Second, it identifies the impact dimensions of the network concept. Third, it presents a framework for investigating 'network impact.' Thereafter, it discusses network performance concepts and outcome indicators. Finally, it links the empirical puzzle to a network typology.

Network Definitions: Contested Meanings

Network definitions reflect the researcher's worldview. The public administration literature emphasises 'networks' as flexible governance arrangements separate from hierarchies and markets (see Powell 1991). In reality, networks support different kinds of governance and collude with hierarchies and markets to coordinate funders' quasi-markets for contracts (Whitehead 2007). In the main, definitions are 'network-centric' and less interested in the citizens or communities networks serve. Some examples are illustrative: 'The concept of network draws attention to the interaction of many separate but interdependent organisations which act in a self-interested manner but nevertheless coordinate their actions through inter-dependencies of resources and interests' (Hanf and O'Toole 1992: 169). 'Networks occur when organizations or individuals begin to embrace collaborative processes, engage in joint decision-making and begin to act as a coherent entity' (Milward and Provan 2004: 8). 'Networks are structures of interdependence involving multiple organizations or parts thereof, where one unit is not merely the formal subordinates of the others in some larger hierarchical arrangement' (O'Toole 1997: 45). Arguably, investigation will be limited if network definitions disconnect from politics and the area-based factors, as 'the possibilities we discern for effective action will be a function of the models we use for understanding reality' (Hanf and O'Toole 1992: 167).

The definitional language is also partial. It marginalises concepts of conflict, collusion, and competition and assumes that cooperation is voluntary, not coercive, and that it does not harm third parties or produce negative unintended consequences (Bish 1978: 23). The term 'collaborative advantage', for example, applies more to network or participant level than 'for society as a whole' (Huxham 1996: 14; Provan et al. 2007). Assuming that collaboration is attainable may also limit problem understanding and ignore the conflicts that arise amongst national administrators and frontline operators with different objectives, responsibilities and accountabilities. As McGuire and Agranoff (2011: 265) claim, the network literature is more about 'how networks can be successfully managed' than overcoming collaborative obstacles. Yet what is the network concept for, if disconnected from local realities and impact-avoidant?

Multi-Dimensional Network Impact

Examining network conceptions from a network-impact perspective requires caution, since definitions of network impact are neither universally agreed in the literature nor well grounded in real-world situations or the policy and historical context, instead relying on researchers' interpretation.

A simple definition of 'network' is 'public policy-making and implementation through a web of relationships between government, business and civil society actors' (Klijn et al. 2010: 1064). Nonetheless, networks have a 'multidimensional nature' (John 1998: 90). Moreover, the term 'network impact' implies a meaning beyond internal network dynamics. In other words, networks have multidimensional impacts, as will be discussed below, from seven perspectives: symbolic, socio-economic, political, structural, processes, behavioural, and area-based.

The Symbolic Impact

The symbolic impact of networks often conflicts with reality. Broadly, networks may be valued more for what they represent than for what happens or is resolved through them. In public policy, networks represent an orderly approach to overcoming problems through constructive joint action, which takes effort, organisation and planning. Since officials use networks to legitimise their ideas, networks are not necessarily long-standing. They also signal as a coping mechanism for dealing with social and economic crises on the presumption that actors will work together, unify a fragmented policy community and steer social change. This applies in areas of decline or regions that lag behind due to lack of investment, industry closure from former recessions, high unemployment, poverty, lack of jobs, early school leaving, unaffordable housing and ethnic ghettoisation (Giguère 2008: 40–1). But without analysing network properties, collective practices and network environment the communicative language of 'network' is illusory. As Edelman (1964) noted: 'systematic research suggests that the most cherished forms of popular participation in government are largely symbolic, but also that many of the public programs universally taught and believed to benefit a mass public in fact benefit relatively small groups' (1964: 4).

Take the arbitrarily used phrase 'network society,' as used here: '"Governance" and "network management" emerge as new responses to the "network society" in which we live' (Hajer and Wagenaar 2003: 4). The vision implies a managed social shift, from individualistic culture (which accepts inequality, a high regard for individual achievement and large differentials of authority, salary, and status) to collectivist culture (which prioritises collective responsibility and less regard for status, power, or wealth) (see Hofstede 2001). This transformation requires equitable citizenship, a government commitment to eradicating unemployment, and markets without exclusion tendencies or a dearth of economic opportunity (Castells and Cardoso 2005). And yet networks in British culture would seem to assist a 'market society' to pursue the good life through power-dependent public–private transactions and individualistic consumption.

Arguably, the 'network society' eroded in a crisis of governability starting in the late 1970s, when social partnership between the state, trade unions and interest groups were dismantled, and industrial and employment relations made way for a neo-liberal flexible market (Crouch 2003: 113). New roles emerged. In the current climate, politicians and advisers shape market-orientated policy, public servants manage reforms and local authorities regulate, subsidise and are customers and partners of the private sector and voluntary sector, both of whom compete for government contracts to deliver local public services (Moore et al. 1989: 50–1; Hill and Hupe 2009: 86–7). Thus, networks for policymakers symbolise a counter to the negative outcomes of neo-liberal fundamentalism and the service fragmentation resulting from funding competitions, organisational self-interests, professional enclaves and labour individualisation (Perri 6 et al. 2002).

While many academics remain optimistic about governing network forms, others question networks' capacity to solve problems cooperatively (Teisman and Klijn 2002: 204). Network websites and image-building brochures validate networks externally, but there may be less internal collaboration taking place in practice than these imply. As Davies (2011) argues, positive network discourse 'fostering an entrepreneurial, reflexive and communicative sociability' may hide a coercive disposition or resistance, thus, it is worth investigating 'how the dialectic plays out' (2011: 5, 153).

The Socio-economic Impact

The socio-economic impact of networks links to the management of unemployment in the wider policy context. However, social and economic policy fields have ideologically divergent responses to unemployment and tend to keep separate networks to avoid conflict. Thus to assume that networks integrate socio-economic interests would be misleading. Sectors emphasising social injustices, social welfare measures and economic disparities, for example, have minimal contact with capitalist representatives who emphasise the role of individual effort and self-reliance in boosting economic growth and wealth creation, underpinned by state aid for industry and supply side interventions. The social and economic rift is historical and resonates in contemporary unemployment responses (Peters 2001: 24–6). Classical economists asserted the natural law of markets for self-regulation, and the importance of personal responsibility, freedom of exchange and creativity. The approach benefited rich landowners who governed parish networks instrumentally to coordinate preferred options irrespective of welfare consequences, and profited from labourers' low wages, which were 'topped up' by parish rates (Polanyi 1957). Wage 'top-up' schemes have existed since the 1600s; especially well known is the eighteenth-century Speenhamland allowance. Attitudes towards unemployed people have also changed over the centuries; at first subjecting them to social expulsion, then to 'confinement' in 'houses of correction,' later the workhouses, thereafter the unemployed character is placed under surveillance and made governable, as 'the obligation to work assumes its meaning as both ethical exercise and moral guarantee'(Foucault 1971: 43–64).

These themes reverberate today in the state's 'moral' management of unemployed people, from blaming their worklessness on skills deficiencies to criticising young mothers. The state regulates the lowest incomes through its 'welfare-to-work' and 'welfare-in-work' policies. 'Out-of-work' welfare is a conditional activation policy, whereby unemployed people must attend regular reviews at local state-led Jobcentres to demonstrate the effort made to improve their employability (Peck and Theodore 2000). Failure to do so will result in their welfare benefits being sanctioned.[2] 'In-work welfare', also known as 'working tax credits' compensates wages below subsistence levels and encourages employers' to keep wages low. The EU also uses structural investment funds to influence national policy through economic recovery plans, flexicurity (public policies to deregulate the social obligations of employers combined with social security and wage subsidies to compensate for labour insecurity) and social policies to support household purchasing power. However, local networks limit open debate on matters such as socio-economic needs, ineffective neo-liberal labour market policies, sharing responsibility for unemployment, underemployment and the gaps in the job market. Rather, closed networks locate unemployment within the socio-economic structural status quo and the prevailing political governance of social welfare, with its

---

[2] A 'sanction', states MacInnes et al. (2013: 146), is 'a reduction in or suspension of Jobseeker's Allowance (JSA) on account of a breach of the terms of a jobseeker's agreement'.

bias toward market solutions, growth, contract exchange, income maintenance policies, and retrenchment ideology. Any deviation would be radical or a direct challenge to the prevailing politics and policy.

The Political Impact

The third conception relates to networks' political effect. Incumbent governments ensure that the network infrastructure in local authorities asserts their political authority and promotes policies aligned to central ideological goals that admit or exclude organisations from decision-making and resource-distribution processes (Peters 2001: 152–62). For example, from 2010, the Conservative–Liberal Democrat coalition government allowed the former Labour government's network infrastructure for integrating state–citizen relations in 'local strategic partnerships' (LSPs) to 'wither on the vine' (Lowndes and Pratchett 2012: 29). In the Conservative tradition, the coalition government promoted a 'civil society empowerment' model to front a retrenchment package of community self-help and sub-regional business-led networks called 'local enterprise partnerships' (LEPs), which handled some of the former Regional Development Agencies' (RDAs') activities for economic growth. Like other non-elected bodies, centrally mandated networks disseminate a portfolio of political objectives through a mixed mode of governing (self-governing, co-governing and hierarchical governing) (see Kooiman 2000:146–53). They also have bargaining power through competitive funding rounds, and a degree of independence from mainstream services and locally elected politicians. But as Stone (2005) observes: 'politics shapes policy, and it poses an ever-present risk that decision-making will rest on a perspective, drawn too narrowly from the relationship between decision makers' (2005: 241).

British politics maintain a two-tier network culture, one for strategic effect, the other for community effect. The former plays to the idea of power-over and the latter power-with. Neither model is sufficient, given that the former has weak network democracy, and the barriers to 'community governance' are well known in the latter (see Rai 2008). In the first model, network regimes may form as 'a subset of governance occurring where the relationship between the political and economic elites is the dominant form of politics to which other groups accommodate' (John 2001: 52). Networks advancing the development goals of resource-rich institutions and large private firms sustain high-income professionals, managers, planners, consultants and regeneration officials over many years on the premise of long-term wealth creation. Construction and service industries may benefit, but job creation, especially for youth, is usually overestimated. As Marsh (1992: 167, 198–9) found in a study of networks tackling youth unemployment from 1970 to 1990, a lack of sustained interest in the policy community was related to politics and privilege, as networks favoured outcomes for business needs above others.

In the second model, community networks are not equal to central state institutions and find it difficult to manage long-term local problems with short-term funding. Officials restrict open dialogue and keep 'self-governing networks' at arm's length in case they do 'resist central guidance' (Rhodes 1997: 59; Pierre 1999, 2005; Stone 1989; Bassett 1996). Unemployed people have no right of representation themselves, and resource-poor organisations supporting those furthest away from the labour market lack the political influence, time and resources to participate; hence, it is hard to ascertain what aspects of unemployment are represented, or whether problems pass from one network meeting to the next. As this book reveals, network accountability problems extend to the institutional system of private government itself. For example, a consultancy report on unemployment commissioned by a case network was only available through the Freedom of Information Act (see Chap. 5), and final expenditure of the £80,000 contract remains undisclosed. Collaboration between state and societal actors supposedly increases accountability, yet networks undermine democracy if activities are undisclosed (Kjaer 2004: 200). Resource-dependent network participants survive through political games of conformity, political economy and social capital, all of which essentially mean the tactical acquisition or defence of money, authority and reciprocal social ties for information and resources (Benson (1975: 231). Indeed, local inter-organisational networks 'are an intensely political process' at the 'bargaining and negotiation' level and 'at the level of ideology concerning the values and conceptions of local economic regeneration' (Moore et al. 1989: 80). However, networks with aspirations to share power can gloss over political struggles.

The Structural Impact

The fourth conceptual network effect emphasises the impact of network relational structures, and connectivity between government, institutions, local authorities, businesses and social enterprises, and voluntary and community organisations. The way a network is structured communicates important status signals to the policy community. In structural terms, a network refers to the links between two or more actors within or among organisations (Lin 1999). Networks may be informally structured and spontaneous, or formally structured with written aims, rules or constitution (Gilchrist 2004; Kenis and Provan 2009: 446; Isett et al. 2011: 162–6). Interchangeable terms for network structure include 'coalition', 'consortium', 'forum', 'hub', 'partnership', 'pilot', 'strategic alliance', 'task force' or 'working group.' However, structural insights into a policy domain and its network connectivity are always time-specific and require interpretation (Wasserman and Faust 1994: xii–xiv; Knoke 1994; Bevir and Richards 2009; Provan and Lemaire 2012: 639). The method known as 'social network analysis' (SNA) measures relational and structural patterns of behaviour. Egocentric ties, for example, refers to individuals' direct links to others such as 'anchorage, density, reachability and range', and the role they play in the social interaction process in terms of 'content, directedness, durability, intensity and frequency of interaction'; while 'whole networks' represent all ties, present or absent, 'latent or activated' (Mitchell 1969: 20–9).

Structural qualities relate to structural advantage (the network or participants are always cited in the public domain), integration (the network develops links between participants) and dependency (the extent to which others are reliant on the network inside or outside), and may set the pace for '"rich get richer" relationships' (Provan et al. 2007: 502–3). While policymakers have encouraged cross-boundary working, the state institution or lead agent often controls network structures to facilitate or constrain interrelations, and the functions required to implement policy 'set the parameters of network actions' (Reid and Iqbal 1995: 6; Rhodes 1997: 12).

The Impact of Processes

Thus a fifth dimension of network impact is represented by network processes, namely, 'network governance' and 'network management' (Gage and Mandell 1990; Agranoff and McGuire 2001; O'Toole and Meier 2004). The efficacy of concepts, applied over decades in places where unemployment persists, is questionable. Network governance theory evolved from the 1980s onwards, first emerging during a legitimacy crisis over state hegemony and market and civil society exclusion from policy decision-making (Geddes 2000). Previously, during the 1960s and 1970s, New Right thinkers had perceived large state bureaucracies as being dominated by labour factions and expensive to run. In response, central government began to privatise the role played by the local authorities to enlarge governance, commissioning private sector providers to manage public services, particularly housing, and non-elected agencies to oversee local problems, giving networks and partnerships the means by which to unify interests. Voluntary sector activists and community development workers were already using networks successfully to lobby for social justice outcomes, and address specific equality issues by working for or against local government (Stoker 1991). However, government networking lagged behind, as it 'denies the political nature of governance, and fails to utilize the resources and capacities of local actors' (Kickert et al. 1997: 8). Thus network management literature evolved to assist governance and 'partnership' approaches, as an alternative to hierarchical 'command and control' style management. However, management transition to a 'New Public Governance' has not been as equitable as expected and new hybrid forms of governance may evolve, as Klijn and Kopenjan (2012): 600) suggest.

In theory, networks in the governance sphere 'may spread horizontally, between organisations of equivalent power and stature as well as between organisations that are, in some respects, related vertically' (Hill and Hupe 2009: 67). Network governance is one mode of governance associated with the 'New Public Management' (NPM) (Rhodes 1997: 55–6, 2000 56–7). NPM expands neo-liberal policy in the public sector using 'business-like ways of organising and managing' (Klijn and Kopenjan 2012: 587). The NPM replaces procedural governance (large-scale top-down public bureaucracies and institutions, laws, rules and standardisation) with network governance (for alliance-building, co-production and reducing hierarchies), corporate governance (focusing on management performance, plans and goals) and market governance (steering competition, contracts and cost-efficiency) (Considine 2001: 5, 23–31; 2008: 17). However, NPM constrains democratic pluralism and political dialogue, as only administrators and public managers have the information to make 'political judgements' (Hoppe 2011: 223–9). Moreover, network governance used as a policy instrument draws 'attention to the mechanisms of rule and relationship between governance and the governed' (Kassim and Le Galès 2010: 3).

'Governance networks' supposedly encourage a non-hierarchical open membership of diverse actors for consensus-building and democratic dialogue, accountable to citizens (Sørensen and Torfing 2008b). Kenis and Provan (2009: 446–9) call this approach 'shared or participant governance'. Conversely, 'governing networks' manage a policy domain, usually constituted by an elite closed membership supporting a governing agent, either a 'lead organisation' or 'network administrative organisation'

(Goldsmith and Eggers 2004; Kenis and Provan 2009: 446–9). Arguably, neither approach satisfies citizen needs. For example, governing networks use 'strategic governance' to plan long-term high-profile regeneration schemes and transport links, but rarely deal with current unemployment or create enough jobs to meet demand. Education partnerships supporting school-to-work transition have poor integration with employers, and young apprentices often lack work placements to complete their training. A survey report on Europe's high youth unemployment, involving over 8,000 people (5,300 young people aged 15–29, 2,600 employers and 700 education providers) between 2012 and 2013, from eight European countries, highlights education to employment barriers, in particular, living costs prohibit education enrollment, skills mismatched to demand, and support systems to find stable jobs were unsatisfactory; for example, 'Fewer than half of UK students completed a work placement, compared with 87 percent of those in France' (Mourshed et al. 2014: 10).

Further, 'community governance' was dubbed 'communities in control' under Labour; subsequently, under the coalition government 'localism and community empowerment' were 'not always supported by strong legal, conceptual or statutory frameworks' (DCLG 2008; Painter et al. 2011: 36). Community networks, like street-level bureaucracies, may also have 'personal biases, including the prejudices that subtly permeate the society' (Lipsky 1980: 85). The notion of 'citizen governance' is hardly applicable to unemployed people who lack the means to counter discrimination. Indeed, the existence of welfare partners depoliticises poverty discourse.

Hence scholars question whether networks shift democracy from 'government to governance' or if governance networks shift power back to government (Rhodes 1997, Chapter 3; Pierre 2000: 244; John 2001: 61; Richards and Smith 2002; Davies 2011). A Westminster model of government can still operate through governing networks if policymakers connected to the vertical hierarchy direct non-hierarchies to secure its own agenda-setting and position of power (see Rhodes 1997: 5–7; Richards and Smith 2002; Richards 2008: 138). Decisions are often made outside networks by those who perceive problems and solutions from the perspective of different models of reality and functional boundaries, and who coordinate or 'compel action from privileged power positions' (Hanf and O'Toole 1992: 166–7; Lawless 1989: 10–11). Consequently, networks fail to adjust from traditional ways of working and the powerful still govern through governance networks (Davies 2002; Teisman and Klijn 2002; Keast et al. 2004; Whitehead 2007; Geddes 2008: 100; Mandell and Keast

2008a; Durose 2009). Britain remains highly centralised and successive governments use networks to counter market-induced service fragmentation (Pierre 2000). Even self-regulating networks cannot escape government's privileged recentralisation pathways and funding opportunities for the powerful (Taylor 2003). As Burns et al. (1994: 6) warned, market models would 'damage democracy' and 'promote ways of behaving which are selfish.' Governance can be used in the following ways: first, as a price tag to order entrepreneurial forms of neo-liberal services; and secondly, as a tactic for restructuring power relations (see Swyngedouw 2005: 1997, on Lemke's (2001) elaboration of Foucault). Hence networks may not be as democratic as hoped, may defer to traditional models of governing, and may even be obstructive, as despite the existence of network management there is no public agreement about who has the right to negotiate policy adjustments (Koppenjan and Klijn 2004: 90; Hoppe 2011: 196–8).

The Behavioural Aspect

The sixth concept relates to the behavioural dimension of networks, and shifts analytical focus to participants' beliefs, attitudes, activities and motivations (Rhodes 2007). Rather, the quality, form and role of network relations, as well as their structure, need explaining (Dowding (2001). For Richardson (2000: 1017), 'actor behaviour changes over time' because 'policy fashions' spread like a virus and challenge networks and policy communities. In this way, 'local government is a victim of the latest fashions that spread like wildfire from innovators to imitators. And just when the imitators have caught on, so a new set of pioneers advocate a new idea with as great a fervour as the first' (John 2001: 166). Networks disseminate ideas to change or manage participants' behaviour. Labour's 'joined-up working' policy, for example, expected network governance to influence participants' behaviour towards greater alliance-building, coordination and connectedness. Yet the social capital that professionals gain for themselves, such as job security or economic resources, may not join up functionally at the neighbourhood level (Granovetter 1973). Network outcomes may still be directed towards the members' self-interests, because of participants' 'structural features and of their position in the implementation process as a whole' (Hanf and O'Toole 1992: 169).

Behavioural theories, from anthropology (culture and manners), political science (political actions), psychology (attitudes and opinions) to economics (financial choices), offer perspectives on why decision-making and cooperation are difficult to achieve through networks; or why they produce unintended effects, such as imitation, herding, contagion and informational cascades (see Easley and Kleinberg 2010: 425–42). Likewise, 'the prisoner's dilemma', 'the tragedy of the commons', 'public choice' and 'the logic of collective action' explain free-rider type problems, from profit-maximising behaviour, shirking and rule-breaking to opportunism, and why laws are passed to regulate self-interested behaviour (Ostrom 1990). In contrast, 'desirable' collective action behaviours of 'reciprocity, reputation and trust' (see Ostrom 1998) do not regulate optimal outcomes for citizens, and 'distrust' may be more protective (see Cook et al. 2005: 61).

Distrust of unemployed people, however, impairs the network goal of reducing unemployment, as it leads to employers preferring to appoint from the existing pool of workers. Some network participants and politicians provoke this distrust by polarising moral evaluation and pitting the 'lifestyle choices' of the 'something-for-nothing' benefit recipients against 'hard-working people.' This is perhaps a coping strategy to safeguard the reputations of these networks against accusations of ineffective unemployment handling (Merton 1968). Network identities can also reinforce inequalities through homophily. Richardson (2011: 543) states 'Practitioners' lack of integrated work across services was not primarily the result of a technical lack of "joining up" but a result of deep-rooted oppositions to the theoretical traditions of human behaviour used by other services.' A male-dominated economic development network presiding over one of the case areas studied in this book refers to six dinner events in the minutes, but unemployment had no mention. As Pierre (2000: 245) comments: 'networks could be assumed to cater almost exclusively to the interests of those actors that participate in the network, a scenario which raises questions about the long-term legitimacy of such governance arrangements.' Hence, the coordination and self-monitoring costs of network behaviour to meet or delay certain ends may be directed towards private benefit, rather than the public good. Thus, network behaviours are as important to policy analysis as processes and structures (Hogwood and Gunn 1986: 212–15).

Area-based Impact

The seventh conception is networks for area-based impact, shaping places, influencing local assets, services and social and economic regeneration. Meegan and Mitchell (2001) claim that area-based policy is 'inherently political'. In other words, regeneration requires 'community influence on the evolution of policy', otherwise 'economic restructuring' may lead to massive job losses, as occurred in Liverpool, or few jobs for locals, as occurred in London Docklands (2001: 2184–92). Thus networks reinforce 'urban restructuring', but also the 'reproduction of inequality' (Pacione 1997). Political interests and funding determine networks' geographic reach, from 'city deals' to neighbourhood schemes. Cities attract more funding but growth areas are no guarantee of reduced unemployment. Opportunities created in one area skewer development in another; hence the uneven spatial development within and between local authorities and rural and urban regions (Urry 1990: 203). Economic decline compounds an area's problems; industries create a low-wage culture to maintain profits, businesses relocate or terminate, properties fall into disrepair, and young people with few prospects resort to substance abuse and crime to release boredom and anger. Skilled or qualified people commute to areas with better job prospects (Taylor 2003: 32). Arguably, network theories appear inadequate to counter the impaired prospects of an area.

Putnam (1993) claims that where you live determines your fate, as places where 'civic culture matters' have dense networks, better prospects in education, employment, housing, childcare and greater trust, and social capital. The former Labour government wanted neighbourhood networks to build social capital and counter the perception that overly professionalised networks had diminished horizontal cooperation. In this way, the bridging and bonding of relations (both weak and strong ties) supply access to resources, advice, credit and information 'which communities are able to mobilise to work collaboratively' (Considine 2008: 19). Yet the entry point is inequitable. Workers from poorer neighbourhoods may unnecessarily sacrifice their own families' needs to undertake childcare or cleaning work for people in prosperous communities with demanding lifestyles, who may then draw on a low-pay economy and acquire social capital for themselves. Granovetter's theory (1973) that jobseekers need 'weak ties' to improve job prospects, as opposed to 'strong tie' contacts that only trade limited job information, lacks resilience during periods of economic stagnation. Even 'job coaches' delivering the government's 'Work Programme', introduced in June 2011 to activate long-term unemployed people into work, cannot meet targets, not because staff or clients are idle, but because the jobs are not there, reports Gentleman (2013). By September 2014, 972,000 individuals had completed the Work Programme and 648,000 (70 %) of these returned to Jobcentre Plus (JCP) (DWP 2014: 6). A witness giving evidence to the Work and Pension Committee about the Work Programme's effectiveness claimed employment and skills agencies had worked together in recent years, but continued: 'Because providers are now in direct competition with each other for outcome payments and with Jobcentre Plus for the few available vacancies, this coordinated approach is falling apart and employers are already getting frustrated with a number of multiple approaches from different agencies chasing their vacancies' (WPC 2013: Ev164). Policy problems interconnect, but only a few actors will work together on a segment of the problem, as locally there is no single network coordinating provision; moreover, people access work through different gateways.

A Multifactor Approach to Network Investigation

A multifactor investigative framework approach guides the systematic investigation of the empirical puzzle in question (in this case, networks), the multiple influences supporting or hindering them, and tests theories about the network performance. Hanf (1978) proposes studying networks as a whole unit of analysis, integrating theory, method and empirical research in order to identify 'those factors which make some networks more successful than others in coordinating the activities of component organizations' (1978: 13). This requires problem analysis, considering the realistic policy alternatives available to policymakers, developing a model to describe networks, evaluating network structure, behaviour and performance and accounting for policy or network failure 'in terms of the structural features of policy formulation and implementation' (1978: 11). In other words, we must 'point out key structural variables which impede or facilitate coordination, and therefore represent potential points of leverage for improving network performance' (1978: 13). Kilduff and Tsai (2003: 121) 'expect to see more careful case studies of network ties and less cavalier neglect of the embeddedness of social ties in particular local sites.' Yet few studies collect 'data on multiple networks' or study the problems of 'hierarchy and control' in the local habitat (Provan and Kenis 2008: 230). A study of the civic culture of local economic development, for example, distinguished local factors but omitted a network perspective (Reese and Rosenfeld 2002). Well-known studies focus on relatively stable US service-delivery networks in the health and social care sector (Alter and Hage 1993; Provan and Milward 1995, 2001). Arguably, networks in unemployment policy require much broader investigation.

Framework Components

Fig. 1.1 depicts five factors (independent variables) most likely to influence network performance, in different neighbourhoods; it is a device to simplify research options and guide comparative case studies in one policy

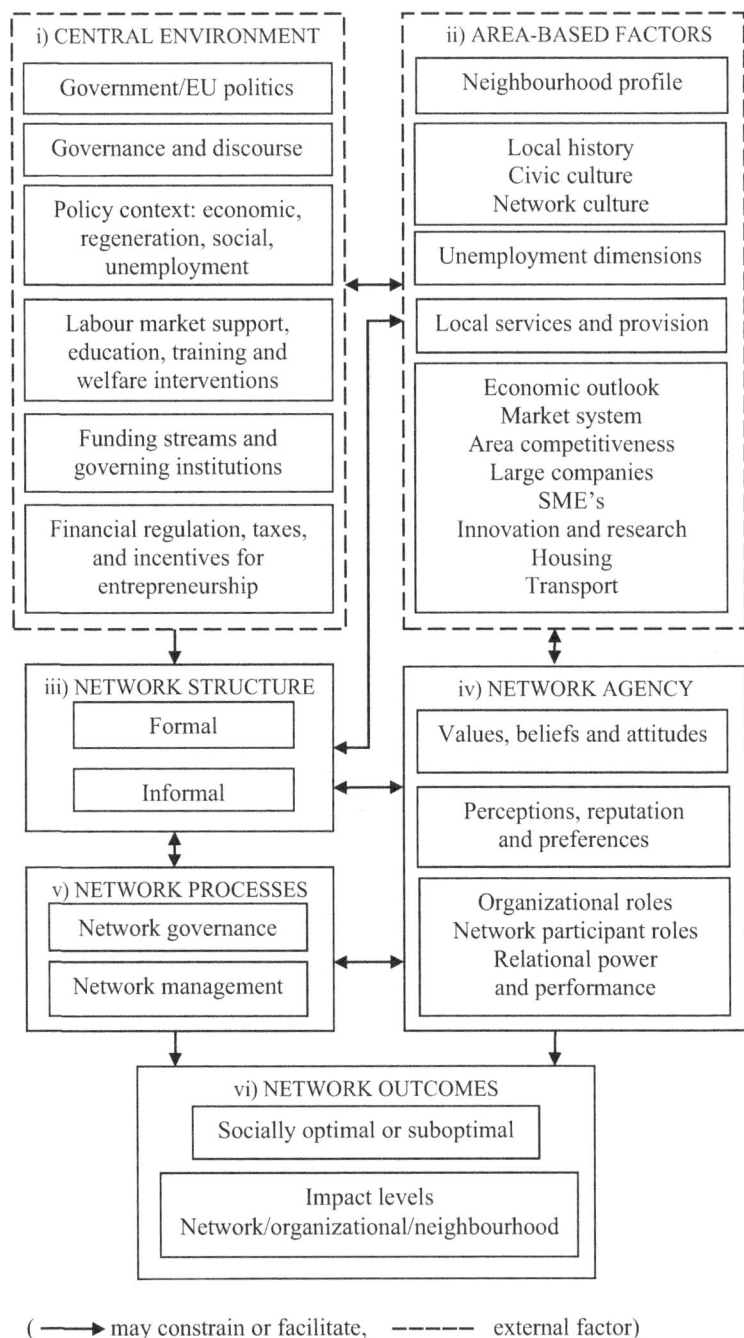

Fig. 1.1 A conceptual framework for investigating network impact

context (unemployment). The aim is to verify how rival and complementary theories facilitate causal effects, and lead to similar or different optimal or sub-optimal network outcomes (dependent variables). The set is neither exhaustive, nor too narrow, as networks involve wide-ranging political, structural, operational, relational and environmental variables. The factors are as follows:

(i) *Central environment*: international, government and EU politics, fiscal levers, social and economic policies and discourse, departments and institutions funding streams, and interventions to reduce unemployment.
(ii) *Network structure*: the variety of formal and informal network types associated with the policy field and relational connectivity.
(iii) *Area-based factors*: neighbourhood demographics, socio-economic situation, local culture, network history, services and assets.
(iv) *Network agency*: organisational/participants' interpretation of problems, influenced by roles, behaviour, beliefs, and attitudes.
(v) *Network processes*: the steering activities used to tackle unemployment, from network governance to network management.

Although factors within the framework have a linear sequence (see arrow direction in Fig. 1.1) in reality, factors overlap. Also, it is not a policy stages model or process map with which to fix a problem, or prescriptive evaluation system; rather, it is descriptive and functions to assess network outcomes.[3] The next section details concepts relating to network performance and outcomes.

Concepts of Network Performance and Outcomes

Scholars suggest that the 'network performance' concept is poorly defined, that theoretical and empirical links between networks and performance are underdeveloped and that network evaluation is 'still in its infancy' (Sydow and Milward 2003: 1; Provan et al. 2007: 509; Kenis and Provan 2009: 441; Meier and O'Toole 2010: 129). The existing studies mainly address network level performance. This book argues that 'network performance' links to the endogenous conditions (factors originating inside networks) and exogenous conditions (factors

---

[3] For further information on the ideal-type, prescriptive and descriptive policymaking models, see Hogwood and Gunn (1986: 42–64), and for the stages model see Hill and Hupe (2009: 115–120).

originating outside of networks) that lead to optimal or sub-optimal network outcomes at three impact levels: neighbourhood, network and organisational/participant. Networks rarely self-evaluate, as making 'collaborators accountable for results may impose new measurement and reporting demands that simply complicate their work and detract from their performance' (Page 2004: 603). Yet formal networks frequently manage or manipulate performance perceptions. Scholars are also wary about evaluating network performance, as qualitative measures are perceived as fallible (Provan and Milward 2001: 414; Provan and Kenis 2008: 247). Provan et al. (2007: 509) state: 'The relative lack of studies examining network effectiveness was somewhat surprising. If we are to understand about networks and network performance, then it is essential that network effectiveness be addressed.' As they see it, 'few have studied network evolution during a sufficiently long period to understand why interorganizational networks might succeed or fail at their mission' (2007: 509). Hanf et al. (1978: 338–9) advised the use of qualitative data, 'judgements and technical assistance' to guide network performance. Likewise, the development of performance indicators is considered a key challenge for future network research (Provan and Milward 2001: 1; McGuire 2002; Mandell and Keast 2008a: 695; Voets et al. 2008: 785; Isett et al. 2011: 163).

Performance Criteria

Kenis and Provan (2009) want network performance studies to 'make fully explicit which performance criteria are being addressed', operationalise performance criteria with realist performance indicators based on potential achievement and 'identify factors that prevent networks performing well on certain criteria' (2009: 442–3). Studies should not standardise assessment; this should depend on the network form, type of inception (voluntary or mandated) or development stage. Instead, the criteria should take into account how networks steer, what they steer and who benefits from their actions, as well as careful contextual interpretation. For example, 'trust' is an aspirational condition in some contexts, but can lead to malpractice. Network conflict supports or hinders communication. Stability supports network formation, but later curtails creativity or information flow; especially in networks emphasising the interests of a few like-minded people (Provan and Milward, 1995, 2001; Turrini et al. 2010: 534). Power diffusion signifies democracy, yet could also reduce networks accountability. The European Employment Strategy, for example, progresses through 'soft law' governance principles and elite decision-making networks, but may undermine 'hard laws' supporting legal rights for citizens and democratic structures, and bypass national interests.

Perceptions of network effectiveness and ineffectiveness may also disconnect from the bigger picture of policy, politics and place. For Goss (2001), effective networks require 'negotiated objectives, shared strategy, pooled resources and mutual "contracts" and need the flexibility to negotiate these locally' (2001: 209). Voets et al. (2008: 784) suggest that network effectiveness assessment is limited and analysis is needed to incorporate production, process and regime dimensions linked to costs. Meier et al. (2006: 167) correlate public managers' networking strategies with educational performance in districts. Klijn et al. (2010: 1081) explore network management strategies (content, connecting and process arrangements) for handling Dutch environmental projects. Sørensen and Torfing (2009: 256) advocate 'operational definitions of network effectivity and democratic anchorage that can be used in empirical case studies of governance networks at multiple levels.' Turrini et al.'s (2010) literature review of service-provision network studies identifies determinants of network effectiveness (structure and functioning) at three levels, client, community and network, but calls for empirical studies with 'specific, reliable and valid measures of success at a network-level' (2010: 547). Danish researchers have studied network responses to National Action Plans (NAPs) for employment in three cities in Denmark, England and France, as well as the Danish local employment council's contribution to effective network governance, but have overlooked the local culture (Torfing 2007a; 2007b; Esmark and Triantafillou 2007; Damgaard and Torfing 2010). Many of the best-known network performance studies rarely consider the influence of politics or place and focus on commissioning environments and service-delivery networks within relatively stable policy domains, such as the health and social care sector. Alter and Hage (1993) measure the performance of health and social care networks in two US districts, in terms of perceived effectiveness, conflict and performance gap. Provan and Milward (1995; 2001) study US mental health service provider networks, using the performance criteria of perceived network effectiveness and ineffectiveness, at three levels: network, organisational participant and client. Cristofoli et al. (2011: 39) examine homecare service networks in three Swiss cities, using network-level 'predictors of network performance', network structure, network functioning and network management at three assessment levels: structure, process and outcomes. Some studies offer guidance to health partnerships for working together and overcoming governance and management barriers, or found shortcomings (Mitchell and Shortell 2000; Perkins et al. 2010; Hunter et al. 2011). Few studies represent ineffective network performance, with some exceptions (see McGuire and Agranoff 2011). Goss (2001: 91–116) considers network problems to stem from constant reform and a lack of joined-up organisational operational systems, which hinders the handling of problems at the local level. Some studies focus on partnership limitation (Geddes 2000; Davies 2007) and inter-organisational coordination limitation (Hanf and Scharpf 1978).

Network Outcomes

The concept 'network outcomes' is distinguishable from 'policy outcomes'. A policy outcome should have an 'observable effect on a target population, consistent with the initial purpose of the policymakers' (Elmore 1979: 603). Studies have linked policy failure to poor implementation and the governance environment (see Bell and Hindmoor 2009: 27–9); especially well known is Pressman and Wildavsky's (1973) classic study of failed implementation in a scheme to hire long-term unemployed minorities in Oakland, California. O'Toole (2011: 117–8) claims Pressman and Wildavsky were unrealistic to base their success criteria on the probability of pluralistic implementation and number of conflict-free decision points, as public managers can be highly networked and new programmes make a difference. Nevertheless, in 2011 Oakland suffered riots and a general strike over unemployment and social and economic inequality. And even 'a policy that is executed need not result in the accomplishments of its objectives' (Lane 1993: 91).

As for network outcomes, 'there is no consensus in the field on which outcomes are of central interest to research' (Oliver and Ebers 1998: 566) or 'constitute successful network outcomes' (Provan and Milward 2001: 417). Even performance management research omits network outcomes, or perceives them as difficult to measure (Page 2004: 591; Walker et al. 2010; Klijn et al. 2010: 1065). In short, network investigation still has fundamental angles to explore—namely the factors shaping network outcomes and the outcomes themselves. Since the outcomes that actors desire cannot all be achieved, the question arises as to which ones can be achieved, and whether citizens, the network or participants benefit. Technically, an outcome sits at the end of a linear performance indicator scale. A service evaluation, for example, retrospectively assesses inputs, outputs and outcomes. In practice, the sequence overlaps, outcomes may be unplanned or unintended and may start a new set of actions. The outcome can be the network form per se, or specific network outcome(s) attributed to purposeful action (or inaction), internal or external goals. For Mandell and Keast (2008b: 729), 'building relationships needs to be recognized as a legitimate outcome, alongside the more tangible outcomes of tasks/activities completed'. Hence, community workers associate optimal outcomes with serendipity and good networking (Gilchrist 2004). Networks also spread amongst elites and reproduce social divisions. Participants can rate members' 'social outcomes' such as influence, trust and reputation, but poor organisations may have limited reputation-building opportunities to prove trustworthiness and judge outcomes (Provan et al. 2009: 875–6, 881–91; Hill and Hupe 2009: 137–8). Undisclosed outcomes may explain why ineffective networks survive, especially if outcomes benefit organisations' reputation, individuals' performances or managers' careers.

Provan and Milward (2001: 415) question whether 'networks really work' as 'good comparative network data that are tied to outcomes are scarce'. As Kenis and Provan (2009: 441) observe, 'even though network organizations may provide excellent services on their own, overall network outcomes may be low'. For example, a study of co-action amongst multi-agencies delivering children's services used outcome-based accountability methods, but found 'limited conclusive evidence of improvement in outcomes or service performance' (Chamberlain et al. 2010: 7). Even EU partnership programmes had 'limited identifiable outcomes', according to Geddes (2000: 787). Unrealistic goals may force networks to concentrate on structure or functions at the network or participant (egocentric) level rather than on outcomes for citizens (Taylor 2003: 18–31; Provan and Kenis 2008: 230; Provan et al. 2007: 505; Provan et al. 2009: 874; Klijn et al. 2010). Although precise measures of network outcomes are lacking, citizens still need to know why networks do not implement better outcomes for their neighbourhoods. As McGuire and Agranoff (2011: 274) state: 'If we are to assert the utility of networks as administrative alternatives that add public value, then outcomes must be included as ultimate dependent variables, in spite of the empirical roadblocks that presently make such an undertaking so challenging.'

Network outcomes: three analytical levels.

To minimize bias, this book uses the following two dimensions of network outcomes to incorporate the 'entire range of variation in the dependent variable' (King et al. 1994: 109):

1) Socially-optimal network outcomes address the neighbourhood level; local employment needs, target local services and improve policy support for unemployed citizens.
2) Sub-socially-optimal network outcomes focus on the network level and organisational/participant level; network processes, system-maintenance and reputation building.

Network outcomes at the optimal end of the range prioritise citizens, and are positive for neighbourhoods. At the sub-optimal end of the range, outcomes weighted towards network functions and participants' interests may fail to achieve neighbourhood optimality. Whether the factors (Fig. 1.1) influence the outcome-range separately or in combination can be analysed at three levels: neighbourhood, network and organisational/participant, assisted by six outcome indicators (Table 1.1): alliance-building, decision-making, problem-solving, network governance, network management, self-interest and associated empirical referents. The term 'outcome indicator' includes intermediate outcomes (such as goals of progress, service improvements or disputes) and completed outcomes (such as reduced unemployment, retrospective funding acquisition or expenditure). Edelman (1964: 207) suggests the use of '"empirical indicators" is inevitably an expression of values'; effectively, a judgement. Caution is required, as other scholars may select different models and indicators. Nevertheless, 'outcome measure' is perhaps too restrictive as a term for describing complex or ambiguous outcomes. Scholars may also advance outcomes-theory and revise or challenge the moral, rational and normative reasoning about the goodness of outcomes, whether desired, produced or prevented. While scholars elevate the content and processes valued by organisational participants, the social obligations of network outcomes and how network management impact neighbourhoods is less clear (see Klijn et al. 2010). Differentiating network outcomes at the three analytical levels, however, is not concerned with rigid interpretation, and overlaps are inevitable, as the following brief description observes.

Table 1.1 Three levels of network outcomes associated with unemployment support

Outcome level	Indicator	Empirical referents (optimal/suboptimal)
Neighbourhood	Alliance-building:	Group representation; local problem representation; community assets; resource-sharing
	Decision-making:	Service cohesion; skills support; economic inward investment; job creation/business starts; new projects/joint schemes
	Problem-solving:	Information advice/services for citizens; benefit take-up; personal support and counselling
Network	Network governance:	Democracy/transparency; joint working; weak ties; strong ties; clique-forming; conflict; leadership; vertical or horizontal coordination
	Network management:	Problem analysis; strategic planning; policy implementation; information-sharing; hierarchical bureaucracy; image-building; network funding
Organisational/ participant	Self-interests:	Self-promotion/reputation-building; power assertion; skills utilisation

Neighbourhood Level

The 'neighbourhood' concept has multiple meanings, across the contexts of governance, service delivery and regeneration (see Sullivan and Taylor: 2007). Geographically, 'neighbourhood' is a small area; usually a ward within a local authority. Neighbourhood profiles include internal dynamics, for example demographics/spatial characteristics, local culture and history and network relations, and external dynamics; the central policies influencing service provision, and targeting disadvantaged areas, economic and community development strategy, jobs and industry, to housing quality and transport, and local politics.

Neighbourhood-level network outcomes focus on place-based needs and interests, assisted by alliance-building, problem-solving and decision-making (Lane 1993: 102). Optimal outcomes might lead to job creation, improvement in referral systems, and access to: better local amenities and resources; information and training provision; community assets; services; signposting events; community healthcare; childcare; counselling; advice on welfare benefits; and help in fighting against discrimination. But who influences unemployment policy if social justice outcomes weaken, and networks distance, unemployed people? Government and local officials want to steer 'neighbourhood governance' outcomes, whereby residents groups within poor neighbourhoods take ownership of their problems, fill service gaps, build local culture, and accrue representational power to participate in decision-making (Lowndes and Sullivan 2008). Yet community engagement networks are not fully formed policy instruments with power to influence labour markets, and neighbourhood strategies are never as inclusive as their mission statements imply. Local people will not be empowered if area gentrification limits prospects for social businesses, and unaffordable housing and high rents displace poor inhabitants to other low-income areas, as without a community-land protection framework network input is negated (Stone 1989). Neither can neighbourhood governance redeem economic problems if decision-makers blame policy failure on residents or external forces, global markets, lack of joined-up government, weak neo-corporatist industrial relations and inflexible institutions (Cope and Goodship 1999: 5; Perri 6 et al. 2002: 34–6). Actors want networks to generate cooperative outcomes that are sensitive to local cultural needs and strengthen civil society, but which may only operate as a 'form of fragmented local crisis management' (Richards et al. 1999: 10–11; Geddes 2000: 797).

Network Level

Network-level outcomes relate to the structures, relations and behaviours that enhance or hinder collective working, assisted by network governance and management processes. The literature values outcomes that enhance 'pluricentric governance based on interdependence, negotiation, and trust' (Sørensen and Torfing 2008b: 3; Rhodes 1997). Likewise, it promotes the transparency of joint working for policy consensus, coordinating plans across cultural divisions, managing institutional complexity, and aligning sector rights, rules, and regulations (deLeon 1994: 234). However, governance instruments (vertical regulation and central planning, contracts and incentives) tie participants to 'fit into the structure within which they have to function' (De Bruijn and Ten Heuvelhof 1997: 122–3). In this sense, networks cannot deliver outcomes that are 'alien to the purpose for which they are organised' (Agranoff and Lindsay 1983: 235). Network managers supposedly counter difficulties by mediating interdependencies between actors with different outcome preferences (Kickert et al. 1997: 10–11). Yet how do network relations represent unemployed people in problem analysis or strategic planning? Does self-referential information-sharing prioritise political image-building above public values and bury the policy problems? In this way, networks deteriorate through misuse, misapplication or 'managerial mistakes in designing or operating [them]' beyond their actual capabilities (Miles and Snow 1992: 53).

Organisation/Participant Level

Organisational/participant-level outcomes satisfy the private interests of individual participants and representative organisations. Different participants value different types of outcomes. Network performance may deteriorate if participants are unable to assert power and influence over network direction, rules or funding, utilise their skills or innovate (Alter and Hage 1993: 258). Outcomes assisting organisations to enhance their reputation for career advancement, job security or self-actualisation increase individualism and fragmentation, but coordination costs may rise (Lane 2000: 205). The Labour government promoted networks intensively after 1997 as a route to local democracy and overcoming social exclusion (Perri 6 1997: 10). Evidently, 'network-rich' managers build 'network capital' to increase their power and learning (Perri 6 et al. 2002: 67). Yet networks of information-sharing experts may lose sight of citizens' needs (Yanow 2004: S14–5). Participation deficit leads to the 'Matthew Effect', whereby poor organisations in poor areas have poor networks, while rich organisations with a broader geographic reach have richer networks, access to funding and policy influence.[4] Meier and O'Toole's (2003) study of educational performance claims managers can use networks to improve policy outcomes, enhance future liaison and better utilise resources by

---

[4] Socio-economic policy studies have found evidence of 'The Matthew Effect', a term attributed to Merton, who claims that individuals or institutions with initial advantage, wealth or status, accumulate advantages and get richer, whereas the disadvantaged become poorer as the cost of acquiring information, knowledge and resources may be too high (see Rigney on Merton 2010: 46).

making 'room for manoeuvring' (2003: 697). However, educationalists have limited integration with employers, who, in turn, have limited capacity for multisector networking. Moreover, few organisations pursue joint outcomes.

STUDYING NETWORKS: A POLITICISED POLICY CONTEXT

This book considers unemployment from the standpoint of public policy failure and network design. Distinguishing network types is an important aspect of the analysis; the network variation is outlined, and a typology is then given to guide the case selection as relevant to the politicised policy environment.

Network Variation: Formal and Informal

Formal networks represent a rational, organised, planned and bureaucratic policy world, although solutions may only be reached by switching structure to informal governance (Giarchi 2001: 67). The network's membership attracts central elites and administrative support to legitimise its network status. Its formal rituals reinforce linkages; these may include meetings, minutes, progress reports, disseminating information and monitoring targets. Such networks often lead a policy sector dealing with multifaceted problems, control a funding stream accountable to government initiatives, or steer the policy interests of authoritative bodies. Sometimes they act like organisations and develop a common vision for integrated planning and management. This creates a power structure referred to as 'network regulation', meaning 'the degree to which a network is constrained in its ability to be self-governing, as opposed to being autonomous' (Alter and Hage 1993: 111). Dependent on public image-building, they are cautious and less conducive to creativity.

Informal networks occur when actors spontaneously interact, or self-organising relationships form, supposedly regulated by trust; they are a legitimate unit of analysis whether interaction leads to formal linkages or not (Sullivan and Skelcher 2002: 5, 42; Alexander 1995: 61). Actors use them for peer support as needs arise, for problem-solving, for learning and experimentation, or to sidestep bureaucracy and reach consensus more quickly. Hence they are associated with personal influence, community structures and social movements. Informality can also be used to manipulate relations, and reduce visibility and accountability (Alexander 1995:

91–115). However, there is no guarantee that formal or informal network types 'will be representative or that certain voices or perspectives will be included' (Isett et al. 2011: 166). Olson (1965) suggests small networks provide an incentive to overcome the free-rider problem, as participants in large-size networks may do little if their actions are more difficult to hold to account or evaluate. Others claim group size is not relevant to individual effectiveness. Rather, it is the group's identity set and lobbying power or its strategies that motivate participation and manipulate interest groups (see Dunleavy 1988: 47–9).

Typology of Networks for the Unemployment Field

Deciding which networks represent the unemployment field is contentious, as typologies change over time. The choice of network may depend on politics, policy funding streams, institutions and organisational interests. Some aspects of Rhodes typology remain relevant; for example, the fact that the '"national government environment" conditions the operations of the policy networks', and that 'self-governing networks' are still considered ideal models for governing (Rhodes and Marsh 1992: 14; Rhodes 1997: 51–2). The empirical network typology below includes structure (size, formal or informal, open or closed membership, origin and geographic focus), relations (e.g., cooperative or hierarchical), interest (policy focus), and resources (staff and level of funding).

- *Contract-exchange networks* disseminate funding and contract rules to service providers.
- *Community networks* support housing or community development personnel with facilitating citizens' input in neighbourhood decision-making on matters such as funding expenditure for 'acorn' grants, small schemes or services.
- *Ego networks* are the personal ties of managers with decisional responsibilities for reducing unemployment, and their 'alters'.
- *Governing networks* convene elite actors to promote government's preferred national and local policy, share information and compete for national funding rounds for grants, projects, or programmes.
- *Inter-organisational networks* involve organisations that shape new ventures or services to support unemployed clients directly or indirectly, or share information ad hoc as needs arise.

- *Peer support networks* operate on a day-to-day basis between like-minded professionals and colleagues.
- *Pilot networks* are short-term experimental programmes targeting neighbourhoods, dependent on cooperation and multi-level governance between state and non-state providers.
- *Sector-specific networks* represent an industry or service sector, such as a chamber of commerce, employers or voluntary sector council.

Nevertheless, the methods for uncovering case networks are not perfect and caution is required; this is because results may be limited or partial, interrelations are often fluid or ambiguous and SNA can only map networks (present or absent) at a given time (see the research design and interview procedure in Appendices 2 and 3). The next chapter considers the relevance of network theoretical foundations to a network impact analysis.

Bibliography

Agranoff, R., & Lindsay, V. A. (1983). Intergovernmental management: Perspectives from human services problem solving at the local level. *Public Administration Review, 43*(3), 227–237.

Agranoff, R., & McGuire, M. (2001). Big questions in public network management research. *Journal of Public Administration Research and Theory, 11*(3), 295–326.

Alexander, E. R. (1995). *How organizations act together.* Luxembourg: Gordon and Breach.

Alter, C., & Hage, J. (1993). *Organizations working together.* Newbury Park, CA: Sage.

Bassett, K. (1996). Partnerships, business elites and urban politics: New forms of governance in an English city? *Urban Studies, 33*(3), 539–555.

Baumberg, B., Bell, K., & Gaffney, D. (2012). *Benefits stigma in Britain.* London: TurnToUs.

Bell, D. N. F., & Blanchflower, D. G. (2013). Underemployment in the UK Revisited. *National Institute Economic Review, 224*(1), F8–F22.

Bell, S., & Hindmoor, A. (2009). *Rethinking governance: The centrality of the state in modern society.* Cambridge: Cambridge University Press.

Benson, J. K. (1975). The interorganizational networks as a political economy. *Administrative Science Quarterly, 20*(2), 229–249.

Bevir, M., & Richards, D. (2009). Decentring policy networks: A theoretical agenda. *Public Administration, 87*(10), 3–14.

Bish, R. (1978). Intergovernmental relations in the United States: Some concepts and implications from a public choice perspective. In K. Hanf & F. W. Scharpf

(Eds.), *Interorganizational policy making: Limits to coordination and central control* (pp. 19–33). Beverly Hills, CA: Sage.

Bogason, P., & Zølner, M. (2007). Methods for network governance research: An introduction. In P. Bogason & M. Zølner (Eds.), *Methods in democratic network governance* (pp. 1–20). Basingstoke: Palgrave Macmillan.

Börzel, T. (2011). Networks: Reified metaphor or governance panacea? *Public Administration, 89*(1), 49–63.

Burns, D., Hambleton, R., & Hoggett, P. (1994). *The politics of decentralisation.* Basingstoke: The Macmillan Press.

Castells, M., & Cardoso, G. (Eds.) (2005). *The network society: From knowledge to policy.* Washington, DC: John Hopkins Centre for Transatlantic Relations.

Chamberlain, T., Golden, S., & Walker, F. (2010). *Implementing outcomes-based accountability in children's services: An overview of the process and impact.* LG Group Research Report. Slough: NFER.

Considine, M. (2001). *Enterprising states, the public management of welfare-to-work.* Cambridge: Cambridge University Press.

Considine, M. (2008). The power of partnership: States and solidarities in the global era. In M. Considine & S. Giguère (Eds.), *The theory and practice of local governance and economic development* (pp. 13–39). Basingstoke: Palgrave Macmillan.

Cook, K. S., Hardin, R., & Levi, M. (2005). *Cooperation without trust.* Russell: Sage Foundation.

Cope, S., & Goodship, J. (1999). Regulating collaborative government: Towards joined-up government? *Public Policy and Administration, 14*(2), 3–16.

Cristofoli, D., Maccio, L., & Pedrazzi, L. (2011). Networks funded by the public sector can and should be evaluated – Ok, but how? *Paper for the 11th Annual Public Management Research Association Conference, June 2–4, 2011 Maxwell School of Management.* Syracuse, US.

Crouch, C. (2003). The state: Economic management and incomes policy. In P. Edwards (Ed.), *Industrial relations in Britain* (2nd ed.). (pp. 105–123). Malden: Blackwell Publishing Ltd.

Damgaard, B., & Torfing, J. (2010). Network governance of active employment policy: The Danish experience. *Journal of European Social Policy, 20*(3), 248–262.

Davies, J. S. (2002). The governance of urban regeneration: A Critique of the 'governing without government' thesis. *Public Administration, 80*(2), 301–322.

Davies, J. S. (2007). The limits of partnership: An exit-action strategy for local democratic inclusion. *Political Studies, 55*(4), 779–800.

Davies, J. S. (2011). *Challenging governance theory: From networks to hegemony.* Bristol: Policy Press.

De Bruijn, J. A., & Ten Heuvelhof, E. F. (1997). Instruments for network management. In W. J. M. Kickert, E.-H. Klijn, & J. F. M. Koppenjan (Eds.),

Managing complex networks, strategies for the public sector (pp. 119–137). London: Sage.
deLeon, L. (1994). Embracing anarchy: Network organizations and interorganizational networks. *Administration Theory & Praxis, 16*(2), 234–253.
Denzau, A. T., & North, D. C. (1994). Shared mental models: Ideologies and institutions. *Kyklos, 47*(1), 3–31.
Department for Communities and Local Government (DCLG). (2008). *Communities in control; Real people, real power. The local government white paper.* Cm7427. Norwich: The Stationery Office.
Department for Work and Pensions (DWP). (2014). Quarterly Work Programme statistics to September 2014. 18th December 2014. Newcastle upon Tyne: DWP.
Dowding, K. (2001). There must be an end to confusion: Policy networks, intellectual fatigue, and the need for political science methods courses in British Universities. *Political Studies, 49*(1), 89–105.
Dowling, B., Powell, M., & Glendinning, C. (2004). Conceptualising successful partnerships. *Health and Social Care in the Community, 12*(4), 309–317.
Dryzek, J. S. (2008). Networks and democratic ideals: Equality, freedom, and communication. In E. Sørensen & J. Torfing (Eds.), *Theories of democratic network governance* (pp. 262–273). Basingstoke: Palgrave Macmillan.
Dunleavy, P. (1988). Group identities and individual influence: Reconstructing the theory of interest groups. *British Journal of Political Science, 18*(1), 21–49.
Durose, C. (2009). Front-line workers and 'local knowledge': Neighbourhood stories in contemporary UK local governance. *Public Administration, 87*(1), 35–49.
Easley, D., & Kleinberg, J. (2010). *Networks, crowds, and markets: Reasoning about a highly connected world.* Cambridge: Cambridge University Press.
Edelman, M. (1964). *The symbolic uses of politics.* Urbana: University of Illinois Press.
Elmore, R. F. (1979). Backward mapping: Implementation research and policy decisions. *Political Science Quarterly, 94*(4), 601–616.
Entwistle, T., Bristow, G., Hines, F., Donaldson, S., & Martin, S. (2007). The dysfunctions of markets, hierarchies and networks in the meta-governance of partnership. *Urban Studies, 44*(1), 63–79.
Esmark, A., & Triantafillou, P. (2007). Document analysis of network typology and network programmes. In P. Bogason & M. Zølner (Eds.), *Methods in democratic network governance* (pp. 99–122). Basingstoke: Palgrave Macmillan.
Fawcett, P., & Daugbjerg, C. (2012). Explaining governance outcomes: Epistemology, network governance and policy network analysis. *Political Studies Review, 10*(2), 195–207.
Foucault, M. (1971). *Madness and civilization: A history of insanity in the age of reason.* London: Tavistock Publications.

Gage, R. W., & Mandell, M. P. (1990). *Strategies for managing. Intergovernmental policies and networks.* New York: Praeger.

Geddes, M. (2000). Tackling social exclusion in the European Union? The limits to the new orthodoxy of local partnership. *International Journal of Urban and Regional Research, 24*(4), 782–800.

Geddes, M. (2008). Government and communities in partnerships in England: The empire strikes back? In M. Considine & S. Giguère (Eds.), *The theory and practice of local governance and economic development* (pp. 100–125). Basingstoke: Palgrave Macmillan.

Gentleman, A. (2013, May 21). Work programme staff struggle to help unemployed when 'jobs aren't there'. *The Guardian.*

Giarchi, G. G. (2001). Caught in the nets: A critical examination of the use of the concept of 'networks' in community development studies. *Community Development Journal, 36*(1), 63–71.

Giguère, S. (2008). The use of partnerships in economic and social policy: Practice ahead of theory. In M. Considine & S. Giguère (Eds.), *The theory and practice of local governance and economic development* (pp. 40–62). Basingstoke: Palgrave Macmillan.

Giguère, S., & Considine, M. (2008). Partnership and public policy: The importance of bridging theory and practice. In M. Considine & S. Giguère (Eds.), *The theory and practice of local governance and economic development* (pp. 1–12). Basingstoke: Palgrave Macmillan.

Gilchrist, A. (2004). *The well-connected community. A networking approach to community development.* Bristol: The Policy Press.

Goldsmith, S., & Eggers, W. D. (2004). *Governing by network. The new shape of the public sector.* Washington, DC: The Brookings Institution Press.

Goss, S. (2001). *Making local governance work.* Basingstoke: Palgrave.

Granovetter, M. (1973). The strength of weak ties. *American Journal of Sociology, 78*(6), 1360–1380.

Hajer, M., & Versteeg, W. (2005). Performing governance through networks. *European Political Science, 4*(3), 340–347.

Hajer, M., & Wagenaar, H. (2003). Introduction. In M. Hajer & H. Wagenaar (Eds.), *Deliberative policy analysis: Understanding governance in the network society* (pp. 1–30). Cambridge: Cambridge University Press.

Hanf, K. I. (1978). Introduction. In K. I. Hanf & F. W. Scharpf (Eds.), *Interorganizational policy making: Limits to coordination and central control* (pp. 1–15). Beverly Hills: Sage.

Hanf, K. I., & O'Toole Jr., L. J. (1992). Revisiting old friends: Networks, implementation structures and the management of interorganizational relations. *European Journal of Political Research, 21*(1–2), 163–180.

Hanf, K. I., & Scharpf, F. W. (Eds.) (1978). *Interorganizational policy making: Limits to coordination and central control.* Beverly Hills: Sage.

Hanf, K. I., Hjern, B., & Porter, D. O. (1978). Local networks of manpower training in the Federal Republic of Germany and Sweden. In K. I. Hanf & F. W. Scharpf (Eds.), *Interorganizational policy making: Limits to coordination and central control* (pp. 303–341). Beverly Hills: Sage.

Hay, C., & Richards, D. (2000). The tangled webs of Westminster and Whitehall: The discourse, strategy and practice of networking within the British core executive. *Public Administration, 78*(1), 1–28.

Hill, M., & Hupe, P. (Eds.) (2009). *Implementing public policy* (2nd ed.). London: Sage.

Hofstede, G. (2001). *Culture's consequences: comparing values, behaviours, institutions, and organizations across nations* (2nd ed.). Sage: Thousand Oaks.

Hogwood, B. W., & Gunn, L. A. (1986). *Policy analysis for the real world*. Oxford: Oxford University Press.

Hoppe, R. (2011). *The governance of problems: Puzzling, powering and participation*. Bristol: Policy Press.

Hunter, D. J., Perkins, N., Bambra, C., Marks, L., Hopkins, T., & Blackman, T. (2011). *Partnership working and the implications for governance: Issues affecting public health partnerships*. Final Report. Southampton: NIHR SDO.

Huxham, C. (1996). Collaboration and collaborative advantage. In C. Huxham (Ed.), *Creating collaborative advantage* (pp. 1–18). London: Sage.

Isett, K. R., Mergel, I. A., LeRoux, K., Mischen, P. A., & Rethemeyer, R. K. (2011). Networks in public Administration scholarship: Understanding where we are and where we need to go. *Journal of Public Administration Research and Theory, 21*(suppl 1), i157–i173.

Jessop, B. (2000). Governance failure. In G. Stoker (Ed.), *The new politics of British local governance* (pp. 11–32). Basingstoke: Macmillan.

John, P. (1998). *Analysing public policy*. London: Continuum.

John, P. (2001). *Local governance in Western Europe*. London: Sage.

John, P., & Cole, A. (2000). Policy networks and local political leadership. In G. Stoker (Ed.), *The new politics of British local governance* (pp. 72–90). Basingstoke: Macmillan.

Kassim, H., & Le Galès, P. (2010). Exploring governance in a multi-level polity: A policy instruments approach. *West European Politics, 33*(1), 1–21.

Keast, R., Mandell, M. P., Brown, K., & Woolcock, G. (2004). Network structures: Working differently and changing expectations. *Public Administration Review, 64*(3), 363–371.

Kenis, P., & Provan, K. G. (2009). Towards an exogenous theory of public network performance. *Public Administration, 87*(3), 440–456.

Kickert, W. J. M., Klijn, E.-H., & Koppenjan, J. F. M. (1997). Introduction: A management perspective on policy networks. In W. J. M. Kickert, E.-H. Klijn, & J. F. M. Koppenjan (Eds.), *Managing complex networks: Strategies for the public sector* (pp. 1–13). London: Sage.

Kilduff, M., & Tsai, W. (2003). *Social networks and organizations.* London: Sage.
King, G., Keohane, R., & Verba, S. (1994). *Designing social enquiry.* Princeton, NJ: Princeton University Press.
Kjaer, A. M. (2004). *Governance.* Cambridge: Polity.
Klijn, E.-H., & Koopenjan, J. (2012). Governance network theory: Past, present and future. *Policy and Politics, 40*(4), 587–606.
Klijn, E.-H., Steijn, B., & Edelenbos, J. (2010). The impact of network management on outcomes in governance networks. *Public Administration, 88*(4), 1063–1082.
Knoke, D. (1994). *Political networks: The structural perspective.* Cambridge: Cambridge University Press.
Kooiman, J. (2000). Societal governance: Levels, models, and orders of social-political interaction. In J. Pierre (Ed.), *Debating governance* (pp. 138–164). Oxford: Oxford University Press.
Koppenjan, J., & Kljin, E.-H. (2004). *Managing uncertainties in networks: A network approach to problem-solving.* London: Routledge.
Lane, J.-E. (1993). *The public sector: Concepts, models and approaches.* London: Sage.
Lane, J.-E. (2000). *New public management.* London: Routledge.
Lawless, P. (1989). *Britain's inner cities* (2nd ed.). London: Paul Chapman.
Lemke, T. (2001). 'The birth of bio-politics': Michel Foucault's lecture at the Collège de France on neo-liberal governmentality. *Economy and society, 30*(2), 190–207.
Lin, N. (1999). Building a network theory of social capital. *Connections, 22*(1), 28–51.
Link, B. G., & Phelan, J. C. (2001). Conceptualizing stigma. *Annual Review of Sociology, 27,* 363–385.
Lipsky, M. (1980). *Street-level bureaucracy, dilemmas of the individual in public services.* New York: Russell Sage Foundation.
Livingstone, D. W. (2004). *The education-jobs gap: Underemployment or economic democracy* (2nd ed.). Ontario: Garamond Press.
Lowndes, V., & Pratchett, L. (2012). Local governance under the Coalition Government: Austerity, localism and the 'big society'. *Local Government Studies, 38*(1), 21–40.
Lowndes, V., & Sullivan, H. (2008). How low can you go? Rationales and challenges for neighbourhood governance. *Public Administration, 86*(1), 53–74.
Lynn Jr., L. E. (2008). New frontiers of public administration: The practice of theory and the theory of practice. *PS: Political Science and Politics, 41*(01), 3–9.
MacInnes, T., Aldridge, H., Bushe, S., Kenway, P., & Tinson, A. (2013). *Monitoring poverty and social exclusion 2013.* York: Joseph Rowntree Foundation and The New Policy Institute.
Mandell, M. P., & Keast, R. (2008a). Introduction. *Public Management Review, 10*(6), 687–698.

Mandell, M. P., & Keast, R. (2008b). Evaluating the effectiveness of interorganizational relations through networks. *Public Management Review*, *10*(6), 715–731.

Marsh, D. (1992). Youth employment policy 1970-1990: Towards the exclusion of the trade unions. In D. Marsh & R. A. W. Rhodes (Eds.), *Policy networks in British Government* (pp. 167–199). Oxford: Clarendon Press.

Marsh, D., & Furlong, P. (2002). A skin not a sweater: Ontology and epistemology in political science. In D. Marsh & G. Stoker (Eds.), *Theory and methods in political science* (2nd ed.). (pp. 17–41). Basingstoke: Palgrave Macmillan.

Marsh, D., & Smith, G. (2001). There is more than one way to do political science: On different ways to study policy networks. *Political Studies*, *49*(3), 528–541.

May, T. (2001). *Social research: Issues, methods and process*. Buckingham: Open University Press.

McAnulla, S. (2002). Structure and agency. In D. Marsh & G. Stoker (Eds.), *Theory and methods in political science* (2nd ed.). (pp. 271–291). Basingstoke: Palgrave Macmillan.

McGuire, M. (2002). Managing networks: Propositions on what managers do and why they do it. *Public Administration Review*, *62*(5), 599–609.

McGuire, M., & Agranoff, R. (2011). The limitations of public management networks. *Public Administration*, *89*(2), 265–284.

Meegan, R., & Mitchell, A. (2001). 'It's not community round here, it's neighbourhood': Neighbourhood change and cohesion in urban regeneration policies. *Urban Studies*, *38*(12), 2167–2194.

Meier, K. J., & O'Toole Jr., L. J. (2003). Public management and educational performance: The impact of managerial networking. *Public Administration Review*, *63*(6), 689–699.

Meier, K. J., & O'Toole Jr., L. J. (2010). Managerial networking, managing the environment, and programme performance: A summary of findings and an agenda. In R. M. Walker, G. A. Boyne, & G. Brewer (Eds.), *Public management and performance: Research directions* (pp. 127–151). Cambridge: Cambridge University Press.

Meier, K. J., O'Toole Jr., L. J., & Lu, Y. (2006). All that glitters is not gold: Disaggregating networks and the impact on performance. In G. A. Boyne, K. J. Meier Jr., L. J. O'Toole, & R. M. Walker (Eds.), *Public service performance: Perspectives on measurement and management* (pp. 152–170). Cambridge: Cambridge University Press.

Merton, R. K. (1968). *Social theory and social structure* (3rd ed.). New York: Free Press.

Miles, R. E., & Snow, C. C. (1992). Causes of failure in network organizations. *California Management Review*, *34*(4), 53–72.

Milward, H. B., & Provan, K. G. (2004). *The* Public manager's guide to network management. *Working Paper Presented at the 2004 Annual Meeting of the American Political Science Association*. Chicago, September 2–5, 2004.

Mitchell, J. C. (Ed.). (1969). 'The concept and use of social networks', in J. C. Mitchell, Social networks in urban situations: analyses of personal relationships in Central African towns. Manchester: Manchester University Press. (pp. 1–50).

Mitchell, S. M., & Shortell, S. M. (2000). The governance and management of effective community health partnerships: A typology for research, policy, and practice. *The Milbank Quarterly, 78*(2), 241–289.

Moore, C., Richardson, J. J., & Moon, J. (1989). *Local partnership and the unemployment crisis in Britain*. London: Unwin Hyman.

Mourshed, M., Patel, J., & Suder, K. (2014). *Education to employment: Getting Europe's youth into employment*. McKinsey & Company.

O'Toole Jr., L. J. (1997). Treating network seriously: Practical and research-based agendas in public administration. *Public Administration Review, 57*(1), 45–52.

O'Toole Jr., L. J. (2008). Governing outputs and outcomes of governance networks. In E. Sørensen & J. Torfing (Eds.), *Theories of democratic network governance* (pp. 215–230). Basingstoke: Palgrave Macmillan.

O'Toole Jr., L. J. (2011). The EDA in Oakland: A case that catalysed a field. *Public Administration Review, 71*(1), 116–120.

O'Toole Jr., L. J., & Meier, K. J. (2004). Public management in intergovernmental networks: Matching structural networks and managerial networking. *Journal of Public Administration Research and Theory, 14*(4), 469–494.

OECD (2009). *The OECD economic outlook: Tackling the jobs crisis*. Paris: OECD.

Office for National Statistics (ONS). (2012, November). *People in work wanting more hours increases by 1 million since 2008*. Available at http://webarchive.nationalarchives.gov.uk/20160105160709/http://www.ons.gov.uk/ons/dcp171776_289024.pdf

Office for National Statistics (ONS). (2013, June). *Full report: Graduates in the UK labour market 2013. Numbers of underemployed in the labour market*.

Oliver, A. L., & Ebers, M. (1998). Networking network studies: An analysis of conceptual configurations in the study of inter-organizational relationships. *Organization Studies, 19*(4), 549–583.

Olson, M. (1965). *The logic of collective action*. Cambridge, MA: Harvard University Press.

Ostrom, E. (1990). *Governing the commons: The evolution of institutions for collective action*. Cambridge: Cambridge University Press.

Ostrom, E. (1998). A behavioral approach to the rational choice theory of collective action: Presidential address, American Political Science Association, 1997. *The American Political Science Review, 92*(1), 1–22.

Pacione, M. (1997). Urban restructuring and the reproduction of inequality in Britain's cities: An overview. In M. Pacione (Ed.), *Britain's cities: Geographies of division in urban Britain* (pp. 7–60). London: Routledge.

Page, S. (2004). Measuring accountability for results in interagency collaboratives. *Public Administration Review*, 64(5), 591–606.

Painter, J., Orton, A., MacLeod, G., Dominelli, L., & Pande, R. (2011). *Connecting localism and community empowerment: Research review and critical synthesis for the AHRC Connected Communities Programme*.

Papadopoulos, Y. (2007). Problems of democratic accountability in network and multilevel governance. *European Law Journal*, 13(4), 469–486.

Peck, J., & Theodore, N. (2000). Beyond employability. *Cambridge Journal of Economics*, 24(6), 729–749.

Perkins, N., Smith, K., Hunter, D. J., Bambra, C., & Joyce, K. (2010). 'What counts is what works'? New labour and partnerships in public health. *Policy and Politics*, 38(1), 101–117.

Perri 6. (1997). Social exclusion: Time to be optimistic. In I., Christie, & H. Perry, *The wealth and poverty of networks* (pp. 1–24). Demos: London.

Perri 6, Leat, D., Seltzer, K., & Stoker, G. (2002). *Towards holistic governance*. Basingstoke: Palgrave.

Peters, B. G. (2001). *The future of governing* (2nd ed.). Kansas: University Press of Kansas.

Pierre, J. (1999). Models of urban governance the institutional dimension of urban politics. *Urban Affairs Review*, 34(3), 372–396.

Pierre, J. (2000). Conclusions: Governance beyond state strength. In J. Pierre (Ed.), *Debating governance* (pp. 241–246). Oxford: Oxford University Press.

Pierre, J. (2005). Comparative urban governance, uncovering complex causalities. *Urban Affairs Review*, 40(4), 446–462.

Polanyi, K. (1957). *The great transformation*. Boston: Beacon Press.

Powell, W. W. (1991). Neither market nor hierarchy: Network forms of organisation. In G. Thompson, J. Frances, R. Levacic, & J. Mitchell (Eds.), *Markets, hierarchies and networks: The coordination of social life* (pp. 265–276). London: Sage in Association with The Open University.

Pressman, J. L., & Wildavsky, A.B. (1973). *Implementation: How great expectations in Washington are dashed in Oakland or, why it's amazing that Federal programs work at all, this being a saga of the Economic Development Administration as told by two sympathetic observers who seek to build morals on the foundation of ruined hopes*. Berkeley: University of California Press.

Provan, K. G., & Kenis, P. (2008). Modes of network governance: Structure, management, and effectiveness. *Journal of Public Administration Research and Theory*, 18(2), 229–252.

Provan, K. G., & Lemaire, R. H. (2012). Core concepts and key ideas for understanding public sector organizational networks: Using research to inform scholarship and practice. *Public Administration Review*, 72(5), 638–648.

Provan, K. G., & Milward, H. B. (1995). A preliminary theory of network effectiveness: A comparative study of four community mental health systems. *Administrative Science Quarterly, 40*(1), 1–33.
Provan, K. G., & Milward, H. B. (2001). Do networks really work? A framework for evaluating public-sector organizational networks. *Public Administration Review, 61*(4), 414–423.
Provan, K. G., Fish, A. C., & Sydow, J. (2007). Interorganizational networks at the network level: A review of the empirical literature on whole networks. *Journal of Management, 33*(3), 479–516.
Provan, K. G., Huang, K., & Milward, H. B. (2009). The evolution of structural embeddeness and organizational social outcomes in a centrally governed health and human service network. *Journal of Administration Research and Theory, 19*(4), 873–893.
Putnam, R. D. (1993). The prosperous community: Social capital and public life. *The American Prospect, 4*(13), 1–11.
Rai, S. (2008). *Routes and barriers to citizen governance.* York: Joseph Rowntree Foundation.
Reese, L., & Rosenfeld, R. A. (2002). *The civic culture of local economic development.* Thousand Oaks: Sage.
Reid, B., & Iqbal, B. (1995). *Redefining housing practice: Inter-organizational relationships and local housing networks.* Unpublished paper (permission to quote from the authors).
Rhodes, R. A. W. (1997). *Understanding governance: Policy networks, governance, reflexivity and accountability.* Buckingham: Open University Press.
Rhodes, R. A. W. (2000). Governance and public administration. In J. Pierre (Ed.), *Debating governance* (pp. 54–90). Oxford: Oxford University Press.
Rhodes, R. A. W. (2007). Understanding governance: Ten years on. *Organizational Studies, 28*(08), 1243–1264.
Rhodes, R. A. W., & Marsh, D. (1992). Policy networks in Britain: A critique of existing approaches. In D. Marsh & R. A. W. Rhodes (Eds.), *Policy networks in British government* (pp. 1–26). Oxford: Clarendon Press.
Richards, D. (2008). *New labour and the civil service: Reconstituting the Westminster Model.* Basingstoke: Palgrave Macmillan.
Richards, D., & Smith, M. J. (2002). *Governance and public policy in the UK.* Oxford: Oxford University Press.
Richards, S., Barnes, M., Coulson, A., Gaster, L., Leach, B., & Sullivan, H. (1999). *Cross-cutting issues in public policy and public services.* London: Department of Environment Transport and the Regions.
Richardson, J. (2000). Government, interest groups and policy change. *Political Studies, 48*, 1006–1025.
Richardson, L. (2011). Cross-fertilization of governance and governmentality in practical policy making on behaviour change. *Policy and Politics, 39*(4), 533–546.
Rigney, D. (2010). *The Matthew effect: How advantage begets further advantage.* New York: Columbia University Press.

Scharpf, F. W. (1978). Interorganizational policy studies: Issues, concepts and perspectives. In K. I. Hanf & F. W. Scharpf (Eds.), *Interorganizational policy making: Limits to coordination and central control* (pp. 345–370). Beverly Hills, CA: Sage.

Scharpf, F. W. (2000) 'Economic changes, vulnerabilities, and institutional capabilities' in F. W. Scharpf and V. A. Schmidt (eds.), *Welfare and work in the open economy: Volume I: From vulnerability to competitiveness* (pp. 21–124), Oxford: Oxford University Press.

Somers, M., & Block, F. (2005). From poverty to perversity: Ideas, markets, and institutions over 200 years of welfare debate. *American Sociological Review*, 70(2), 260–287.

Sørensen, E., & Torfing, J. (2008a). Theoretical approaches to democratic network governance. In E. Sørensen & J. Torfing (Eds.), *Theories of democratic network governance* (pp. 233–246). Basingstoke: Palgrave Macmillan.

Sørensen, E., & Torfing, J. (2008b). Introduction: Governance networks research: Towards a second generation. In E. Sørensen & J. Torfing (Eds.), *Theories of democratic network governance* (pp. 1–21). Basingstoke: Palgrave Macmillan.

Sørensen, E., & Torfing, J. (2009). Making governance networks effective and democratic through metagovernance. *Public Administration*, 87(2), 234–258.

Stoker, G. (1991). *The politics of local government* (2nd ed.). Basingstoke: Macmillan.

Stone, C. N. (1989). *Regime politics: Governing Atlanta 1946–1988*. Lawrence, KS: University of Kansas Press.

Stone, C. N. (2005). Rethinking the policy-politics connection. *Policy Studies*, 26(3/4), 241–260.

Streeck, W. (1989). Skills and the limits of neo-liberalism: The enterprise of the future as a place of learning. *Work, Employment & Society*, 3(1), 89–104.

Sullivan, H., & Skelcher, C. (2002). *Working across boundaries: Collaboration in public services*. Basingstoke: Palgrave Macmillan.

Sullivan, H., & Taylor, M. (2007). Theories of 'neighbourhood' in urban policy. In I. Smith, E. Lepine, & M. Taylor (Eds.), *Disadvantaged by where you live?: Neighbourhood governance in contemporary urban policy* (pp. 21–42). Bristol: Policy Press.

Swyngedouw, E. (2005). Governance innovation and the citizen: The Janus face of governance-beyond-the-state. *Urban Studies*, 42(11), 1991–2006.

Sydow, J. and Milward, H. B. (2003, June 27). Reviewing the evaluation perspective: On criteria, occasions, procedures, and practices. In *Papier präsentiert auf der 10th International Conference on Multi-Organisational Partnerships, Alliances and Networks*. University of Strathclyde, Glasgow.

Taylor, M. (2003). *Public policy in the community*. Basingstoke: Palgrave Macmillan.

Teisman, G. R., & Klijn, E.-H. (2002). Partnership arrangements: Governmental rhetoric or governance scheme? *Public Administration Review*, 62(2), 197–205.

Therborn, G. (1986). *Why some people are more unemployed than others*. London: Verso.

Torfing, J. (2005). Governance network theory: Towards a second generation. *European Political Science*, *4*(3), 305–315.

Torfing, J. (2007a). A comparative and multi-level analysis of governance networks: A pilot study of employment policy. In P. Bogason & M. Zølner (Eds.), *Methods in democratic network governance* (pp. 21–40). Basingstoke: Palgrave Macmillan.

Torfing, J. (2007b). Empirical findings: Seven network stories. In P. Bogason & M. Zølner (Eds.), *Methods in democratic network governance* (pp. 41–73). Basingstoke: Palgrave Macmillan.

Tuite, M. F. (1972). Toward a theory of joint decision making. In M. Tuite, R. Chisholm, & M. Radnor (Eds.), *Interorganizational decision making* (pp. 1–9). Chicago: Aldine.

Turrini, A., Cristofoli, D., Frosini, F., & Nasi, G. (2010). Networking literature about determinants of network effectiveness. *Public Administration*, *88*(2), 528–550.

Urry, J. (1990). Conclusion: Places and policies. In M. Harloe, C. G. Pickvance, & J. Urry (Eds.), *Place, policy and politics: Do localities matter* (pp. 187–204). London: Unwin Hyman.

Voets, J., Van Dooren, W., & De Rynck, F. (2008). A framework for assessing the performance of policy networks. *Public Management Review*, *10*(6), 773–790.

Walker, R. M., Boyne, G. A., & Brewer, G. (Eds.) (2010). *Public management and performance: Research directions*. Cambridge: Cambridge University Press.

Wasserman, S., & Faust, K. (1994). *Social network analysis: Methods and applications*. Cambridge: Cambridge University Press.

Whitehead, M. (2007). The architecture of partnerships: Urban communities in the shadow of hierarchy. *Policy and Politics*, *35*(1), 3–23.

Work and Pensions Select Committee (WPC) (2013). Can the Work Programme work for all user groups. *House of Commons*. London: The Stationery Office Limited.

Yanow, D. (2004). Translating local knowledge at organizational peripheries. *British Journal of Management*, *15*(S1), S9–S25.

CHAPTER 2

Theoretical Background

There are no exclusive theories or agreed set of values by which to judge network impact. However, if a theory's scope is too narrow, speculative or lacks reality, the impact explanation may be distorted or trivialised in causal terms (Geddes 2003). Chapter 1 claims that network theories appear to elevate managerial culture, emphasising the importance of professional practice, performance and interaction in supporting political discourse, and the contagion effect of ideologies, rather than socio-economic outcomes for localities. Network literature reveres collaboration and trust to stabilise networks, yet 'trust' is sector specific, personal, often short-lived and conducive to the formation of cliques and cannot regulate individualism, or stabilise institutional and political behaviour, labour coordination or the idiosyncrasies that affect economic demand. Hence, this chapter considers how theoretical strands contextualise our understanding about network impact. First, it links network performance to theoretical traditions. Then it reviews the roles of network structure (conditions), agency (actors' behaviour and interactions) and network processes (namely, network governance and network management). The conclusion of the chapter considers theory limitation. The final chapter of the book considers whether the theories are too abstract to explain case network outcomes and form the basis of generalisations, or whether they may be distorted by unforeseen results due to unintended consequences, chance or 'factors outside the scope of the theory' (King et al. 1994: 10, 106).

Network Theory

Geddes (2003) believes theories should do more than 'identify factors that contribute to outcomes'; rather, they should also try 'to explain the relationships among the moving parts of the processes leading to outcomes' (2003: 219). This book aims to explain the impact of factors' 'moving parts', in terms of: (i) local network structuring; (ii) political actions influencing networks; and (iii) deliberative processes for organising and steering network outcomes. Three disciplines, structural sociology, political science and public administration and management, fit the methodology and have the explanatory potential necessary for application to our testing of theories, case descriptions, and analysis of the network conditions that facilitate or limit impact. A brief overview follows.

Structural Sociology

The nineteenth-century founders of social science (for example, Marx, Durkheim and Weber) were not network theorists, but their social and economic theses complement studies of social structure in network analysis. Marx and Weber considered how economic structure divides society, whereas Durkheim examined what holds society together. Marx regarded capitalism as more than an economic system; rather, he saw it as a web of complex social relations originating from material activities influencing material relations and symbolic discourse about material realities. Marxists might emphasise the contribution of networks to structural inequality, the shaping of material interests and commoditisation. A Durkheimian account might see networks as serving social solidarity; uniting the divisions of labour and protecting individuals against state monopoly. For Weberians, any class or group can use social structure to increase power and status and unify social action. Yet neo-liberal competition strategies discourage network solidarity, and power dispersal is not always equitable and increases management surveillance. Hence, anarchists might view networks as structures which challenge economic rule.

Structural perspectives of social relations emerged during the 1930s and were applied to the field of sociometry, subsequently developed by anthropologists at Manchester University and further utilised in a broad range of disciplines from economics to epidemiology.[1] Social network

---

[1] The sociological network tradition is reviewed in Scott (1991: 7–16), Mizruchi and Galaskiewicz (1994: 230–1) and Kilduff and Tsai (2003: 3–4, 35–65). Also see Wasserman

analysis (SNA) methods evolved to map network boundaries and relational positions (affiliations and ties) at the centre or periphery of a network. Sociologically the number of contacts an actor has made (out-degree) and received (in-degree) indicates their tie-strength, reputation and power, but also how information flows in the network. Mapping policy relations adds precision to narratives about community power (Hunter 1953; Dahl 1961), strong or weak ties embedded in economic behaviour (Granovetter 1973), and resource dependencies (Benson 1975). For example, a website promoting an employment strategy group implies that officials are collaborating, but the frequency of actors' interactions maybe weaker than appearances suggest. The issue of whether these ties are reciprocal, or if the actor have no ties (and are 'isolates') is 'the basic building block [...] of any network study' (Provan and Milward 1995: 10). However, critics suggest that structural studies under-theorise the 'causal role of ideals, beliefs and values', and lose sight of 'culture, agency and process' (Emirbayer and Goodwin 1994: 1446–7; Skelcher and Sullivan 2008: 762). Moreover, SNA methods are time-consuming, their findings can be predictable and time-specific, and the quality of relations requires interpretation (Dowding 1995). This is possibly why few network studies use SNA to demonstrate the areas in which public service relations might be strengthened (with exceptions; see Provan and Milward 1995; 2001; Cotterill and King 2007: 5). Nevertheless, structural sociology is central to understanding social and economic exchanges in network association.

Political Science

Political science highlights network tensions between a) central and local relations, b) centralised and decentralised policy, and c) inequitable and equitable resource distribution (Taylor 2003: 43–5). On the one hand, theorists of power in policymaking highlight state-centred networks (Hay and Richards 2000: 5). These scholars claim UK networks sustain an 'asymmetric power model' of government that reproduces structural inequality and unequal plurality in politics and institutions; this 'top-down view of democracy' is levered by 'force, legitimacy, state bureaucracy, tax-raising powers and legislation' (Richards 2008: 54–5). Additionally, the 'elitist' model places the business community in partnership with government,

and Faust (1994: 4–17), and for a literature review, see Berry et al. (2004) and Klijn and Koppenjan (2012).

insulated from pressure groups and citizens. On the other hand, pluralist and democratic theorists expect governance networks to regain control of local governance and assist consensual politics for 'governing without Government' (Rhodes 1997: 59; Stoker 2004). In this model, multi-actors can mediate policymaking through decentred, democratic, non-hierarchical network governance (Sørensen and Torfing 2008; Bevir 2010). Yet the 'consensual' model is usually operated under central guidance, led by a government partner in proximity to state institutions. Moreover, network governance is image-managed through websites or glossy reports to demonstrate success to politicians or aggrandise producers' careers (Wolman and Page 2002: 492; Davies 2011: 2). Other critics associate network governance with democratic accountability problems, whereby governance failure is inevitable (Jessop 2000; Papadopoulos 2007). Yet how can networks address neighbourhood problems, if their area-based policies are not designed to tackle the structural causes of employment exclusion or social injustice (Dietz 2002; Taylor 2003: 31–2)?

Public Administration/Public Management

Networks support public administration/public management operations and professionals in a governance context so as to negotiate and implement policymaking goals and systems (Considine 2001; Sydow and Milward 2003: 1–5; Goldsmith and Eggers, 2004: 157–9; Kenis and Provan 2009: 442). However, the tension between 'top-down' and 'bottom-up' implementation approaches remains as public managers retain their primary decision-making roles (Hill and Hupe 2009: 42–56). During the 1990s, Dutch scholars established public sector network management theories for coordinating stakeholders' interests in policy networks, but they overlooked the politics and power dimensions (Kickert et al. 1997; Le Galès 2001: 171; Berry et al. 2004). Despite reams of 'partnership' advice, all fraternities play governance games; there is no levelling the playing field. Rhodes (1999: xv) claims that 'any account of the management of networks is necessarily about politics'. For Davies (2007: 780), state managerialism colonised networks, re-emphasised the role of the centre and eroded prospects for democracy, as the costs of challenging network processes or of exiting those networks may be too high. Evidently, public administration depoliticises networks to defend public spending cuts and a revolving door of interventions based on inadequate macroeconomics and work effort theories.

Structure, Agency and Processes

Aspects of the structure/agency debate frame unemployment narratives. Some view unemployment as a problem of 'economic change' (attributing it to structure), others to 'feckless, immoral individuals unwilling to work' (attributing it to agency) (McAnulla 2002: 274). Network processes also assist to decrease political struggle (agency), and increase instrumental rationality (structure) for managing and quantifying individual actions, marketisation, and profits (Blokland on Weber 2006: 30–1). The danger is that scholars' preoccupation with smoothing antagonisms and governance solutions for social harmony between society's bodies and groups ignores the issue of governance for economically excluded citizens. The network-centric impact perspectives below offer insights.

The Impact of Structure

Governments' mandated network structures influence structural contexts and the 'material conditions which define the range of actions available to actors' (McAnulla 2002: 271). The actors who represent network structure also shape new structures or reproduce existing structure, including: administrative and governance systems; bureaucratic 'command structures'; policy and programme design; funding streams; contracts; institutions; organisational constitution; legal systems; regulations, and quality controls. Structural problems beyond the network, however, may impair policy implementation and stem from a lack of 'joined-up government' and central departments' dysfunctional policy and processes (Davies 2009; 2011: 67; Richards 2008: 115). Keast et al. (2004: 364) suggest that '… unless policy-makers have a full understanding of what it means to work through network structures, they will continue to develop traditional policies and management techniques that mitigate against the positive attributes of networked arrangements'. Moreover, imposing network structure (members, size, and form of organisation) may not be the right coordination strategy to meet the policy needs, and may prove detrimental to network behaviour (Scharpf 1978: 353, 362, 364; Pollitt 2003: 58), such that:

> Once an inter-organizational strategy has been pursued, a network of relations is created that may constrain actors' subsequent behaviour. (Mizruchi and Galaskiewicz 1994: 231)

Networks fail if structures, tasks and interactions do not match the policy and institutional requirements, as '[f]ancier top-down structures' that try to correct 'maladapted steering patterns' may be inappropriate and increase coordination costs (Hanf and O'Toole 1992: 167, 173; see also Kickert and Koppenjan 1997: 51 in reference to Scharpf 1978: 363). Networks often lack the structural authority to share responsibilities; this problem is often compounded by their varying 'world views, goals and organizational procedures' (Hanf and O'Toole 1992: 172; Davies 2004: 31). Instead, they offer structural advantage, greater influence and better access to the advisers, information and statistical data necessary to secure interests, resources and outcomes in return for money, information, support or clients (Lin 1999: 34; Galaskiewicz 1985: 284, 295; Marsden 2005). Role-equivalent actors with powerful resources occupy central positions in networks, yet connectivity may be hollow; reciprocal or bridging ties may be less than expected and actual decision-making may occur with closer organisational associates outside the network (Heinz et al. 1990: 32). Network structure creates a governance market of narrow objectives and ad hoc returns for citizens whose real needs, more often, extend beyond what the networks can deliver.

The Impact of Agency

Communication structures shape human behaviour, but intervention requires individual power to influence outcome preferences and make things happen (Mayhew 1980: 357). However, uncertainty surrounds the outcomes of agency for citizens.

First, 'trust' is assumed to influence goal-directed or serendipitous network behaviour across the political, cultural, social and economic strata and, in turn, to influence policy implementation, reputational perceptions, connectivity, information dissemination, resource exchange and decision exchange (Kilduff and Tsai 2003: 6–8, 95–6; Klijn and Edelenbos 2008: 206). Trust inside networks, however, is inequitable, yet seldom discussed, or studied only within a single organisation (Provan and Kenis 2008: 237–8; Wong and Boh 2010). Van den Brink and Benschop (2014) studied the adverse impact of trust and gendered networking practices in academic career progression, with implications beyond this setting. Indeed, trust diffuses through homogenous networks, steering norms amongst like-minded people in order to reduce risk or avoid social dilemmas. Ostrom (1998: 12), links behavioural foundations to 'the trust that individuals have in others; the investment others make in trustworthy reputations; and the probability that participants will use reciprocity norms'. Another study suggests actual reciprocal ties are less important than has been assumed, as trustworthiness accrues through prior informal relations and occupational status (Wong and Boh 2010). Trustworthiness may also lead to 'in-group favouritism' (Begley et al. 2010: 288) or complacency, and organisational and economic malfeasances (Granovetter 1985: 491–2). Conversely, distrust increases 'formal control mechanisms' and coordination costs, and can mean that work tasks take longer to develop (see transaction cost economics theory in Williamson 1979: 235, 246, and 261; Knoke 2001: 151; Alter and Hage 1993: 78). Distrust between neighbourhoods impacts interrelations and resource allocation. Likewise, trust can be short term or depend on reputation.

Hence network agency is also seen as a way to enhance reputation (Graddy and Chen 2006: 543). Earned or assigned reputation is an ancient trading mechanism associated with good or poor performance, labour allocation, previous contacts, collaboration or defection, resource power and reputed interests. Reputation indicators include grants and contracts awarded, targets achieved and the number of employees with status and higher salaries who emulate funding bodies' performance management needs (Rouse and Smith 2002: 46–7). Consequently, high-reputation organisations may avoid working with organisational/participants with low reputations. Yanow (2004: S11) asks why managers are 'so little interested in the work of those they manage, especially when that work entails local knowledge that has strategic implications for their organisation. It would seem 'the further from the centre, the less power and status one is perceived to have, and the less legitimacy is accorded to one's knowledge' (Yanow 2004: S22). Less well-known groups may be effective, but neighbourhoods depend on the reputational strength of local organisations to represent needs and attract funds. Thus, reputational advantage require empirical investigation as the 'processes of a collaboration are as central to leading its activities as are the participants associated with it' (Huxham and Vangen 2000: 1165–6; Considine 2001: 1).

The Impact of Processes

Network processes (network governance and network management) steer structure and agency towards desirable goals. Urban governance processes assume networks pursue collective goals assisted by co-production in policymaking and 'close linkages among norms and values, institutions, political objectives, and policy outcomes' (Pierre 2005: 449). The governance approach aims to devolve decision-making to multi-level stakeholders in a local policy domain, but Wilson (2003: 321; 2004: 12) claims that the outcome of this is more usually 'multi-level dialogue' than multi-level governance. Governance is not a commitment to effective action, and the concept's assumptions require examination. Kickert et al.'s (1997b: 7–10) three overlapping governance models provide a starting point: these are 'rational central rule', 'multi-actor', and 'network', the facets of which highlight processes of coordination, alliance-building and connectivity, respectively.

Rational central rule (coordination)—this model uses traditional managerial approaches to steer governance and bring more actors into the 'jurisdiction of one central coordinating authority' (1997b: 7–11). But central policies can be awkward for local governance, and work against civil society inclusion or make assumptions about the problem's dimensions in advance of local diagnoses. Possible outcomes may include an increase in planning to justify expenditure, the creation of feasibility studies to legitimise or delay actions, an increase of perverse bureaucracy and coordination processes, the production of publicity for programme goals, and increased officialdom over network autonomy and diversity.

Multi-actor (alliance-building)—this model involves alliance-building between actors who tackle local issues. Through self-governance and self-determination, actors hold back adverse action by the government whilst lobbying the latter for local resources. Kickert et al. (1997b: 9) suggest that localised governance is too radical. With no centralisation of decision-making, alliances may not be able to 'calculate rationally the optimum mix from a field of competing values' (Warren 1967: 412). Outcomes might be conflict between the local and central levels of governance, group infighting and the formation of cliques, or numerous agencies driving new services and opportunities through ad hoc planning or service duplication.

Network (connectivity)—this model encourages collective working amongst equals for resource exchange and problem-solving (Alter and Hage 1993: 169). Outcomes depend on the contact and information flow between local leaders and those actors who represent unemployed people and steer funds to meet their needs. This requires 'discretion to solve public problems in creative ways' (Page 2004: 591). Actors may cooperate, yet connectedness can reduce smaller agencies' autonomy and some agencies may monopolise. This is because even shared governance tends to gravitate to the level of a proxy authority (lead organisation or network administrative organisation).

Thus, overcoming governance uncertainties and applying the implementation detail may depend on network management approaches, as used by government and EU programmes within local authorities for national, regional and local policy coordination (see Painter et al. 1997; Bache 2000; Johnson 2004: 84). Whereas a classic management style sets and controls well-defined goals, network management is: '... aimed at coordinating strategies of actors with different goals and preferences with regard to a certain problem or policy measure within an existing network of interorganizational relations' (Kickert et al. 1997b: 10, original emphasis). Arguably, network management theory lags behind the public management practice of 'government by network' and local provision continues to tackle shortfalls in service coordination and community involvement; in other words, it is 'business as usual' (Taylor 2000a, 2000b: 1019; Goldsmith and Eggers 2004: 22). British community network literature has a negligible interest in network management (Skelcher et al. 1996; Gilchrist 2004). Clearly, any public or private sector actor can fill the role of a network manager or administrator; however, they would need to have the authority to work across organisational boundaries to reconcile cultural differences and varying institutional rules. This might depend on the choice of network management styles; Kickert et al. (1997c: 181–91) distinguishes these as 'instrumentalism', 'interaction' and 'institutionalism'. Specifically, they emphasise goal-directing, games and rules, respectively.

The first of these styles, instrumentalism (or goal-directing), uses persuasion or bargaining strategies to ensure network members take a particular course of action, for example by 'influencing network behaviour toward the implementation of public programmes' (O'Toole 1997: 119). This could involve 'agency aggrandizement or [a prioritisation of] public relations rather than in policy results' (Meier and O'Toole 2003: 697), especially if the 'specialism' and authority of programme goals 'distracts from the maintenance of cohesion and coordination in the government service' (Stahl 1960: 35–7). Instrumentalisation of networks alienates certain groups while others exercise their policymaking influence, and makes pluralism, local democracy, and self-regulating network governance difficult to achieve (Sørensen and Torfing 2008).

The second style, interaction (games), refers to the games that lead to joint actions, such as the making of meeting arrangements, coordination of events, solution-brokering and conflict mediation (Kickert and Koppenjan 1997: 47, 52–4). Interaction also occurs spontaneously amongst participants who recognise mutual benefits and dependencies when opportunities arise. Collective interaction in pluralistic contexts must be appropriate, as participants who pursue self-interested outcomes cannot be easily coordinated (Scharpf 1978: 352). One study suggested that potential network managers were 'into game management more than network structuring', as they operated within 'structures largely dictated elsewhere' (Painter et al. 1997: 230). Participants assess the benefits of interaction according to their own interests, as networks representing diverse belief systems are likely to involve games with conflict.

The third style, institutionalism (or rules), projects norms and procedures in such a way as to convey the image of governance, authority and trustworthiness. The ruling group filters policy and funding interests to comply with system maintenance at several administrative levels, supposedly to reflect the public interest, as opposed to maximising the actors' own values. Johnson's (2004) study of network management in UK urban regeneration suggests that the institutionalised network structure dominating network processes was inadequate for implementing the necessary policies. Hierarchical structures controlling the policy environment are less open to opposition or change, except through incremental improvement and learning. Network management requires fast decision-making to attract opportunities, but new problems may still arise as policy results are not predictable (De Bruijn and Ten Heuvelhof 1997: 133).

Conclusion

Theoretical traditions elevate professional network practice, and privilege network-level and participant-level impact but are weak in modelling impact for neighbourhoods. Network theories tend to ignore powerful contextual elements that shape outcomes, from politics through policy to local culture. Structure/agency dimensions add more insight into network responses to societal problems (McAnulla 2002: 290–1). Evidently, large network structures can pay lip service to all goals and reflect political power, but coordination costs increase and network perceptions and morale can deteriorate, as 'complex systems are difficult to make effective' (Hanf and O'Toole 1992: 173; Alter and Hage 1993: 211–26; Provan and Milward 2001: 418). Consequently, network agency is like a weak trade union equivalent which enables providers to negotiate a piece of the pie and policymakers and managers to negotiate the status quo for their own job security. The 'networked society' sets the unemployed citizen apart with no voice or full citizenship to protect him or her against perverse sanctions, and with no bargaining rights in places with weak local economies and poor labour regulation. Network processes advocate cooperation and collaboration, yet the impulses of British society are individualistic, pluralistic and exploitative, and governance is deeply flawed. Likewise, networks for trust building has great appeal for people already employed, but 'trust' is hardly an economic solution without a legally binding right to a useful remunerative job. Theoretical charms are not strategies for supporting unemployed people to join the local labour force. Nor can they overcome structural inequality in the absence of an effective profit orientated economy. The next chapter broadens the enquiry to include the unemployment perspectives and provide more insight into the empirical puzzle.

Bibliography

Alter, C., & Hage, J. (1993). *Organizations working together*. Newbury Park, CA: Sage.
Bache, I. (2000). Government within governance: Network steering in Yorkshire and the humber. *Public Administration, 78*(3), 575–592.
Begley, T. M., Khatri, N., & Tsang, E. W. K. (2010). Networks and cronyism: A social exchange analysis. *Asia Pacific Journal of Management, 27*(2), 281–297.
Benson, J. K. (1975). The interorganizational networks as a political economy. *Administrative Science Quarterly, 20*(2), 229–249.
Berry, F. S., Brower, R. S., Choi, S. O., Xinfang Goa, W., Jang, H., Kwan, M., et al. (2004). Three traditions of network research: What the public management research agenda can learn from other research communities. *Public Administration Review, 64*(5), 539–552.
Bevir, M. (2010). *Democratic governance*. Princeton: Princeton University.
Blokland, H. (2006). *Modernization and its political consequences*. New Haven and London: Yale University Press.
Considine, M. (2001). *Enterprising states, the public management of welfare-to-work*. Cambridge: Cambridge University Press.
Cotterill, S. and King, S. (2007). Public sector partnerships to deliver local e-Government: A social network study. In *International Conference on Electronic Government* (pp. 240–251). Springer Berlin Heidelberg.
Dahl, R. A. (1961). *Who governs?* New Haven: Yale University Press.
Davies, J. S. (2004). Can't hedgehogs be foxes too? reply to Clarence N. Stone. *Journal of Urban Affairs, 26*(1), 27–33.

Davies, J. S. (2007). The limits of partnership: An exit-action strategy for local democratic inclusion. *Political Studies, 55*(4), 779–800.

Davies, J. S. (2009). The limits of joined-up government: Towards a political analysis. *Public Administration, 87*(1), 80–96.

Davies, J. S. (2011). *Challenging governance theory: From networks to hegemony.* Bristol: Policy Press.

De Bruijn, J. A., & Ten Heuvelhof, E. F. (1997). Instruments for network management. In W. J. M. Kickert, E.-H. Klijn, & J. F. M. Koppenjan (Eds.), *Managing complex networks, strategies for the public sector* (pp. 119–137). London: Sage.

Dietz, R. D. (2002). The estimation of neighborhood effects in the social sciences: An interdisciplinary approach. *Social Science Research, 31*(4), 539–575.

Dowding, K. (1995). Model or metaphor? A critical review of the policy network approach. *Political Studies Association, 43*(1), 136–158.

Emirbayer, M., & Goodwin, J. (1994). Network analysis, culture, and the problem of agency. *The American Journal of Sociology, 99*(6), 1411–1454.

Galaskiewicz, J. (1985). Interorganizational relations. *Annual Review of Sociology, 1,* 281–304.

Geddes, B. (2003). *Paradigms and sand castles: Theory building and research design in comparative politics.* Ann Arbor: University of Michigan Press.

Gilchrist, A. (2004). *The well-connected community. A networking approach to community development.* Bristol: The Policy Press.

Goldsmith, S., & Eggers, W. D. (2004). *Governing by network. The new shape of the public sector.* Washington, DC: The Brookings Institution Press.

Graddy, E. A., & Chen, B. (2006). Influences on the size and scope of networks for social service delivery. *Journal of Public Administration Research and Theory, 16*(4), 533–552.

Granovetter, M. (1973). The strength of weak ties. *American Journal of Sociology, 78*(6), 1360–1380.

Granovetter, M. (1985). Economic action and social structure: The problem of embeddedness. *American Journal of Sociology, 91*(3), 481–510.

Hanf, K. I., & O'Toole Jr., L. J. (1992). Revisiting old friends: Networks, implementation structures and the management of interorganizational relations. *European Journal of Political Research, 21*(1–2), 163–180.

Hay, C., & Richards, D. (2000). The tangled webs of Westminster and Whitehall: The discourse, strategy and practice of networking within the British core executive. *Public Administration, 78*(1), 1–28.

Heinz, J. P., Laumann, E. O., Salisbury, R. H., & Nelson, R. L. (1990). Inner circles or hollow cores? Elite networks in national policy systems'. *The Journal of Politics, 52*(2), 356–90.

Hill, M., & Hupe, P. (Eds.) (2009). *Implementing public policy* (2nd ed.). London: Sage.
Hunter, F. (1953). *Community power structure: A study of decision-makers.* Chapel Hill: University of North Carolina Press.
Huxham, C., & Vangen, S. (2000). Leadership in the shaping and implementation of collaboration agendas: How things happen in a (not quite) joined-up world. *The Academy of Management Journal, 43*(6), 1159–1175.
Jessop, B. (2000). Governance failure. In G. Stoker (Ed.), *The new politics of British local governance* (pp. 11–32). Basingstoke: Macmillan.
Johnson, C. L. (2004). *A comparative study of "joined-up" working in three regeneration programme case studies.* Doctoral Thesis, Birmingham: Aston University.
Keast, R., Mandell, M. P., Brown, K., & Woolcock, G. (2004). Network structures: Working differently and changing expectations. *Public Administration Review, 64*(3), 363–371.
Kenis, P., & Provan, K. G. (2009). Towards an exogenous theory of public network performance. *Public Administration, 87*(3), 440–456.
Kickert, W. J. M., & Koppenjan, J. F. M. (1997). Public management and network management: An overview. In W. J. M. Kickert, E.-H. Klijn, & J. F. M. Koppenjan (Eds.), *Managing complex networks: Strategies for the public sector* (pp. 35–61). London: Sage.
Kickert, W. J. M., Klijn, E.-H., & Koppenjan, J. F. M. (Eds.) (1997a). *Managing complex networks: Strategies for the public sector.* London: Sage.
Kickert, W. J. M., Klijn, E.-H., & Koppenjan, J. F. M. (1997b). Introduction: A management perspective on policy networks. In W. J. M. Kickert, E.-H. Klijn, & J. F. M. Koppenjan (Eds.), *Managing complex networks: Strategies for the public sector* (pp. 1–13). London: Sage.
Kickert, W. J. M., Klijn, E.-H., & Koppenjan, J. F. M. (1997c). Managing networks in the public sector: Findings and reflection. In W. J. M. Kickert, E.-H. Klijn, & J. F. M. Koppenjan (Eds.), *Managing complex networks: Strategies for the public sector* (pp. 166–191). London: Sage.
Kilduff, M., & Tsai, W. (2003). *Social networks and organizations.* London: Sage.
King, G., Keohane, R., & Verba, S. (1994). *Designing social enquiry.* Princeton, NJ: Princeton University Press.
Klijn, E.-H., & Edelenbos, J. (2008). Meta-governance as network management. In E. Sørensen & J. Torfing (Eds.), *Theories of democratic network governance* (pp. 199–214). Basingstoke: Palgrave Macmillan.
Klijn, E.-H., & Koopenjan, J. (2012). Governance network theory: Past, present and future. *Policy and Politics, 40*(4), 587–606.
Knoke, D. (2001). *Changing organizations: Business networks in the new political economy.* Colorado: Westview Press.

Le Galès, P. (2001). Urban governance and policy networks: On the urban political boundedness of policy networks: A French case study. *Public Administration*, 79(1), 167–184.

Lin, N. (1999). Building a network theory of social capital. *Connections*, 22(1), 28–51.

Marsden, P. V. (2005). Recent developments in network measurement. In P. J. Carrington, J. Scott, & S. Wasserman (Eds.), *Models and methods in social network analysis* (pp. 8–30). New York: Cambridge University Press.

Mayhew, B. H. (1980). Structuralism versus individualism: Part 1, shadowboxing in the dark. *Social Forces*, 59(2), 335–375.

McAnulla, S. (2002). Structure and agency. In D. Marsh & G. Stoker (Eds.), *Theory and methods in political science* (2nd ed.). (pp. 271–291). Basingstoke: Palgrave Macmillan.

Meier, K. J., & O'Toole Jr., L. J. (2003). Public management and educational performance: The impact of managerial networking. *Public Administration Review*, 63(6), 689–699.

Mizruchi, M. S., & Galaskiewicz, J. (1994). Networks of interorganizational relations. In S. Wasserman & J. Galaskiewicz (Eds.), *Advances in social network analysis, research in the social and behavioural sciences* (pp. 230–253). Thousand Oaks, CA: Sage.

O'Toole Jr., L. J. (1997). Implementing public innovations in network settings. *Administration and Society*, 29(2), 115–138.

Ostrom, E. (1998). A behavioral approach to the rational choice theory of collective action: Presidential address, American Political Science Association, 1997. *The American Political Science Review*, 92(1), 1–22.

Page, S. (2004). Measuring accountability for results in interagency collaboratives. *Public Administration Review*, 64(5), 591–606.

Painter, C., Isaac-Henry, K., & Rouse, J. (1997). Local authorities and non-elected agencies: Strategic responses and organizational networks. *Public Administration*, 75(2), 225–245.

Papadopoulos, Y. (2007). Problems of democratic accountability in network and multilevel governance. *European Law Journal*, 13(4), 469–486.

Pierre, J. (2005). Comparative urban governance, uncovering complex causalities. *Urban Affairs Review*, 40(4), 446–462.

Pollitt, C. (2003). *The essential public manager*. Maidenhead and Philadelphia: Open University Press/McGraw Hill.

Provan, K. G., & Kenis, P. (2008). Modes of network governance: Structure, management, and effectiveness. *Journal of Public Administration Research and Theory*, 18(2), 229–252.

Provan, K. G., & Milward, H. B. (1995). A preliminary theory of network effectiveness: A comparative study of four community mental health systems. *Administrative Science Quarterly*, 40(1), 1–33.

Provan, K. G., & Milward, H. B. (2001). Do networks really work? A framework for evaluating public-sector organizational networks. *Public Administration Review*, 61(4), 414–423.

Rhodes, R. A. W. (1997). *Understanding governance: Policy networks, governance, reflexivity and accountability*. Buckingham: Open University Press.

Rhodes, R. A. W. (1999). *Control and power in local-central government relations* (2nd ed.). Aldershot: Ashgate.

Richards, D. (2008). *New labour and the civil service: Reconstituting the Westminster Model*. Basingstoke: Palgrave Macmillan.

Rouse, J., & Smith, G. (2002). Evaluating new labour's accountability reforms. In M. Powell (Ed.), *Evaluating new labour's welfare reforms* (pp. 39–60). Bristol: The Policy Press.

Scharpf, F. W. (1978). Interorganizational policy studies: Issues, concepts and perspectives. In K. I. Hanf & F. W. Scharpf (Eds.), *Interorganizational policy making: Limits to coordination and central control* (pp. 345–370). Beverly Hills, CA: Sage.

Scott, J. (1991). *Social network analysis: A handbook*. London: Sage.

Skelcher, C., & Sullivan, H. (2008). Theory-driven approaches to analysing collaborative performance. *Public Management Review*, 10(6), 751–771.

Skelcher, C., McCabe, A., Lowndes, V., & Nanton, P. (1996). *Community networks in urban regeneration*. Bristol: The Policy Press in Association with Joseph Rowntree Foundation.

Sørensen, E., & Torfing, J. (2008). Introduction: Governance networks research: Towards a second generation. In E. Sørensen & J. Torfing (Eds.), *Theories of democratic network governance* (pp. 1–21). Basingstoke: Palgrave Macmillan.

Stahl, G. O. (1960). More on the network of authority. *Public Administration Review*, 20(1), 35–37.

Stoker, G. (2004). *Transforming local governance*. Basingstoke: Palgrave Macmillan.

Sydow, J. and Milward, H. B. (2003, June 27). Reviewing the evaluation perspective: On criteria, occasions, procedures, and practices. In *Papier präsentiert auf der 10th International Conference on Multi-Organisational Partnerships, Alliances and Networks*. University of Strathclyde, Glasgow.

Taylor, M. (2000a). *Top down meets bottom up: Neighbourhood management*. York: Joseph Rowntree Foundation.

Taylor, M. (2000b). Communities in the lead: Power, organisational capacity and social capital. *Urban Studies*, 37(5–6), 1019–1035.

Taylor, M. (2003). *Public policy in the community*. Basingstoke: Palgrave Macmillan.

Van den Brink, M., & Benschop, Y. (2014). Gender in academic networking: The role of gatekeepers in professorial recruitment. *Journal of Management Studies*, 51(3), 460–492.

Warren, R. L. (1967). The interorganizational field as a focus for investigation. *Administrative Science Quarterly*, *12*(3), 396–419.

Wasserman, S., & Faust, K. (1994). *Social network analysis: Methods and applications*. Cambridge: Cambridge University Press.

Williamson, O. E. (1979). Transaction-cost economics: The governance of contractual relations. *Journal of Law and Economics*, *22*(2), 233–261.

Wilson, D. (2003). Unravelling control Freakery: Redefining central-local government relations. *British Journal of Politics and International Relations*, *5*(3), 317–346.

Wilson, D. (2004). New patterns of central-local government relations. In G. Stoker & D. Wilson (Eds.), *British local government into the 21st century* (pp. 9–24). Basingstoke: Palgrave Macmillan.

Wolman, H., & Page, E. (2002). Policy transfer among local governments: An information-theory approach. *Governance: An International Journal of Policy, Administration, and Institutions*, *15*(4), 477–501.

Wong, S. S., & Boh, W. F. (2010). Leveraging the ties of others to build a reputation for trustworthiness among peers. *Academy of Management Journal*, *53*(1), 129–148.

Yanow, D. (2004). Translating local knowledge at organizational peripheries. *British Journal of Management*, *15*(S1), S9–S25.

CHAPTER 3

Unemployment Policy Context

During Britain's economic boom years (1997–2008), financial markets and house prices soared and the unemployment rate fell to 4.8 %. The Labour government at that time attributed their unemployment policy success to the implementation of a National Minimum Wage, the New Deal scheme and their introduction of Working Tax Credits (Gregg and Wadsworth 2011). However, despite these statistics and network interventions, neighbourhood unemployment continued to rise in certain local authorities during this period. Even the Labour government's Social Exclusion Unit report, *Jobs and Enterprise in Deprived Areas* (SEU 2004), reported on a lack of progress in closing the gap between the best and the worst-affected areas:

> Looking at wards, eight out of 10 areas saw worklessness fall from 1998–2001. But levels of relative inequality between the best and worst-off places grew during this time as wards with the highest levels of worklessness saw the smallest improvements. (SEU 2004: 34)

Moreover, the report stated that nine out of ten neighbourhoods in the worst 10 % for worklessness in 2001 'were also in the worst 10 % in 2003' (SEU 2004: 34). This book draws on case studies in some of the 'worst 10 %' wards and tracks the rise in worklessness since 2004. A recent study of worklessness suggests the private sector cannot generate enough jobs to replace the public sector jobs axed during the last recession.[1] Moreover, the next decade will only produce 'a marginal narrowing in the gap between the best and worst parts of the country' (Beatty et al. 2011: 23). In response to area imbalance, networks in the mixed economy support two divergent capitalist approaches: neoclassical or Keynesian (see Hasluck 1987). The former subsidises free-market conditions for growth and coordinates competition amongst capitalists to progress technological advances. The latter tries to counter inequitable growth and welfare dependencies by bettering local people's life chances through skill enhancement or better job-search information. However, networks in places with weak local economies or specialised industries may be limited to supporting the flow out of unemployment without the help of employment reforms, such as by trading social goods where these needs are not met in the marketplace. The purpose of this chapter is to keep in mind the empirical puzzle and consider whether networks can realistically address the wide range of problems in the unemployment policy field. The first section explains the causes of unemployment and nine unemployment dimensions requiring network responses. The second surveys UK unemployment policy, from macro-economic adjustments, EU influence and employment-boosting schemes, to the barriers to full employment. The third section considers how roles and responsibilities in the unemployment policy field constrain network performance. The chapter's conclusion suggests that without alternative policy directions networks will continue to repackage the same or similar policies and struggle to alleviate unemployment in neighbourhoods. As Hasluck (1987: 17–18) suggests, local labour market processes are not well understood, and require a theoretical framework for analysing the operations and consequences of local employment initiatives. This chapter explains the ongoing problems that a framework would need to address. A short introduction of the basic definitions follows.

Defining and Measuring Unemployment

Unemployment refers to the state of being involuntarily or voluntarily jobless. Involuntary unemployment occurs when people who are willing and able to work cannot find a job. High unemployment in countries, regions or neighbourhoods suggests economic inefficiency and social mismanagement. This is especially so in cases where capitalist economies use

---

[1] A recession refers to two or more consecutive quarters of negative growth. The 2008 recession lasted for six quarters.

unemployment as a control factor over profits, inflation and wage bargaining. Some voluntarily unemployed people are neither seeking work nor eligible to claim benefits. Others are waiting for jobs, or have stopped applying for jobs and have become discouraged; some are unpaid carers with family responsibilities, are incarcerated, or have retired early.

Measures of unemployment began in the early twentieth century during the Great Depression, but were never exact since the records included only those who paid national insurance. Even today, the full scale of unemployment is unknown since official figures only count people claiming unemployment-related benefits (Green 1998: 96).[2] Separate figures count the economically inactive, such as students, or people on incapacity benefits (IB; the Employment and Support Alliance or ESA has been gradually replacing IB since October 2008) who are not in work as a consequence of ill health or severe disability. Decreasing unemployment rates is not a measure of economic well-being if numbers of casual or part-time 'underemployed' workers who want to work full-time increase (see Appendix 1). Likewise, an increase in people registering as self-employed may not reflect an enterprise boom, as for some it may be a means of surviving on a low income, acquiring credit or receiving 'in-work' benefits, thereby avoiding the stigma of unemployment if permanent positions are unavailable.[3] Making unemployment benefits difficult to obtain also masks the real situation; indeed, since 2008 the number of people experiencing hardship has risen (DWP 2013a: 23). Employers may also stipulate low pay and of short hours to avoid the threshold of auto-enrolment in pensions. Under the coalition government, benefit sanctions escalated following new rules implemented in October 2012. These also apply to people who refuse zero hours contracts; hence, the recent increase in 'disposable' staff with involuntary insecure temporary contracts (working for ad hoc hours,

---

[2] A government definition of an unemployed person is someone of working age looking for a job and in receipt of the Jobseeker's Allowance. However, the actual unemployment level is estimated to be far higher than the unemployment rate. Likewise, it is difficult to make country comparisons of unemployment rates as measurement criteria differ between statistical institutions: The International Labour Organization, The OECD, and The Statistical Office of the European Communities (Eurostat).

[3] Between 2008 and 2015 self-employment increased by 730,000, from 3.8 million to 4.5 million; hence, the rising rate of employment is due, in part, to the increase in those declaring themselves self-employed (ONS 2016: 5, 45). The trend is unabated, as by June 2016 4.7 million people are self-employed (ONS 2016a). Moreover, average incomes from self-employment fell by 22 % between 2008/2009 and 2012/2013.

not jobs), reliant on 'in-work' benefits due to low rates of pay. Between December 2012 and December 2013, the decision to apply a Jobseeker's Allowance benefit sanction was made 1.03 million times; initially, it was not disclosed to the recipients whether these sanctions would be applied beyond the minimum 4-week sanction (DWP 2014: 13). In 2014, parliamentarians debated the physical and mental suffering of benefit recipients wrongly sanctioned due to poor communication of the rules, lack of common-sense reasoning, perverse decision-making and serious administrative incompetence. The opposition party feared that Department for Work and Pensions (DWP) staff would apply trivial sanctions in order to meet their own performance targets, or to mete out punishment (Hansard HC Deb. 2014: col.1056–82). A Citizens Advice Bureau (CAB) press release stated that the culture of 'sanction first and ask questions later' was devastating lives, while appeals waste public money (the cost of these to date is currently £7 million since October 2012) and lead to delays in people receiving the benefits to which they are entitled (CAB 2014). Moreover, time spent on tribunals, finding food to survive, dealing with rent arrears or the threat of homelessness distracts people from job-hunting and pushes them into debt and health problems. The benefit sanction appeals procedure has since been reformed and people can submit early evidence to counter benefit sanctions. However, sanctions are still causing havoc and local network responses are uneven or indifferent.

False reporting is also commonplace, such that error or spin can 'massage the message'. Esther McVey, Minister of Employment claimed in a statement made to the House of Commons on 14 October 2013 that, since the general election in 2010, 'unemployment is down by 400,000' (Hansard HC Deb. 2013: col.427 and 431). After a subsequent complaint by a member of the public that this claim was inaccurate, the UK Statistics Authority investigated it and in a letter dated 17 March 2014, stated that the figure for this reduction published by the National Office for Statistics was nearer 7000 (see UKSA 2014). McVey (Hansard HC Deb, 2013: col.416) also highlighted the national drop in people aged from 16 to 24 years of age who are not in employment, education or training (NEETs), although MacInnes et al. (2013: 66–7) claim the number of NEETs who are care leavers has risen sharply. From October to December 2015, the number of NEETS (aged from 16 to 24) increased by 5,000 from the previous quarter to 853,000 (ONS 2016b). As Wolf (2011: 38–9) points out, NEETs churn in and out of work and struggle to find sustainable employment. In other words, the presentation of unemployment-related facts and statistics requires nuanced understanding.

Causes of Unemployment

For centuries, societies have created the problem of unemployment for themselves and have then found it difficult to resolve. Unemployment is a choice for a country, not a binding option. Its causation has been attributed to: the misuse of human potential; lack of motivation to find work; bad harvests; rising food prices; decreased demand for goods; over-production; technological advances; distributive injustices; the demise of traditional industries; capitalist control of wages; land prices; skills mismatch; uneven geographic employment opportunities; the demands of global and domestic markets; a lack of universal standards; and society's weak response to the problem (Garraty 1978). Personal motivation is not sufficient to avoid unemployment. Other factors may include: where you live; class background; age, gender and ethnicity; your education and skills; career preference; caring responsibilities; access to networks; government policy; public infrastructure; economic climate; oil prices and energy costs; private investment and risk taking; number of vacancies in the job pool; and even bad luck or timing. Hence, the capacity of networks to improve the unemployment situation requires realistic consideration, since high national unemployment is symptomatic of periodic general (Shaikh 1978). The 'great recession' in 2008–9 and the Eurozone crisis, for example, aggravated Europe's high unemployment rate following the collapse of financial institutional networks that had been handling inflated asset prices, unregulated securities and derivatives, and had also been supplying poor people in EU countries and the USA with credit and mortgages they were unable to repay. The high indebtedness of households slows growth, as reduced savings and expenditure impact on business investments and market confidence. Bell and Hindmoor (2009: 82) argue that the 'failure of governments across the world to anticipate and prevent the crisis from occurring ... reflected a failure of political will rather than governing ability'. For example, during the financial crisis, the UK government intervened in the deregulated globalised financial market by using public funds to salvage and nationalise several major banks and preserve their remuneration and bonus structure despite their poor performance. Recently, public debt has distracted attention from the problems experienced within the private sector and domestic market.

Public spending increased under Labour; thereafter, the coalition government imposed public spending cuts. Fiscal austerity measures, lack of domestic credit flow and investment in public services, poor industrial growth and low exports can prolong periods of high unemployment. By late 2011, Britain had 2.70 million unemployed people, with 463,000 job vacancies and rising numbers of long-term unemployed.[4] By late-2016, unemployment fell to 1.60 million, but long-term sickness remains over 2 million. Job vacancies decreased from 763,000 (Nov-Jan) to 741,000 (May-Jul) (ONS 2016a). Many vacancies advertised are normal job churn, or the underemployed and involuntary part-time workers are ready to fill them (Bell and Blanchflower 2013). Anecdotal reports suggest some Jobcentres advertise job scams, 'commission only' sales positions, or jobs replicated with 'zero-hour' contracts. 'Flexible jobs' designed to meet employers' needs are not necessarily family-friendly, and may lead to the exploitation of young people and those with few qualifications in the labour market. From June 2014, a statutory procedure gave employees the right to request flexible working hours; however, this right is not absolute and decision outcomes are based on the business's needs. In the period September to November 2011, youth unemployment (aged 16–24) peaked at 22.5 %, and fell to 13.6 % in February to April 2016 (ONS 2016c); however, the rate began increasing before the 2008/2009 'credit crunch' recession, during which time job-centred policies were lacking (WPC 2012: 61; Mourshed et al. 2014: 16; HM Treasury: 2014: 55). In the EU-28 countries, youth unemployment decreased in January 2015, leaving 4.889 million young people (under 25) unemployed, including 3.281 million in the Eurozone area (Eurostat 2015). But job quality decreased, and those at risk of poverty and social exclusion increased in one-third of EU member-states, including Britain. For instance, 'the stepping-stone function of temporary employment (whereby workers on temporary contracts move up to a permanent contract) has diminished notably' (EC 2014: 5). Yet economic downturns aside, neighbourhood unemployment is a long-standing problem.

Unemployment Dimensions: Past and Present

At least nine unemployment dimensions require a network response: historic, type of unemployment, economic, social, political, institutional, neighbourhood and spacial, individual, and group (see Table 3.1). Yet a single network could not coordinate them all, and several dimensions relate to 'social exclusion' (Percy-Smith 2000: 8–11).[5] The remainder of this section considers each dimension in turn. It begins with a brief historical account of unemployment under the capitalist system; however, the literature review is selective due to space limitations.[6]

---

[4] Long-term unemployment refers to the number of people unemployed for one year or more.
[5] See debates and reservations on the social exclusion/inclusion agenda in Davies (2005).
[6] This section draws on the history of unemployment literature (see Garraty 1978), but omits the history of capitalist development, from the structural conditions framed in medieval Europe, to twentieth-century perspectives, including feminist, eco-environmental, and Gramscian.

Table 3.1 Unemployment-related themes affecting network functioning

Unemployment dimension	Interpretation
Historic	Poor record of unemployed representation
	Problem-solving based on blaming unemployed people, rather than a problem of industry.
Type of unemployment	Frictional/seasonal/structural/cyclical/long-term/technological.
Economic	National and international trends
	National inflation/growth/balance of payments
	Economic theories disputed
	Recessions (fiscal and monetary responses)
	Industrial relations/stakeholder involvement
	Business conditions, growth, competition, regulation, planning, investment.
	Long-term unemployment
	Local economic development/job creation
Social	Social exclusion (socio-economic disadvantage)
	Multiple cross-cutting issues
	Skills under-utilised/barriers
	Low income
	Weak social networks
Political	National and international trends
	Attitudes towards welfare support
	Perceptions/representation of unemployment
	Pro-poor policies/job creation/job guarantee
Institutional	NPM and privatisation goals
	Rules and activation strategies
	Public Employment Services
	Funding streams and information flow
	Inflexible short fixed-term contracts
Neighbourhood and spatial	Historic and cultural factors
	Demographics/area reputation
	Local labour market
	Skills mismatch
	Housing/poor local services
	Transportation
	Regeneration and local governance
	Concentrations of at risk groups
Individual	Educational achievement and skills level
	Low self-esteem/confidence
	Mental and physical ill health
	Welfare benefits, debt counselling, needs assessment
	Access to jobs (quality, pay and durability) and training information and advice
Group	Cohort discrimination: youth, disabled, gender, lone parents, family obligations, carers, ethnic minorities, graduates and older people

The Historical Dimension

Understanding network performance requires an awareness of how past attitudes may influence the future and close off positive resolution. Arguably, four narratives of economic thinking have recurred in a historical trajectory and influence present-day networks. First, societies (pre-capitalist and capitalist) were not equally affected by the causes and symptoms of work loss nor were resolutions shared equally. Second, governing networks have maintained the status quo as a social control device to prevent unrest, rather than address the structural weaknesses of the capitalist system. Third, networks have experienced difficulties in implementing, overseeing and administrating work schemes and welfare distribution for centuries. Fourth, public elites and business fraternity have blamed unemployed people for not finding work, even during periods when jobs were scarce. Notably, high morals, namely upward and downward social comparisons imposed on human struggle, still resounds in policymaking today, keeps the realities of unemployment out of the discourse, and runs through the brief historical outline to follow.

Classical Economics

The 'classical economists', a term coined by Karl Marx to refer to eighteenth- and early nineteenth-century production theorists (Ricardo, Smith, Malthus, Say and Mill) had little conception of unemployment. They believed labour was plentiful, overpopulation was the cause of working-class poverty and that 'idleness was a personal not a social problem, and that hunger would compel everyone to work' (Garraty 1978: 73). Further, they viewed sporadic work or stoppage as a temporary self-correcting condition caused by over-production, and by the fixed and circulating capital that disrupted trade when consumers' interests changed. However, Adam Smith's classic work, *Inquiry into the Nature and Causes of the Wealth of Nations* (1776), does not purport 'free markets' to mean unfettered capitalism, as modern neoclassical economists assume (see Tabb 1999: 44–52). For example, Smith's economic sentiments go beyond labour as a commodity. He distrusted market monopolies and large transnational corporations, the opulence of religious institutions, power, and privilege in government. Rather, he advocated a clear role for government as the supplier of public provision; from education for all, to building and maintaining transport infrastructure. Being class-conscious, Smith describes the plight of overstretched working-class labourers and apprentices in terms of their lack of promotion. He also objected to trade guilds meeting to lower labourers' wages, and favoured domestic local economies over international free trade. Özler (2012) suggests Smith's fixation with market independence was a defence strategy to counter feelings of dependency towards his mother and benefactors, and a longing for his father who had died before he was born. As such: 'In the denial of the reality of his personal dependence, he conceived of market as institutions that would lead to independence and hence to prosperity' (2012: 27).

Nevertheless, abuses of power and wealth were long-standing. The division of labour into specialisations and development of new technologies increased poverty, and subsistence measures, financed through a local rates levy for cash, goods or clothing were insensitive to the working class. Parish boards implemented the 'Act for the Relief of the Poor' (1601) on an ad hoc basis, depending on parochial attitudes towards the 'deserving' and 'un-deserving' poor. Individuals in possession of a parish settlement had the legal right to relief, but this inhibited the free movement of labour; however, landowners often employed casual labour from neighbouring parishes without having to increase the parish rates. Between 1750 and 1850, land enclosures intensified for sheep rearing, and the open-field system of small-scale farming was abolished. By late 1794, work poverty and seasonal unemployment were widespread. Food prices increased due to poor harvests and overseas wars disrupted trade on a regular basis, especially for cloth workers. In 1795, Justices of the Peace in the Berkshire parish of Speenhamland met with the initial aim of setting a minimum wage; instead, they subsidised low wages from the rates, relative to family size or the price of bread. Variants of the system were used by employers in other counties to sustain low wages and temporary workers. This still resonates today in welfare provision and the tax credit system for people on low incomes (Michie and Wilkinson 1994: 24; Grover and Stewart 2002). Smaller ratepayers (self-employed or reliant on family labour), who subsidised the low-waged landless labourers employed by rich landowners, became poor at the expense of their rich neighbours, and were forced to seek work in cities.

In 1815, following the end of the Napoleonic Wars, poor relief payments reached nearly 1 million, and a period of unrest and protest ensued (Mann 2012: 521). Victorian reformers (inspired by Bentham's

Utilitarian doctrine of social planning, moral self-righteousness, and self-improvement through hard work, and material self-interest) believed poor relief undermined the work ethic; they made a virtue of lower- and middle-class aspirations for social prosperity (Girling 1997: 122–33). Consequently, the Poor Law Amendment Act of 1834 replaced the parochial Speenhamland system of outdoor relief (direct payments to poor families) with centralised administration of indoor relief (payments to workhouses) as a deterrent and test of destitution. Welfare administration reforms, from the appointment of parish overseers to the introduction of a London-based Poor Law Commission and the involvement of state bureaucrats, led to popular protests (Somers and Block 2005: 267). Despite perceptions of citizens abusing outdoor relief (such as intentionally increasing family size to increase relief payments), the potential for corruption in the management of indoor relief contracts was immense. Local business providers kept inadequate records and audits performed by the 'Guardians of the Poor' lacked rigour.[7] Moreover, forcing large numbers of frequently unemployed labourers to work for aid in workhouses or in counter-cyclical parish-based employment schemes was futile, as such methods could not occupy people with varying physical strength and skills at short notice or produce goods to compete with private industry in case it created unemployment elsewhere. Nevertheless, optimism for capitalism and laissez-faire markets grew under the pretext that men and women would flourish through enterprise in a democratic labour market and counter a government ruled by corrupt amateurs and aristocratic landowning elites. Themes of liberty, property and equality inspired support for capitalism; however, the commoditisation of social relations instigated by the Industrial Revolution, and the impact of three recessions on poor people in the nineteenth century, stirred interest in the social economy. Consequently, thinkers began to view competition and wealth inequality problematic.

Socialist/Marxian Economics
The ideas of socialist theorists, utopian socialists and social reformers evolved during the early nineteenth century as a response to the exploitation of economic growth by capitalists and the government, and to the erosion of

---

[7] A board of guardians, comprising town dignitaries and politicians, administrated the system of poor relief through central funds and local taxation.

compassion towards the working classes. For example, Robert Owen (1771–1858), a Welsh pioneer of mutual-benefit societies, advocated working for each other through social planning and cooperation rather than competition. However, his idealist self-sufficient community in America, known as 'New Harmony', failed in 1827 due to disputes over property, the clash of group interests versus those of individuals, and poor management. Henry George (1839–1897), an American economist, claimed that rural landowners forced people off the land and into cities, but that high land prices in cities left little for wage distribution. Therefore he blamed unemployment on landownership in *Progress and Poverty* (1879), and advocated a single redistributive tax. Charles Booth (1840–1916) (English businessman, social scientist and parliamentary candidate), did not consider the morality of economic systems; instead, he surveyed the life and labour of East London's poor people between 1886 and 1903, and found that 35 % of them lived in abject poverty. John A. Hobson (1858–1940), an economist, believed under-consumption and excessive savings without profitable investment caused unemployment, and proposed the funding of social provision by taxing the savings of the wealthy. In contrast, Karl Marx (1818–1883), the socialist economist author of *Das Kapital* (1867), believed capitalism's cyclic 'success' depended on the existence of a reserve of involuntary unemployed people to stabilise wages in depressions when demand dropped, and be exploited in boom periods. People do not revolt, as some Marxists would argue, because the subordinated class hope for upward mobility through hard work, and their false consciousness cannot recognise the real economic situation. Capitalism improved living standards for multitudes, but tensions remained between social institutions (legal, welfare, and cultural) that governed norms (such as money and ownership) and the wealth industries that created the best conditions in which to accumulate profits. Rather, two cultures pull apart within economics: one rooted in moral philosophy, such as that of John Stuart Mill (1806–1873); and one rooted in science-based methods that views unemployment as a result of insufficient job-searching, physical fitness and individual character, rather than as a symptom of industrial and class oppression.

Structural Approaches/Employment Initiatives/Keynesian Economics

Public employment services became central to unemployment policy, offering a job-matching function for employers, and job-search assistance for prospective employees. Herbert Asquith's Liberal government (1908–15) established labour exchanges in 1909, following William

Beveridge's recommendations to Winston Churchill, the Home Secretary at the time. Compulsory national health insurance and unemployment benefit for wage earners in certain industries followed in 1911. A network of labour colonies, work camps, training farms and seasonal work systems was also established (see Field 2009). Socialist reformers Sidney and Beatrice Webb proposed a 10-year public works plan to be activated during an economic depression caused by business cycles, or held back during periods of buoyancy; nevertheless, government rejected relief programmes. However, during the First World War (1914–18), government controlled agriculture, railways and the iron and steel industries, and made industrialists regulate production and wages. In the post-war period, the free market economy returned and thousands of unemployed ex-veterans demanded work or maintenance, but their hopes would soon be dashed, as from 1921 onwards, a series of American bank failures caused an increasing economic crisis. By 1929, a global market depression had taken hold, and with rising unemployment overseas, the Conservative government (1924–29) halted the emigrant training programme supporting unemployed people to seek agricultural work in Australia or Canada. With the next general election looming (5 June 1929), the Inter-Departmental Committee of Officials on Unemployment' examined the opposition party's proposals for reducing unemployment. These included The British Liberal Party's election manifesto pamphlet, 'We Can Conquer Unemployment', and the 'Melchett-Tillett report' on industrial reorganisation and industrial relations (both published March 1929), (see Cabinet 1929). While government ministries recognised the limited prospects for men and women finding work in depressed areas, the Treasury did not endorse state-assisted intervention or relief works (Garraty 1978: 207). Instead, they confined solutions to training, the creation of Juvenile Unemployment Centres, migration and the market, as 'It is to the normal working of ordinary industry that the employable population must look for livelihood' (Cabinet 1929: 108). The second Labour Government 1929–31, establishes national central policies and county councils; whereby the 'Local Government Act 1929' replaces the 1834 Poor Law and the old poor law board of guardians who had administered poor relief in England and Wales. By 1932, UK unemployment was estimated at 3.4 million, with one in six out of work, and one man in four having been unemployed for more than a year. Yet, gross domestic product (GDP) was in recovery and employment participation increasing. However, working-age population was also increasing. But political followers of various schools of economic thought, did not attribute interwar unemployment to the economic systems cyclical, structural and regional incapacity to absorb surplus labour. Rather, government imposed welfare cuts and sanctions on people searching for non-existent jobs (see policy background in Garside 2002). Field (2009) estimates 200,000 unemployed men received training in the Ministry of Labour work camps from 1929 to just before the start of the Second World War; attendance at compulsory 'instructional centres' for unemployed men became voluntary from 1932, but mandatory again following the 1934 Unemployment Assistance Act, which centralised and separated the control of insurance benefits from the 'dole'. This popular shorthand for a small unemployment allowance originates from the medieval religious practice of doling out charitable gifts to unemployed people; for example, a 'dole of bread'. Like the workhouse test of 1834, training was a 'work test', and those who declined to participate risked losing their 'dole'. (The three-month training programmes provided some basic adult education, and work (from road-building and land drainage to forestry) to 'recondition' the bodies of long-term-unemployed young working-class men from 'distressed areas', who were undernourished and deemed too 'soft' for heavy manual work). However, many perceived the Labour Party, trade unions and trade councils as offering little support to unemployed people, as the training initiatives were for non-existent jobs. The National Unemployment Workers' Movement, organised by members of the Communist Party of Great Britain, likened the training provision to 'slave camps', and coordinated national protests against pitiful unemployment allowances and severe means testing.

The American response to unemployment was far-reaching. In 1933, President Franklin Roosevelt combatted unemployment using an emergency public works programme called the 'New Deal'. This led to the American government's prioritisation of meeting its 'social welfare/right-to-work' obligations, rather than balancing the federal budget (see Harvey 2013: 40–4; 2014: 146–79). Roosevelt's concern was to occupy millions of previously unemployed skilled and unskilled people in direct government jobs, with the aim of transforming the American landscape. This scheme continued into the early 1940s and its major programmes occupied manual workers in labour-intensive civic works in transport infrastructure (road-building, street works, airport construction, bridges, house-building); public buildings and improvements (hospitals, parks, schools, playgrounds, swimming pools); public services (utility systems, dams, sewage). Further, teachers, researchers, nurses and librarians provided essential goods and services, while artists, theatres and musicians spread cultural democracy (2014: 149–57; Garraty 1978: 206). Unemployment local councils and social movements were established, but growing interest in communism created paranoia amongst opposition leaders who discredited the New Deal as anti-capitalist, communist and even fascist. However, Roosevelt claimed the New Deal went beyond political labels and was a simple, practical response to an overwhelming need. Campaigns to eradicate communist thinkers from the American profile arose once more in the 1950s, when thousands of workers, government officials and people in the entertainment industry lost their jobs and careers on the basis that they were communist sympathisers.

In the 1930s, the economist John Maynard Keynes, following Hobson and others, theorised that thrift led to unemployment. Keynes' (1936) seminal book, *The General Theory of Employment, Interest and Money*, advised capitalist governments to balance the economy during periods of economic depression by creating a multiplier effect, as easy credit and borrowing money to finance public works programmes, suppliers and workers would make up for economic shortfalls in demand, expand output and repay the budget deficit. Critics of the Keynesian approach believed overproduction and a surplus of goods would create a drop in demand and lead to job losses and inflation. Instead, they advocated monetary policies such as high interest rates to stimulate investment, and reduced wages to depress prices and stimulate spending. Moreover, although Taylorism, Fordism and regulation theories advocated mass consumption to counter market instability, and stabilise employment, when demand for labour fell, supply chains rarely shared losses and 'union proposals for dealing with unemployment were primarily self-protective' (Garraty 1978: 190–1). Moreover, Keynesian theories were designed to apply to a patriarchal society based on stable families and the subservience and domestication of women, in which unemployment was considered a male problem.

Neo-Liberal Economics/Conservatism

Unemployment remained at levels between 10 and 20 % until 1939 and the outbreak of World War II, at which point many women entered the workforce for the first time, undertaking manual work reserved previously for men. Following the 1944 White Paper *Employment Policy*, a Labour government (1945–51) was elected on the promise of full employment. The government's use of Keynesian approaches, alongside informal industrial and trade union relations, managed unemployment at acceptable levels until the late 1950s; thereafter, international economic pressures accelerated economic decline, including the breakdown of the Bretton-Woods exchange rate system and the OPEC oil-price hike of 1973–75. Further, national production trends were changing and a lack of competitiveness may have undermined labour markets. Thus, British industry embarked on restructuring and rationalisation, with spatial consequences on the towns and cities where firms were concentrated (Hasluck 1987: 83–91).

The Labour Party split over incomes policy and trade union disputes fuelled political unrest in the late 1970s. Failure to improve industrial relations and economic planning assisted Margaret Thatcher's Conservative party to sweep into power promoting neo-liberal policies, private sector 'partnership' and enterprise. The Thatcher government (1979–90), influenced by American politics and 'New Right' intellectuals, Milton Friedman and Friedrich Hayek, believed that Keynesianism, social engineering, collectivism and EU social policy directives interfered with labour market regulation (Richards and Smith 2002: 94–5, see Box 5.1). In contrast, a deregulated 'free market' economy and a small rise in unemployment would lower inflation, control borrowing, spending and interest rates, and restrain wages and trade union powers. Moreover, 'rolling back' the state to retrench welfare and enterprise public services would lead to economic success (Taylor-Gooby and Daguerre 2002: 46). This might be called a form of privatising Keynesian policy, as Crouch (2008) suggests. Thatcher's government was prepared to let go of the coal industries, for example, and live with the resulting mass unemployment (Judge and Dickson 1991: 27). Thus Therborn (1986: 134) claims Britain to be the 'most clear-cut case where unemployment had a directly political cause'.

In 1979–80, the second oil shock doubled prices and was followed by a global economic downturn, and the self-adjusting market theory began to lose its explanatory power. High unemployment also persisted in many European countries, hence the European community targeted employment policy, and structural funding for employment programmes in economically disadvantaged groups and regions (Goetschy 1999).

A UK recession in 1981 led to high unemployment, a property slump and, later, to the Poll Tax protests. In response, the Conservatives sold off nationalised industries to reduce state borrowing and reduced unemployment benefits, thereby creating a homelessness problem. Further, cheap labour programmes absorbed hundreds of thousands of long-term unemployed people. The first, in 1982, was the 'Community Programme' of public works, which occupied mostly white (94 %) males (70 %) in a 'netherworld of pretend jobs and training' according to Harris in *The Guardian* (26 April 2006), and Hasluck (1987: 156–8). A second, 'The Enterprise Allowance Scheme', a welfare-through-enterprise programme intended to kick-start an 'enterprise culture', had by 1983 provided 325,000 unemployed people with a business start-up allowance for up to a year. By 1985, unemployment had reached 3.3 million. In 1988, the Education Act stressed the importance of skills for work, and replaced the Manpower Services Commission (MSC) (established in 1973 and involving the Trades Union Congress, Confederation of British Industry, local government representatives and educationalists) with the Training and Enterprise Council (TEC) an employee-led non-elected quango (Ramsden et al. 2007). Differences within the MSC, accountabilities to several departments and special schemes had strained consensus; moreover, 'individual trade unions, youth organisations, and voluntary associations concerned with the issue were, in effect outside the network' (Marsh 1992: 178–9). TECs eroded local governance further, by undertaking community consultations aside from local government (Painter et al. 1997: 228, 237).

By the late 1980s, unemployment was linked to underinvestment and a decline in the engineering sectors, from metal and vehicle manufacturing to chemicals (Rowthorn 1991: 52; Hasluck 1987: 74–9).[8] In 1990, Britain joined the European Exchange Rate Mechanism at an overvalued exchange rate, forcing British firms to reduce costs, and working conditions eroded. Michael Heseltine at the Department of Trade and Industry (DTI), believed underinvestment in business sector training stalled growth, therefore in 1993, Business Link (BL) established business support, delivered by local advice centres in partnership with TECs, local authorities, chambers of commerce and local enterprise agencies. However, British business associations had historically been autonomous from the state, and reforms to this were contentious. State intrusion also increased, in the forms of means testing for welfare benefits and sanctions (Taylor-Gooby and Daguerre 2002: 42). Thereafter, the 'neo-liberal economic discourse' of New Labour governments from 1997 (Fairclough 2000: 63) has attacked welfare recipients, as have those of subsequent governments.

[8] Between 1955 and 1983, Britain had the 'greatest percentage decline of manufacturing employment of any Western country' (see a country comparison in Rowthorn (1991: 38), and Table 2.2: 39).

Contemporary Perceptions: Unemployment Blaming

In recent decades, unemployment understanding has corresponded to: (i) institutional responses; (ii) measures and statistics; and (iii) language reinforcing attitudes. Governments try to skewer unemployment statistics (Moore et al. 1989: 4–7). This is done by, for example: cutting benefit eligibility; emphasising numbers of employed people or raising the school-leaving age; forcing foreign workers to leave, or moving people onto 'welfare-to-work' programmes; unpaid work experience; self-employment schemes, or other benefit streams. In 2014, the UK Statistics Authority (UKSA) cautioned the Department for Work and Pensions (DWP) for using weak numerical evidence in a press statement claiming causal links between benefit capping and moving people into work. Another blaming approach emphasises 'the *unemployability* of the jobless', citing their lack of effort in job-searching, their need for skills improvement and displaying a lack of realism about the jobs and salaries available to them (Kitson et al. 2000: 632, italics in the original; Mitchell and Muysken 2008: 4). Politicians, popular opinion, the media and even academia assume that welfare encourages unemployment duration and free-riders, while the 'moral hazard' concept stigmatises the unemployment subsistence process and legitimises welfare deterrents; meanwhile, decreasing consumption and demand increases poverty, and stereotyping leads to 'labelling, status loss, and discrimination' (Link and Phelan 2001: 380).[9] Scapegoating of unemployed people makes society the innocent victim, protects public figures from feelings of ineptness, and is easier than tackling the labour structure (see Douglas 1995: 55–67; Somers and Block 2005).

Types of Unemployment

Different types of unemployment require different network responses. *Cyclical or general unemployment* is a phenomenon linked to boom-and-bust patterns in the business cycle, whereby a percentage of the workforce bears all the losses, especially new entrants and older workers. Citizens rarely support work-sharing options, wage fluctuations or wage insurance, as they fear exploitation, confrontation over profit margins or higher taxes

---

[9] For decades, unemployed people have used similar expressions to describe their experience, such as 'treated like the scum of the earth' or perceived as 'a down-and-out'. Their perceived 'outsider' stigma separates them from society (Silver 1994: 570). As the wealth gap grows, society needs more welfare to maintain sociability and the poor carry the blame (Spicker 1988: 39).

(Garraty 1978: 261–2). *Frictional unemployment* occurs when people are between jobs or entering the workforce after leaving college; however, when unemployment is high, people try to retain their jobs or stay in education. *Seasonal unemployment* relates to sporadic employment in sectors that operate at certain times of the year, particularly, tourism, agriculture and construction. *Structural unemployment* occurs in regions or localities when key sectors decline or a major industrial plant closes, leaving communities without capital or direction. People who cannot retrain, change occupation or relocate to find work become discouraged and retreat from the labour market (MacKay 1998: 54–5). *Technological unemployment* occurs when industrial production is automated to reduce costs and jobs diminish for unskilled people; hence, the paradox of 'jobless growth' (Murray and Forstater 2013). Finally, reducing *long-term unemployment* following recession and during economic recovery is difficult, as 'growth by itself is not enough'; moreover, employers often conflate all unemployed people with the long-term unemployed, and then disregard their potential even for rudimentary jobs (Hogarth et al. 2003). Networks can make clear the types of local unemployment, and challenge employers' recruitment practices, but resolution requires political will, as the following dimensions set out.

Economic Dimension

During periods of economic crisis, the Bank of England handles monetary policy and the Treasury handles fiscal intervention as it sees fit, as well as national policies to control growth, inflation and unemployment, and the balance of payments. As recessions begin, national unemployment rates rise as demand falls, and inappropriate or poorly managed policy is costly if direct and indirect tax revenues fall. When this occurs, governments have to borrow more to cover additional welfare benefit payments, but austerity measures reduce purchasing power. Regional and sub-regional policies lever economic welfare; massive corporate subsidies to boost supply and demand, coordinate economic development, attract capital and investment, stimulate local enterprise and create favourable conditions for trade and business. Ways of achieving the latter include improvement of industrial relations and working conditions, encouraging growth to support national income, levering funds to disadvantaged areas, promoting international trade, simplifying planning rules, improving access to cash flow, finance and job creation and the introduction of low corporate taxes.[10]

---

[10] Britain's corporate tax rate is currently the lowest rate in the G7 group and was the fourth lowest in the G20 in 2014.

The macro-economic effects of unemployment on the micro-economy determine expenditure and credit decisions at the household level and the situation for the local labour market and firms in low demand welfare dependent areas. National policies seek to influence local economic development and attract inward investment for the following: area, transport and telecommunication improvements, property-led regeneration, business incubators and small-business start-up loans, or financial incentives for skills training. Developing a local social economy may involve stakeholders in marginal activities, from small business start-ups through community businesses and cooperatives to credit unions. However, poorly executed partnerships between 'institutions of economic development', and continuous reforms of the required occupational qualifications and of regulation, discourage employers' engagement (Ramsden et al. 2007: 235–7; Wolf 2011: 56). As a result, the local labour market is ill-prepared for economic crises.

The Social Dimension

The social impact of unemployment perversely creates employment through the provision of intervention programmes, and sector support roles from housing, education and training policy to social services and health provision.[11] Economic failure and social structuring create unemployment, but 'few welfare state regimes ... see employment as a part of social policy' (Roller 2005: 42). The EU and governments highlight the importance of 'social inclusion', or 'participation' 'and community cohesion', yet downplay socio-economic disadvantage. Unemployment benefit does not address the self-identity and social needs provided by work; moreover, job creation rarely 'trickles down' to the long term-unemployed (MacKay 1998: 64; Crisp et al. 2014: 76). Networks have a role to play in: (i) improving access to core services and local employers; (ii) representing local issues upwards; (iii) supporting community and voluntary infrastructure and investment; (iv) strengthening people's networks through job fairs, information and learning opportunities; and (v) utilising local skills through volunteering. Civic work should not be overestimated, however, and welfare benefit rulings can oppress local network participation, for example, in Local Exchange Trading Schemes (Croall 1997: 74–7). Moreover, small social economies may even perpetuate inequalities (Hudson 2009: 50).

---

[11] The sociological literature documents the social impact of unemployment; see definitive accounts in the report, *Understanding Workless People and Communities: A Literature Review* (Ritchie et al. 2005).

The Political Dimension

Some scholars are resolute that 'unemployment policies have to be understood in essentially political terms' (Moore et al. 1989: 10–11). However, EU member states have yet to coordinate their political ideologies and policies to stem unemployment during financial crises. Centre-left policies emphasise social justice, economic and social equality, and welfare reform, while right-wing polices protect individual freedom and private enterprise (Castles and McKinlay 1979: 175). In Britain, welfare policies since Thatcher's Conservative government have shifted to the right, emphasising the individual's contract with society to get a job, education, and skills training. This is reflected in the policy of 'activating' benefit recipients to prepare for work and participate in work programmes, as advocated by global neo-liberal institutions such as the OECD (Organisation for Economic Cooperation and Development). The approach is interventionist and involves tough sanctions, the contracting-out of services to the private or voluntary sector to deliver supply-side policies, and making services for getting people job-ready and into the labour market more personalised and efficient; it often directs them towards menial and routine low-paid jobs (Considine and Lewis 2003: 48; Peck 2004: 205). This approach distracts attention from demand-side policies and labour market barriers (Beatty et al. 2009: 26).

Institutional Dimension

Institutions close to the state represent the processes for governing unemployment support and for stabilising social assistance, policy, planning, employability skills and training strategies, employment advice and funding allocation. These institutions manage the policy field and beneficiaries through their roles in overseeing advisory services, contractual management, administration and monitoring tasks. Beatty et al. (2009: 26), suggest that institutions create barriers to work – from the benefits system, housing, and childcare provision to transport facilities. Consequently, local organisations can perceive institutions as being unresponsive to clients' needs and inflexible. For example, Jobcentre Plus's (JCP) rigid business culture regards clients as a 'cost-unit' based on job outcomes and sometimes deters cooperative working. Institutional contracts, rules and regulations govern pluralism and define which policy norms providers should endorse; but the fear of non-compliance or not meeting targets creates a climate of fear and deters constructive criticism at the front-line of service delivery as if targets are not met, non-compliance means exit. Institutions structure behavior, but stakeholders require flexibility and autonomy throughout the delivery chain so as to adjust policy at design, implementation, and delivery stages (OECD 2009: 13–14).

The Neighbourhood and Spatial Dimension

Unemployment persists in areas where there are regional income and job disparities, industrial decline, geographic or transportation disadvantages, and which would require environmental improvements and large-scale jobs investment to make neighbourhoods attractive to live and conduct business in (Lawless, 1989; Beatty et al. 2009: 14). One-fifth of the people claiming 'out of work' benefits in England in May 2007 lived in the 10 % most rundown neighbourhoods (DCLG and DWP 2007: 12). Regeneration policy, however, rarely integrates with 'social regeneration' (Ginsburg 1999: 55).

Citizens in low unemployment areas enjoy better health and education and less crime, whereas those in high unemployment neighbourhoods endure multiple disadvantages (Green 1998: 106–10; UKCES 2010). Social and geographic segregation reinforce differentiation, such as by social class and lifestyle, income and ethnicity. One study compared poor neighbourhoods with socially mixed areas and suggested that 'higher levels of unemployment in the deprived areas may be related to stigmatisation and the low reputation of potential workers from these areas and low educational achievement' (Atkinson and Kintrea 2001: 2292–3). Likewise, media stereotypes of 'work-shy' neighbourhoods overlook their strong work ethic (Fletcher 2008b: 112). Residents in wards with 'concentrated worklessness' have less chance of leaving poverty, which is the reason why Labour wanted networks to coordinate approaches to deal with such issues (SEU 2004: 14–5). Nevertheless, Esther McVey, Minister of Employment between 2013 and 2015, supported the view that long-term unemployed people in Hull who could not find success through the government's Work Programme scheme should seek work elsewhere in the country (Hansard HC Deb. 2013: col. 434). Low income means low housing ownership and a social housing shortage; in some regions high rents and moving costs stop people from relocating to find work (Crowley and Cominetti 2014:10–14, 35–36). Problems compound, as neighbourhoods with poor housing investment or stock mismanagement and environmental degradation experience higher rates of crime, drug misuse and youth unemployment. Hence, unemployment is as much a matter of social geography as it is a supply or demand problem.

The Individual Dimension

McLaughlin (1992) states that it is the 'same people who are always at risk of unemployment in an inefficient labour market founded on structural inequalities of locality, sex, race, disability, and age' (1992: xiii). Low ambition may result if peers lack jobs or qualifications, and some families have negative perceptions of the education system as a result of their members' experiences of humiliation or bullying at school. People made redundant in areas of employment decline have difficulty competing. The longer a person is unemployed the less they are considered employable, as their confidence levels drop, skills become outdated and employers discriminate against them (Green 1998: 115). This particularly applies to people with literacy and numeracy difficulties, a prison record, drug and alcohol addictions, homelessness, or having responsibilities as a carer (UKCES 2010). A report of 14–19 vocational education provision suggests children who experience learning difficulties are steered towards vocational courses that do not lead to higher education or jobs (Wolf 2011). Lone parents are pressurised to return to work whereas mothers in two-parent households are not (Percy-Smith 2000: 19). Countries participating in the OECD acknowledge the link between job stress and mental health, yet inactivity also induces health problems (Warr 1987: 207). 2.04 million people in the UK aged 16–64 are not looking for work due to long-term sickness (ONS 2016a). Poor support provision is also a cause of stress, and anecdotal reports suggest providers of back-to-work schemes pressurise clients to take low-paid agency work in order to meet 'pay on results' contracts and bonus thresholds. Others argue that the minimum wage legislation deters employers from appointing youth, who often enter the workforce on pay below this threshold. Recurrent unemployment is commonplace; two out of five recipients who find work are back on Jobseekers' Allowance within 6 months. Making work pay through tax credits has not been successful for housing benefit recipients, who 'still face extremely weak work incentives' (Brewer and Shephard 2004: 46). The recent austerity in Britain's welfare state has created food insecurity and increased dependency on food banks. Perhaps food chains will formalise a welfare state partnership in future years. Meanwhile, without the right to work and plentiful sustainable jobs, people's health and job prospects will forever erode in the shadow of Whitehall reforms.

The Group Dimension

Unemployment affects some groups more than others. Despite discrimination law, labour market prejudice affects ethnic minorities, disabled people, unqualified school leavers, carers, lone parents, people over 50, ex-offenders, as well as the long-term unemployed who are perceived to have outdated skills (UKCES 2010). While these cohorts face barriers to work even as the economic climate improves, '… it is less clear that this vulnerability can be resolved by training alone' (Hasluck (1987: 153). Groups may also perceive local authority recruitment practice negatively if the workforce does not reflect diversity (CO 2003: 45). Ethnic minority groups have made significant gains in accessing university in the past two decades, yet these have not translated into job opportunities, and institutional racism, gender, ethnic/religious intolerance and class politics continue to constrain their social mobility (see CoDE 2013). The Office for National Statistics (ONS) notes 49 % of all young black men aged 16–24 are unemployed, and African-Caribbean graduates are three times more likely to be unemployed than their white counterparts. Yet a recent 'full report' on graduates in the UK labour market ignored discrimination factors (see ONS 2013). A BBC *Panorama* programme suggests the plight of young qualified black men who are desperate to work in the UK is worse than in South Africa (BBC 2013). The white working-class population are also stigmatised and devalued. Amongst the disability cohort, people with mental health problems have the highest rate of unemployment, yet the psychological support needed to re-enter the job market is underestimated (Ritchie et al. 2005: 2). Thus, how networks represent local groups will be evident in network outcomes.

UK Unemployment Policy

This section outlines policy responses to unemployment, including macroeconomic instruments, EU policies, the full-employment concept and policy interventions: the latter include labour market activation, skills and training, incentives to attract firms, support for business enterprise and job creation. Thereafter, it surveys the governance infrastructure supporting policy roles.

Policy Contexts: Unemployment, Employment and Welfare

Unemployment policy can be defined as government actions to control or reduce unemployment and maintain the capitalist system. Policy measures range from macro-economics, welfare benefits and welfare-to-work employment incentives to supply-and-demand interventions, including programmes to remove potential workers from the market. Employment policy, on the other hand, tackles in-work conditions, tax distribution and aspects of EU law, including equal rights, the responsibilities of workers and employers, health and safety, and the free movement of labour. Welfare policy aims to reduce citizens' dependency on the state and: (i) motivate unemployed people to find work; (ii) engage them in a moral-behavioural contract (rights and responsibilities); and (iii) apportion justice to job opportunities. Critics suggest the policy's focus on the individual self-reliance necessary to find work, overcome welfare dependency and to engage in lifelong learning, supply-side training and employment programmes, overlooks unemployed people's literacy and transportation needs, as well as factors such as labour market inequalities and economic conditions, and the shortage of jobs in some areas (Crisp 2008: 174; Fletcher 2008b: 99). Moreover, local authorities and trade unions are not prominent in debating Britain's welfare state; rather it is the Treasury, separate from the DWP, that plays the greater role in social and welfare policymaking (Taylor-Gooby and Daguerre 2002: 50).

Managing unemployment is contentious, as the theories and experiments informing the politicised policy spectrum can be fallible. Policies on the political right of the spectrum aim to minimise government intervention and optimise self-reliance and entrepreneurialism. In modern times, the free market doctrine of the opinionated economist Milton Friedman influenced President Reagan and Prime Minister Thatcher's administrations to privatise public services (Friedman 1953: 3–16). Friedman believed the strength of labour unions and wage settlements over the minimum wage had made the 'natural rate of unemployment' (based on a self-correcting market where involuntary unemployment does not exist) higher than it ought to be (Tabb 1999). Consequently, according to this worldview, 'job flexibility' became a political asset and the orthodox solution for correcting unemployment in the marketplace; whereby fixed-term and casual job contracts would regulate and adjust real wages of low-skilled workers downwards. But 'flexibility' is not a full employment policy. Global financial institutions and transnational advisers (the World Bank, International Monetary Fund (IMF), and the OECD) have also advised countries to align their economic policy reforms to the NAIRU (non-accelerating inflation rate of unemployment). This approach links unemployment to changes in inflation, productivity and supply-side shocks, and requires labour market flexibility to reduce labour regulation as well as supply-side intervention to make people more 'employable', but pays less attention to wage controls, demand deficiencies and the social structures of work. Although labour flexibility corresponds with social and cultural liberalisation from family constraints, it has led to an erosion in labour conditions, and gender wage inequality and unaffordable childcare remain problematic.

Policies on the left of the spectrum expect government to steer self-interest and demand deficiencies in the business cycle be adjusted so as to promote a more equitable society; one supported by labour rights, unionisation, wage-bargaining and market regulation. The British economist John Maynard Keynes advocated raising the government deficit during market depressions to fill the demand gap. In this worldview, a laissez-faire economy is impossible. Capitalism needs government to support market functions, from social institutions and industrial subsidisation to fiscal and monetary policy (Triantafillou 2011). Unemployment is the product of the 'strategies of capital', not an isolated market law (Amin 1997: 15). Moreover, some scholars question the validity of unemployment theories guiding public policymaking; particularly, the Beveridge Curve (which links unemployment rate and job vacancy rate), the Phillips Curve (the cause-and-effect relationship between the rate of unemployment and inflationary pressures), and the 'natural rate' of unemployment hypotheses (Garraty 1978). Some claim the NAIRU concept is a myth perpetuated to avoid full employment (Mitchell and Muysken 2008: 78, 116; Storm and Naastepad 2012; Farmer 2013). Others emphasise the influence of policy conditions on levels of employment: for example, World Bank neo-liberal policy; weak socio-economic settlements between private actors' accountabilities and the government's role as lender and insurer of last resort; the lack of harmonised taxes and depleted revenues; 'the fatal logic of supplysidism and monetarism'; poor growth across Europe; and the addiction to capital accumulation (Leaman 2012: 182).

Macro-economic Policy

Macro-economic policy entails fiscal and monetary actions to stabilise imperfect markets and manage economic downturns. Governments' fiscal policies target budget arrangements for taxation and income distribution, spending and borrowing, national and regional business cycle and international trade. The central bank manages monetary policy and operates independently to control growth and inflation through money supply, credit flow, interest and exchange rates, bonds and equities and so forth. Bulpitt (1987: 173–5) suggests the 'monetarism' that emerged during the mid-1970s was a political tactic to rebuild central autonomy at a distance from certain decisions and groups; respectively, income policy and trade unions. Gamble (1987) links Britain's poor recovery from recession to adversarial politics, such that 'governments have failed to create either a stable environment for decision-making in private and nationalised industry, or to sustain long-term policies aimed at expansion' (1987: 17). Institutions organise fiscal and monetary policy separately, and socio-economic settlements depend on whether unemployment is a political or economic priority, on whether there is a commitment to higher public expenditure or lower taxation, and on the extent of concern over inflation and the public deficit (Greive Smith 1994: 268). Economic recovery from recession and frictional or structural unemployment involves long periods of adjustment, job creation, training, and screening and hiring new entrants; hence, the hysteresis theory suggests that the long-term unemployed lose out because employers perceive their employability skills as depleted. Therefore, the objectives for dealing with unemployment are: (i) economic management through regulation, finance or public production; and (ii) social justice, managed through redistribution of income and in-kind transfers such as free employment advice and training (Barr 1998: 99).

The opinion of economists divides between Keynesian and neo-classical approaches. By the 1950s, a 'golden age' of near full employment and low inflation was attributed to Keynesian management, a mixed economy, and fortuitous post-war conditions; these included the investment and expansion taking place at that time, the availability of cheap raw materials and energy, improved inter-state relations, better class relations and labour force inclusion (Gamble 1987: 8–10). However the Keynesian fiscal stimuli of higher public spending, government borrowing, low taxes to stimulate demand, fixed exchange rates, and budgets to regulate income and wealth inequalities were not sustained. International and domestic macro-economic instability and two oil price hikes during the late 1960s and 1970s raised production costs, and industrial productivity and manufacturing investment fell, profits eroded and trade liberalisation reduced export competitiveness, while technology and service sectors grew. Labour relations between employers and unions failed, and weak income policy led to unrest

(Hasluck 1987: 81–6). Neo-classical economists argue that fiscal policies in the long term lead to higher unemployment, inflation and taxes, and to reduced private sector growth. Yet the switch to monetarist policies, supply-side policy, welfare reform and in-work benefits has not been effective in reducing neighbourhood unemployment. The macro-economic neo-liberal agenda offers no solution for weak economies. Consequently, the EU has tried to maintain an influential role in promoting alternative economic policy in the belief that 'the struggle against unemployment requires more transnational coordination and agency' (Lahusen et al. 2010: 180).

European Policy Influence on Britain's Employment Policy

The impact of EU policy initiatives and structural funds on member states' employment policy and economic and social welfare regimes is not long lasting, as 'most of the mechanisms available to tackle labour market issues are still with national governments rather than in Brussels' (Walsh 2009: 11; Lahusen et al. 2010: 194–5). Although networks disseminated EU funding policy, there is no integrated European unemployment policy, as member states resist structural reform, fear burden sharing, the free movement of labour and tax harmonization, and bargain to retain decentralised national economic policy in support of different types of capitalism and welfare infrastructure. Hall and Soskice's (2001) seminal text, *Varieties of Capitalism*, offers a typology of production systems in various countries, from 'coordinated market economies' (social) to 'non-coordinated market economies' (liberal); both of which are capable of similar economic performance but differ in terms of the impact of well-being and institutional networks. Positive country attributes are associated with the former, social type of economy. This includes Germany and other Scandinavian countries that have legacies of guild systems and trade associations, lower unemployment levels, social solidarity, coordination of labour relations, integrated apprenticeship training, wage equilibrium and shorter working hours. The latter liberal type is associated with negative aspects of production in countries like the UK and USA, which have weak industrial coordination, unstable networks, erratic apprenticeship training, a self-funded higher education system, a labour force for short-term gains, diminishing labour costs for inequitable growth and high wage differentials. A comparative study of Britain, Germany and Switzerland suggests that in Britain, unemployment is framed as a 'technical problem', encouraging individuals to resolve their own needs, rather than as a political debate about capital and labour (Cinalli and Füglister 2010: 74). The lack of direct action in Britain means policy actors 'do not need to acknowledge the unemployed due to the limited challenges that they pose' (2010: 87). Decisions made at EU executive-level dominate agenda-setting, from EU hard law (employment legislation and regulatory structure) to soft law directives (symbolic motives, from 'equality for all' to 'economic inclusiveness'), but principals' preferences, which are delegated to decision-making agents through domestic-level networks, structural funds and policy implementation, can weaken under political pressures (Young 2010: 56–7). This has been illustrated by the recent Greece crisis. Reputedly, Britain is already an EU outlier, because it has structurally insulated its political system from wage negotiation, trade unions and lobby groups (Hassel 2006: 141, 249–51). Furthermore, Britain's majority vote to exit the EU, following the Conservative Party's referendum in June 2016, and subsequent political and economic turmoil leaves the poorest in no better state since 'exit campaigners' rhetorical employment policy appears largely dependent on the impact of the National Living Wage and immigration control. Germany's legal structure supports labour relations, but critics claim its economic policy is too competitive. Without a common industrial policy backing the universal rights of citizens, the unemployment levels within member states will vary considerably.

The Role of European Employment Policy and Governance

Although Britain is set to leave the EU, future policy must learn from history. From the 1970s on, EU employment law improved working conditions, including gender equality and parental leave, but Europe's unemployment problems worsened. Jacques Delors, EU President, appealed in the White Paper *Growth, Competitiveness and Employment* (EC 1993) for the European community to take 'action on jobs', as looking at countries' comparative performance demonstrated that 'growth is not in itself a solution to unemployment' (1993:10–16). Delors wanted to halve Europe's unemployment rate and create 15 million jobs in five years, to use social dialogue for 'a genuine right to initial or ongoing training throughout one's lifetime' and to link ecology to economics, as market prices and industrial productivity had ignored pollution costs (1993: 10–17, 100, 146). His vision incorporated a single market currency within one macro-economic framework to increase growth opportunities alongside trans-European networks to advance physical infrastructure and faster technologies. EU member states rejected the costs of the White Paper's far-reaching social objectives, although its policy themes reverberate today. EU summits from 1994 onwards have mainly focused on strengthening the decision-making infrastructure and supporting national policy coordination, multi-level 'civil society' governance, national level activation policies and programmes to train the long-term unemployed (for an account of the genesis of EU employment strategy, see Goetschy 1999). The European Council's Amsterdam Treaty (1997) and the Luxembourg 'Jobs Summit' held in November 1997 sought to institutionalise 'European Employment Strategy', using the 'open method of coordination' (OMC) reaffirmed under the European Employment Pact (1999) (EC 1999). Copeland and ter Haar 2013: 29) claim the process is open to bias, as active members' can 'cherry-pick' interests or sway opinion to suit national aspects or interpret objectives differently. European governance does not meet its own 'competition policy' standards, in that executive networks monopolise or privilege governmental actors, squeeze out less powerful representatives and restrict network competition, while regulatory institutions set the rules. Scharpf (2001) claims the 'heroic self-image' of the EU Commission in the White Paper (CEC 2001) on governance is inflated, and demonstrates 'a disturbing lack of understanding of the preconditions of successful multilevel governance in Europe' (2001: para 5). Indeed, partnership proposals involving a comprehensive set of actors, 'civil society' networks, and EU decision-makers and legislators would be unfeasible in practice. For example, roles and accountabilities between local/regional authorities and social partners in formulating EU-mandated National Action Plans (NAPs) for employment (from NGOs, trade unions and private sector representatives to faith groups), are ambiguous and less than transparent at national level. EU co-funding programmes use local networks to cascade and coordinate 'European Employment Strategy' goals, but Britain's NAP, subsumed into the annual National Reform Programme, is largely unknown (see, for example, HM Treasury 2014). Nevertheless, 'the Commission is deeply involved in the processes of defining, selecting and managing programs at the regional and local level—which makes for extremely cumbersome bureaucratic procedures and often wreaks havoc with the integrity of administrative institutions and practices at national and subnational levels' (Scharpf 2001: 6.2).

From 2010, the European Commission semester, supported by a 'Stability and Growth Pact', strengthened its surveillance of member states' annual budgetary and economic activities. It announced that medium-term budget plans must now demonstrate preventative and corrective fiscal actions to reduce excessive deficits and annual government borrowing, or penalties will apply. The Commission makes country-specific recommendations in 'national reform programmes' (NRP) to address gaps and weaknesses and meet five goals: employment, research and development, climate change, education, and fighting poverty and social exclusion. An Annual Growth Strategy guides the process for integrating macro-economic, micro-economic and employment policies for growth and jobs in the coming year. In 2014, for example, priorities for economic policy coordination are: (i) growth-friendly fiscal consolidation; (ii) restoring lending to the economy; (iii) promoting growth and competitiveness; (iv) tacking unemployment and social consequences of the crisis; and (v) modernising public administration. The latest EU funding to help local areas stimulate growth is the European Structural and Investment Fund (ESIF) (this combines European Regional Development Funding (ERDF), the European Social Fund (ESF) and the European Agricultural Fund for Rural Development). From 2013, the UK government assigned Local Enterprise Partnerships (LEPs) to design multi-year Strategic Economic Plans, and handle England's ESIF funding allocation and expenditure of £5.3 billion for the period 2014–20, along with economic, environmental, rural and social partners. LEPs will disseminate information and support partners to compete and build projects for delivery from March 2015, subject to LEP, government and EU approval. As with previous funding rounds, programmes focus on innovation, 'community-led local development' aligned to EU priorities (see above) and managerial rules, regulatory clauses and monitoring. However, the overarching agenda is questionable, as 'growth' has a poor record of leading to equitable distribution and the support of social needs. Thus, we might ask whether 'full-employment' policies have survived or been abandoned.

The Road to a Full Employment Economy, Route Abandoned?

Governments tolerate unemployment at fixed levels, rather than intervene to cure it. As Moore et al. (1989: 9–11) observe, 'the government's strategies have not really been employment related'. Over the decades, networks promoting 'full employment' policies did not explore how significant numbers of unemployed people might be employed. The attention deficit is unsurprising, as mainstream economics dismiss full employment on the grounds of inflation control. Pro-full-employment scholars, however, claim macro-economic policy is flawed, serves the wealthy and leads to wage inequality (see Pollin 2008; Storm and Naastepad 2012; Murray and Forstater 2013; Harvey 2014). Networks, therefore, would need to replace conventional policies that have run their course with unconventional policies for good local jobs. Simpson et al. (2009: 960) argue that employment policy would need to create 1.1 million new jobs and target neighbourhood-level labour markets and job deficits to bring every ethnic group in every locality up to the average employment rate for England and Wales. To this end, networks would need to engage with the following full employment perspectives: human rights, international growth, domestic perspectives, economic perspectives, structural perspectives and policy interventions, as outlined next.

The Human Rights View

To enable citizens to prosper through work, a full employment policy would need to implement the right to a decent job for everyone that wants one (see footnote 1). In 2011, the UN Human Rights Council issued guidance on human rights and business, and the European Commission requested that member states implement a response and produce action plans. The British plan is more of an aspirational statement. It claims that human rights safeguard personal freedoms, contribute to economic development and are good for business; yet it ignores unemployed people's economic exclusion (FCO 2013). Economic inclusion requires legislation, and the political representation of unemployed people in labour relations and job planning, but these aspects are underdeveloped.

The International Growth View

For decades, transnational institutions (World Bank, IMF and the OECD) have projected a neo-liberal version of 'full employment' minus a jobs plan; the onus is on entrepreneurs to create growth, wealth and jobs and utilise the free market for cheap labour and resources. Global growth in China, India, Brazil, South Africa and Mexico has increased the middle classes' spending power; however, pro-poor growth and full employment capacity diminished. The United Nations International Labour Organisation (ILO) claims global unemployment reached nearly 202 million in 2013, and is expected to rise further. Growth appears to be uneconomic for the underemployed and unemployed, and the growth (traditional economics) versus development (sustainable economics) debate continues.

Herman Daly (2008: 512–6), an ecological economist and advocate of 'development without growth', claims the World Bank's Commission on Growth and Development publication, *Growth Report: Strategies for Sustained Growth and Inclusive Development* (2008), makes growth assumptions, lacks clear thinking, downplays negative aspects and voices of dissent, and fixates on GDP as opposed to alternative indexes, such as human well-being measures. The Delors White Paper of 1993 had already questioned 'whether an increasing part of the measured economic growth figures does not deal with illusionary instead of real growth progress and whether the traditional economic concepts (e.g. GDP as traditionally conceived) may be losing their relevance for future policy design'(EC 1993: 146). Planning for growth, competitiveness and employment can never be 'balanced' or sustainable without recourse to poverty reduction measures. Nevertheless, pro-poor policies, social protection measures and equitable growth profiles to assess the link between employment-intensive and productivity-intensive sectors are still in their infancy (Hull 2009). In May 2011, the OECD launched its Better Life Initiative to compare the well-being of people and nations beyond GDP measures, using broad dimensions; from housing conditions and job quality (income, security and unemployment), to democracy and work-life balance (see OECD 2014). However, 'well-being' indicators lack a universal or European human rights perspective of full employment; moreover, countries have not adopted the ILO's resolution for decent work (employment, rights, protection and dialogue) and unemployment policy has not converged in Europe (Mitchell and Muysken 2008; Ernst and Berg 2009: 42–3; Lahusen et al. 2010).[12] International bodies, the European Commission and nation states fear losing competitiveness; hence, 'growth and innovation' policies dominate network agendas. These policies may not be the right course of action to effect inclusive change, without directly addressing demand-deficient markets or creating enough jobs to absorb long-term unemployed people (see Hasluck 1987: 105–8).

---

[12] In 2015, the UN General assembly set out principals for decent work. Global leaders will replace Millennium development goals with action plans for decent work as cited in 'The 2030 Agenda for Sustainable Development' (UN 2015). However, it does not consider the right to acquire a job.

The Domestic View

At the domestic level, 'full employment' relates to the 'number of hours per week each household would need to work to achieve [a] customary standard of life' (Kitson et al. 2000: 636). Networks react to 'the way work is organised and the possibilities of cooperation' between employment, domestic and family life, and welfare structure, not only between the state and market (2000: 640). This version of full employment emphasises the principles of work quality and the need for an abundance of decent jobs. It requires networks with political authority to plan 'full employment', influence the deregulated marketplace and address widespread entry-level job shortages; work programmes in 'depressed local labour markets', on the other hand, attempt to 'raise employability without raising employment' (see Peck and Theodore 2000: 739). Without government input to achieve a full employment economy, the fate of the unemployed rests on economic bubbles, and gains in the reduction of unemployment reverse when the economy turns.

The Economic View

Economists in neo-liberal Britain, New Zealand, Australia and USA understand 'full employment' as the lowest level of involuntary unemployment possible, relative to targets in the capitalist system at a given time. From the 1940s onwards, William Beveridge considered a 3 % unemployment margin acceptable for Britain's labour force to account for seasonal and frictional unemployment (Gardiner 2000: 673). In Britain and the USA today, an acceptable unemployment rate is 4–5 %. NAIRU supporters claim such unemployment levels are necessary as a trade-off by which to control inflation and wage bargaining, but rates vary depending on the country's conditions. Nevertheless, as Philpott (1997: 4) argues, '…the belief that there can never again be jobs for all who seek them seems hard to sustain'.

Structural View

A structural account of 'full employment' concerns the reproduction of the capitalist order through the 'supply of labour, the containment of inflation through the maintenance of a downward pressure on wages […] linked to neo-liberal ideas about the cultural basis of free markets and the patriarchal family' (Grover and Stewart 2002: 3). Structural barriers to employment impact oppressed groups, such as youth, women, ethnic minorities, older people and disabled people. For example, women supply labour on equal terms with men, yet undertake the larger proportion of the care responsibilities that 'affect access to jobs and earnings over a life course' (Gardiner 2000: 679–81). Private-sector family care is expensive and renders full-time jobs impractical, while training for better-paid jobs is not always feasible (2000: 680–3). Hence, working-class women will be forever in low-paid part-time undervalued employment. Feminist economic literature advises a structural shift from a 'full-time employment pathway' (modelled on individual self-sufficient workers without care responsibilities) and a 'gendered employment/care pathway', to a 'shared employment/care pathway' (2000: 685). Then again, the barriers to employment pathways for all cohorts should be considered.

The Policy View

Marx doubted full employment is feasible for technical reasons. Keynes and Friedman thought it possible, though through different policy approaches. Keynes advocated government intervention and Friedman regarded unemployment as a problem created by labour obstacles. Labour's Employment White Paper, *Building Britain's Recovery: Achieving Full Employment* (DWP 2009), aimed to get 'eight out of every ten people of working age' into employment, and support 'new high-skilled job creation in innovative sectors for the future' (2009: 11). However, technological advances (e.g., artificial intelligence, automation, computerisation, digitalisation, drone technology, green science and robotics) have resulted in the diminution of labour and government and industry have not considered how job creation might assist society as a whole in a changing labour market (Marchant et al. 2014). The Conservative Party manifesto scarcely mentioned job creation in the 2000s. Rather, it called for 'welfare reform' to absorb benefits claimants into the labour market through sanctions and work programmes, and immigration control (Conservative Party 2008). These proposals were taken forward by the former coalition government, whose Chancellor of the Exchequer, George Osborne, claimed in a speech at Tilbury Port that a 'modern approach to full employment means backing business', and that 'you can't abolish boom and bust' (Osborne 2014). However, his desire 'to have more people working than any of the countries in the G7 group', contains an inbuilt contradiction. Without a jobs protection policy, capitalism's periodic general would stymie full employment. Nevertheless, Osborne implemented a four-way policy schema. First, tax cuts, rate capping and the Employer

Allowance scheme to encourage businesses to invest and hire more workers. Second, low-paid workers (public and private sector) gained a tax threshold increase to make 'work pay', and working tax credits (top-up wage) continued. A year later, however, Osborne in his summer budget speech wants businesses to pay higher wages as the 'in-work' benefits are no longer affordable (Osborne 2015). Third, unemployed claimants have been placed under 'ordeal mechanisms' to find work, including compulsory daily attendance at the Jobcentre and keeping benefit payments low. Fourth, benefit payments to the long-term unemployed will be stopped if they refuse to join 'Help to Work', an unpaid community placement scheme (launched in April 2014) involving work experience, 'whether it is making meals for the elderly, clearing up litter, or working for a local charity'. Notably, the 'trailblazer' pilot to the scheme left the majority still in receipt of benefits after 'treatment' (see DWP 2013b).

Trade unions and some industries oppose 'workfare' schemes because they alter the purpose of welfare benefits, from a subsistence payment while looking for real jobs to enforced cheap labour, usually to subsidise rich companies and charities with low-skilled jobs lacking living wages. Osborne (2014) claims to reject artificial job creation 'on borrowed money', but also implies the parasitic nature of benefit claimants, and states: 'It's certainly not fair to taxpayers like you, who get up, go out to work, pay your taxes and pay for those benefits'. This applies especially to claimants who cannot speak English, and are 'not even attempting to learn it' (2014). Nonetheless, ESOL (English for speakers of other languages) courses have long been oversubscribed and underfunded (see Chap. 5). As Crisis, a national charity for single homeless people, noted: 'We have observed the gradual erosion of Skills for Life and particularly ESOL provision over the past few years' (Crisis 2010: 4). Political leaders know that unemployed people do not start recessions, that they lack a public voice and that unemployment experience impairs mental and physical health and places social services and the public health bill under pressure. Likewise, the stereotyping of unemployed people is perhaps one of the reasons employers would rather not appoint them (Hogarth et al. 2003: ix). Conversely, selective industries receive massive taxpayer-funded subsidies and positive promotion. In-work benefits to offset poverty are another indirect subsidy to firms. If the minimum wage rises, low-wage casual workers' hours are likely to reduce. As it stands, the ungovernability of conjunctional unemployment (reflecting the external fluctuation of supply and demand), and structural unemployment (based on internal actions, including 'social protection, wage setting and labour force mobility') suggest 'full employment' limitation (Triantafillou 2011: 570).

Policy Interventions

Central institutions ensure governing networks implement or support policy interventions designed to address local labour market conditions, such as Active Labour Market Policies (ALMPs), skills and training, incentives to attract firms, support for business enterprise and job creation. Interventions, however, are restricted and 'heavily influenced by external forces and events in the rest of the economy', whereby 'the root cause of decline is not of local origin' (Hasluck 1987: 47, 87, 142–3; Sullivan and Taylor 2007). Post-recessions, money circulation depletes, supply and demand for goods and services reduces, companies and investors hoard profits or savings, and banks stem credit flow to households and businesses. Nevertheless, the usual interventions focus on piecemeal solutions that have never directly addressed the real employment conditions. An overview follows.

Activation Policy

ALMPs 'activate' unemployed people in OECD countries to participate in labour market training and improve their employability in exchange for social benefits. They provide or promote labour market programmes for unemployed persons, and employment services that address frictional unemployment when jobs are available and workforce skills mismatch, but they do not provide jobs. Hanf et al. (1978: 303–41) explored the coordination limits of German and Swedish ALMPs in case studies of local implementation networks that assisted in the areas of: job training strategy; private sector development and efficiency; expenditure on employment services; employment measures for disabled people; training; subsidised employment; youth measures; and unemployment compensation. Notably, they found that the networks failed to tackle the shortage of decent jobs or define a labour-controlled social policy. The study cautions: 'Without more fully articulated theories of how collective and individual interests can be simultaneously satisfied the provision of vital social services may continue to be haphazard, inequitable and ineffective' (Hanf et al. 1978: 332). Decades later, neither countries that support the

'Nordic welfare model' and spend more on ALMPs, nor neo-liberal non-coordinated market economies like Britain that spend less on them, have overcome a jobs dearth. Even Sweden's job guarantee scheme, introduced in 2009, failed to combat high youth unemployment.

From New Deal to the Work Programme

In the late 1990s, public employment services modernised round 'Welfare to Work' strategies and supply-side policies (Finn 2000). In 1998, Labour used a £5 billion windfall tax on privatised utilities to fund the 'New Deal', a compulsory programme for the eligible unemployed based on President Bill Clinton's welfare reform schemes implemented in America in 1996 (HM Treasury 2007: 104; Taylor-Gooby and Daguerre 2002: 21). The New Deal transformed welfare entitlements into a social contract of conditional rights and obligations. As Peck and Theodore (2000: 729) suggest: 'The causes of unemployment are therefore conceived in individualistic and behavioural terms: the old problems of demand deficiency and job shortage have been dismissed'. Measures included making work preferable to benefits, tax incentives, welfare retrenchment and placing responsibilities on job-ready individuals to find work. Personal advisers customised employment support for eligible jobseekers, which included action plans, job-search counselling, employability skills training (CV writing, interview techniques, English literacy), job matching, and signposting to job fairs.[13] Some of these services were contracted out to providers to help jobseekers get into both state and non-state sectors, targeting long-term unemployed people through 'back-to-work' schemes, job brokerage, job clubs and information one-stop shops. Finally, local services were expected to cooperate in networks and keep abreast of changes in service provision, department and institutional restructuring, welfare policy reform, funding streams and successive 'flagship' programmes. However, JCP personal advisers struggled to extend community provision, cope with their client workload, meet pay-performance targets and 'respond imaginatively and flexibly to the needs of deprived communities' (see Fletcher 2008a: 578; McNeil 2009: 37). Media reports suggested that New Deal clients attending back-to-work courses had 'demoralising' experiences (BBC Radio 5 2009, f5 April), with the exception of getting their benefit claims processed: 'people really don't feel that they are necessarily getting the kind of personal support that they had been promised' (BBC Radio 4 2009, 13 August).

---

[13] In 2005–06, JCP employed 9,300 advisers at a cost of £238 million for salaries.

In 2010, the Conservative-Liberal Democrat coalition government abolished New Deals, Flexible New Deals, the £1 billion Future Jobs Fund, a minister for young citizens and youth engagement, Employment Zones and the Working Neighbourhood Fund (WNF). Instead, they awarded large profit-making companies 'payment by results' contracts to administer the £5 billion 'Work Programme' (part funded by the ESF) over a five-year period from June 2011. Its manifesto (DWP 2012: 2) states: 'There are still around 5 million people of working age receiving out-of-work benefits, around half of whom receive IB [Incapacity Benefit]. At around 2.5 million, unemployment is still much too high. Youth unemployment is a particular concern, and the proportion of individuals living in workless households is one of the highest in the EU'. Moreover, it relates previous 'Welfare to Work' programme problems to fragmentation, over-specified interventions and poor incentives to deliver strong results (2012: 2). As part of the programme, 40 contracts were issued to 18 providers. It involved staged payments to support long-term unemployed, or those at risk of becoming so, into sustainable work (providers may claim up to £13,700 per 'customer', depending on their situation). Supplementary incentives are also possible, such as payment incentives to providers for reducing reoffending. From August 2011 a 'skill conditionality' mandate was added to the welfare contract to make all benefit recipients on Jobseeker's Allowance (JSA) and ESA recipients in the Work-Related Activity Group in England attend training should their JCP adviser consider they have a barrier to employment through skills deficiency in English or maths, IT ESOL or occupational training. Clients aged 19 and over may be referred for skills assessment with a JCP adviser, or for work-based programmes delivered by a market of college or training providers.

By December 2013, the Work Programme had achieved 48,000 sustained job outcome payments (see the DWP website). Providers engage local supply chains in partnership with local experts and organisations to assist outcomes; however, commercial confidentiality restrains governance and network accountability is blurred. Hence, the National Audit Office (NAO) and Parliamentary Committee of Public Accounts (CoPA) investigated the following: the lack of transparency in sharing results between Work Programme prime providers; including sub-national data and expenditure; mismanagement in the contracting out of public services to private sector firms; and whether these providers appointed former senior civil servants (see CoPA 2013, 2014a). One firm manages all DWP contracts, and the results are not shared amongst providers; hence, there is a perception that government controls the message (CoPA 2014a). During Labour's administration, the media frequently reported fraudulent claims made by contractors to meet targets (BBC News UK 2012, 23 March), poor-quality services, and the expulsion of smaller providers in the supply chain who lacked the cash flow to handle high-risk subcontracts. Moreover, the contractor operating a digitalised job-matching service called 'Universal Jobmatch', which has replaced Employer Direct and Employer Direct Online for companies and the JCP job website and search facility for job-seekers, has been criticised by the media in 2014 for advertising fraudulent jobs and distorting job vacancy statistics.

Skills and Training

Skills for labour have shifted focus in recent decades, from an emphasis on workplace duties overseen by collective 'craft unions', to a focus on individual workers with portable skill sets and competencies overseen by managers. The skills fixation began in the 1980s, alongside a 'virtuous cycle' of macro-economic policies and OECD measures which was designed to build an 'Active Society' instead of a welfare-dependent society (Bednarzik and Sorrentino 2012: 11). The cycle aimed to 'provide a predictable noninflationary environment', along with structural policies to make 'labour and product markets more flexible', and human resource policies to reduce structural unemployment through 'training and improved job matching' (2012: 11). It pushed unemployed people 'into a relentless succession of training programmes designed to address deficiencies in skills and character' (Mitchell and Muysken 2008:4). Many unemployed claimants go back to the queue for jobs after completing employability training (for example, in maths, English, IT, ESOL, customer service, business administration, health and social care) or into unpaid work experience. Indeed, 'Welfare to workfare' or 'supply-side fundamentalism', does not guarantee work or 'access to the specific type of job for which they are qualified' (Peck and Theodore 2000: 729; Keep and Mayhew 2010; Hasluck 2011: ix, 34–36). In practice, skills worldviews and skills strategy are fraught with contradictions (Appelby and Bathmaker 2006: 714).

On the one hand, basic education is a way to fight human oppression and inequality, as people lacking skills in literacy and mathematics encounter more poverty, unemployment or reduced employment prospects. UNESCO (the United Nations Educational, Scientific and Cultural Organisation), for example, promotes 'lifelong learning' for moral and intellectual solidarity, beyond political and economic objectives. The OECD, on the other hand, want people to invest in education throughout their lives, because skills become obsolete in the 'knowledge economy', and learning new skills helps to retain or gain employment and contribute to economic growth and global competiveness (Bednarzik and Sorrentino 2012: 13). Lanning and Lawton (2012: 2) suggest 'the promised "knowledge economy" remains an impartial account' and that the tendency to accept assumptions behind the transformative agenda of supply-side policy, such as education for economic security, work-based needs, commercial values, and career development, rather than to apply critical thinking skills, should be questioned.

Evidently, flexible choice and part-time learning modules have widened participation in further and higher education, but funding priorities and peoples' socio-economic background and situation still restrict educational attainment. Demand for high-level skills has increased, while that for low-skilled jobs has decreased, but half of UK jobs do not require post-secondary education and a third of employers offer no staff training; consequently, overeducated people compete for low-paid jobs at the bottom of an hourglass economy (Hasluck 2011). The OECD identifies that 'a substantial proportion of young people in all the countries surveyed are overqualified in their jobs and/or not working in jobs corresponding with the subject area that they studied' (OECD-ILO 2014: 8–9). The numbers of graduates increased from 17 % in 1992 to 38 % in 2013 (ONS 2013: 3). Graduates working in non-graduate jobs represent 47 % of those who graduated recently and 34 % of those who have been out of full-time education for more than five years (2013: 13–4).

Consequently, scholars warn against 'relying too heavily on improvements in the supply of skills to solve economic and social problems' (Crouch et al. 2004: vii; Keep 2007; Keep and Mayhew 2010: 572). Lord Leitch's review of skills, *Prosperity for All in the Global Economy: World Class Skills* (Leitch 2006) stated in the UK the skills of working-age people were improving, but were unlikely to meet the requirements for a world-class skills base by 2020. Lord Leith's recommendations included national targets to address the stock of low-skilled people without qualifications, investing in intermediary skills, increasing the number of adults with degrees, and using the Employment and Skills Boards to support regional economic strategies. Yet 'skills drives' create a bottleneck without high levels of demand.

Indeed, countries with closed or market-orientated economies experience skills surplus or shortages. For example, Cuba has too many medics and teachers, while India has too few qualified plumbers because of the low wage structure. Skills planning requires liaison between industry and educationalists, and labour market analysis; but matching national, regional or local skills supply to demand is difficult without a skill-forecasting dashboard, labour signalling is not accurate and the range of skills certification befuddles industry and employers. Beyond 'licence to practise' occupations there is no robust pairing between training and occupations, and recruitment practices involve many variables, from experience to personality.

Business groups claim a high-level skills shortage is stalling productivity growth, but self-financing students already subsidise industry indirectly. Employers' engagement with the apprenticeship system is often limited or piecemeal, work-related learning lacks take-up, and work experience placements are bureaucratic. The Wolf Report (2011) claims UK apprenticeship progression routes are inadequate compared with those in place in other countries; moreover, the pull of further education and claims of severe skills shortage distract from the lack of jobs and apprenticeship places across skill levels (2011: 25, 51, 89). In Austria, Germany and Switzerland, apprenticeship frameworks link employment with citizenship and are embedded financially, institutionally, and through governance partnerships at sectoral and local levels between unions, employers and the state. But these countries have not eradicated unemployment, as markets alone cannot accommodate all abilities, or create enough apprenticeship places, sustainable jobs or jobs for graduates. Undoubtedly, young people who lack careers guidance, basic skills, academic qualifications and work experience will struggle to find even low-paid unsustainable work, especially with additional pressures, such as leaving care, criminal convictions, and long-term unemployment.

National apprenticeship programmes have expanded rapidly since 2008/2009, alongside youth traineeships since 2013/2014 for unpaid work experience to reduce youth unemployment. Apprenticeships peaked in 2010/2011 for those under the age of 19 but have since declined; moreover, while 19–24 and 25+ starts have increased, and achievements increased for the latter, they have declined for under 19s and 19–24-year-olds in 2012/2013 (SFA 2014: 9). The government apprenticeship framework is not without criticism as funding criteria regularly change or are abolished, depending on politics, national demand or departmental restructure. Hence, local authorities called for a skills funding devolution. For example, the Train to Gain scheme closed in 2010; Access to Apprenticeships, introduced in May 2011, closed to new starts in December 2013; Advanced Learning Loans, introduced in August 2013 for 24+ studying level 3 and above, closed in March 2014; the Wage Incentive Scheme, paying employers £2275 to appoint an 18–24 year-old person claiming JSA for six months, closed in August 2014; and funding eligibility for Apprenticeship Grants to employers of 16–24 year-olds will focus on firms with fewer than 50 employers from 2015 (SFA 2014: 12–13).

Government also funds large profit-making firms to operate apprenticeships and workplace learning.[14] For example, in 2012/2013, the Skills Funding Agency (SFA) awarded the global food chain McDonald's Restaurants Ltd. £6,946,108 for adult skills training and £6,299,163 for 16–18 apprenticeships to train thousands of people in London, and the supermarket Tesco Stores Ltd. in the central eastern region received £3,639,310 for adult skills training and £465,112 for 16–18 apprenticeships in 2013/2014. The Virgin Media Group requested an increase in the government's apprenticeship support, and received more than £1 million between 2012 and 2014 for adult skills and apprenticeships. While public sector reports and websites showcase employers' successful engagement with apprentice funding models, some firms associate job creation schemes with bad publicity and labour exploitation to reduce the unemployment count. Apprentices aged 16–18 and those in the first year of level 2 or 3 are paid £2.73 per hour, while the remainder are entitled to the minimum wage, and unpaid traineeships present no cost to the employer. Anecdotal reports suggest apprentices struggle to meet travel costs and lack information about travel assistance or available funding.

A skills shortage in construction work and the lack of affordable housing is perhaps why northern firms travel hundreds of miles with apprentices on round trips to undertake contract work in southern England, as anecdotal evidence suggests. Lawless (1989: 107) states: 'Unless there are sufficient local jobs, however, those receiving training may simply displace others in the labour market that leads to a reshuffling rather than a reduction, in unemployment'. In 2010–11, the hairdressing and beauty sector created a skills surplus; 94,420 people trained and 18,016 job vacancies (see Table 5 in Gardiner and Wilson 2012a: 12). Some fields require prolonged vocational

---

[14] (See funding allocations at the FE data library at https://www.gov.uk/government/statistical-data-sets/fe-data-libraryapprenticeships–2).

training and deter individuals who lack the finance to invest (Finegold and Soskice 1988). Reportedly, vacancies are difficult to fill in civil, structural and mechanical engineering; consequently, skills scarcity pushes salaries higher and reduces development affordability in less profitable regions. Unemployed jobseekers with vocational qualifications have also suffered from being portrayed as skills-deficient. One partnership between JCP and the National Skills Academy for Environmental Technologies refers unemployed plumbing and heating engineers and electricians to up-skill courses in environmental technologies, yet there is no national plan to utilise skills on an ongoing basis and make Britain energy efficient. The government's stop-start 'Green Deal' funding for home energy efficiency improvements is oversubscribed or is restricted to selected cities, and 'green' jobs supplied through local authorities have not been pursued.

For decades, the belief has been that central environment 'should act as the sole architect of the E&T [employment and training] system, directing the actions of the other stakeholders' (Keep 2007: 173). Hence, the public-funded UK Commission for Employment and Skills (UKCES) calls for employer-led partnerships with training institutions and trade unions (2011).[15] Inevitably, employers may not value training frameworks if they lack consistency or are not transferable; then again, shifting skills governance from government to employers on a voluntary basis is unrealistic. Thus, from April 2017, the Conservative government will collect an apprenticeship levy from eligible public and private sector employers to raise £3billion, achieve 3 million apprenticeship places by 2020, and establish an employer-led Institute for Apprenticeships to set standards and influence local skills commissioning.

Incentives to Attract Firms

Since the 1960s, governments have inserted private sector principles into spatially selective schemes as a way to invigorate underdeveloped areas, and create jobs. Incentives such as from capital investment, property development or tax breaks to attract firms to areas of decline are not always effective, for the following reasons: they are not directly designed

---

[15] UKCES is the successor to the Sector Skills Development Agency from 2008. It oversees 19 Sector Skills Councils and links to National Skills Academies that rolled out incrementally from September 2006, through a competitive bidding process representing 10 sectors. However, UKCES funding will be withdrawn in 2016/17.

to overcome the specific problem of unemployment; local economic plans scratch the surface of employment need; and outcomes are often wildly overestimated or a gamble. Chapter 4 explains the detail through the lens of urban regeneration policy.

Support for Business Enterprise

Since 1983, successive governments have provided business start-up programmes and enterprise allowance schemes for unemployed people.[16] Small UK firms create and destroy the largest number of jobs: new businesses are vulnerable to financial losses as they expand, and dropout is high in the first year of trading, nevertheless, participation in business support services for new and expanding enterprises is low (DBIS 2011: 32; Huggins and Williams 2009: 28–31). To boost enterprise, national programmes target specific places or cohorts. The local authority also maintains a level of training provision and information: small-medium enterprise grants; cheap loans, and venture capital funds; tax breaks for companies that employ young people; microfinance investment; international trade assistance and business hubs; and small commercial units for business start-ups. Social enterprise, community business, and cooperatives and development trusts also have support and encouragement, but are still restricted by market conditions. Specialist providers pursue funding to deliver business outreach, advice and coaching using a community fieldwork approach, beyond the standard enterprise agency service. LEPs have also driven localised business support, and the Chamber of Commerce is another source of advice and mentoring. Between April 2011 and September 2015, 152,880 unemployed people claiming JSA transferred to the New Enterprise Allowance Mentor starts, featuring business mentoring support, and 76,960 people transferred to the New Enterprise Allowance Business starts and access to business start-up loans of up to £1000 (DWP 2015: 2–3). While firms may create jobs and revenue in the future, there is no evidence that small businesses reverse unemployment trends in low demand areas.

---

[16] The Enterprise Allowance Scheme was launched by the Conservative government in August 1983, and replaced by the Business Start-Up Scheme in April 1991. The latter was administrated by TECs until March 1995. The Labour government launched self-employment options under the 2002 New Deal work programme and a Six Month Offer in April 2009 for those claiming JSA for six months. These schemes closed in 2011, only to be replaced by the coalition government's Work Programme in June 2011.

Job Creation

In capitalist societies, 'continuous political support for job creation programs has been fragile' (Janoski 1990: 259). Creating jobs in areas of low growth is immensely challenging, and neighbourhoods adjacent to high-growth areas may continue to have high unemployment. Moore et al. (1989: 30) recognise this conundrum, as 'Even when local urban labour markets are "job rich"—that is, having a greater number of jobs than the local labour force—local unemployment could remain high because of mismatches between the unemployed and jobs in term of skills, training and the influx of non-residents'. Networks that cite 'job creation' as a policy goal may find the private sector is unable, or unwilling, to create sustainable jobs for unemployed people, and job targets do not protect citizens from bad jobs that particularly affect women, ethnic minorities and people with less formal education. Governing networks have little interest in job conditions or pay-setting policy, yet job quality matters. Precarious jobs with low pay, little responsibility, reward, skill or career pathway lead to job churn and repeated unemployment claims; all of which contribute to social and economic problems. In 2006, over half of men and a third of women had returned to make a new claim within six months and this pattern was shown to be constant in over a decade, and a quarter of vacancies administrated by JCP were temporary jobs (SSAC 2007: 9–10). Indeed, 40 % of repeat JSA claimants could only find temporary work (see CoPA 2007: 5, 7, 9, Ev 2; Leitch 2006; Hasluck 2011: 35). Full employment, however, requires a whole framework approach supporting job quality, and affordable housing, childcare, food and transport.

The Committee of Public Account's (CoPA) enquiry into progress delivering local growth initiatives outlines concerns about the over-optimistic job outcomes predicted, and lack of coordination between government departments. It is 'unclear who in central government is responsible and accountable for the success of local growth policies' (CoPA 2014b: 13). Responsibilities for boosting job creation under the previous Labour government were equally ambiguous. Successive governments also seem unable to design schemes to support youth into employment. In 2013, the school participation age rose from 16 to 17, and 18 from 2015, thereby increasing pressure on local authorities to ensure the cohort remain in education or training. The 'youth contract wage incentive payments' scheme, for example, pays employers £2275 for taking on a young person for six months' work, but it only spent an estimated 4 % of the total funding available in the first year of operation to May 2013 (Crowley and Comnetti 2014: 20). Moreover, only 35 % of the NEETs enrolled progressed to a positive destination (into further education or training) (2014: 20). (With a target of 160,000 work placements and only 4690 recruits, the DWP abolished the three-year scheme earlier than planned. A Local Government Association (LGA) press release stated that 'Almost all young people could either be in work or learning by 2020 if local authorities were able to lead on youth employment schemes' (LGA 2014). According to ONS data (April-June 2016), this would mean engaging 843,000 NEETs in England. Although apprenticeship places increased 15 % in one year for 16–17 year olds, from 41,738 in March 2013 to 49,228 in March 2014 (DfE 2014), how can employers in places with slow growth and job shortages connect young people to jobs with prospects? Innumerable short-term interventions have been trialled, from public works programmes (for human contribution, not profitable production), work placements and apprenticeships, to support for disadvantaged groups on housing estates. From 2017, for example, a new Work and Health programme will replace the Work Programme, although at mid-2016, the voluntary sector await the details. Yet as Crowley and Comnetti (2014: 29) state, 'This is not a short-term problem, but is rather the result of the UK's highly entrenched pattern of economic disadvantage, with many areas yet to recover from a long-term decline in their key industries'. The problem for networks is that attention focuses on 'schemes' to stimulate artificially the local economy, not equalities.

Governance Structures: Roles and Responsibilities

This section outlines governance structures supporting unemployment policy in England that were in place while the book's fieldwork was undertaken, and is as up to date as space allows. The policy field is subject to continuous reform, complexity and short-term partnerships. Fig. 3.1 depicts stakeholders directly or indirectly supporting the policy environment.[17] As Goldsmith and Eggers (2004: 8) observe:

---

[17] In a study of France's and Sweden's European Employment Strategy and Britain's NAPs in the period 1998–2004, the term 'stakeholder' rather than 'social partner' is said to be a British trait; the former is passive, whereas the latter is more conducive to policy design (see Esmark and Triantafillou 2007: 113).

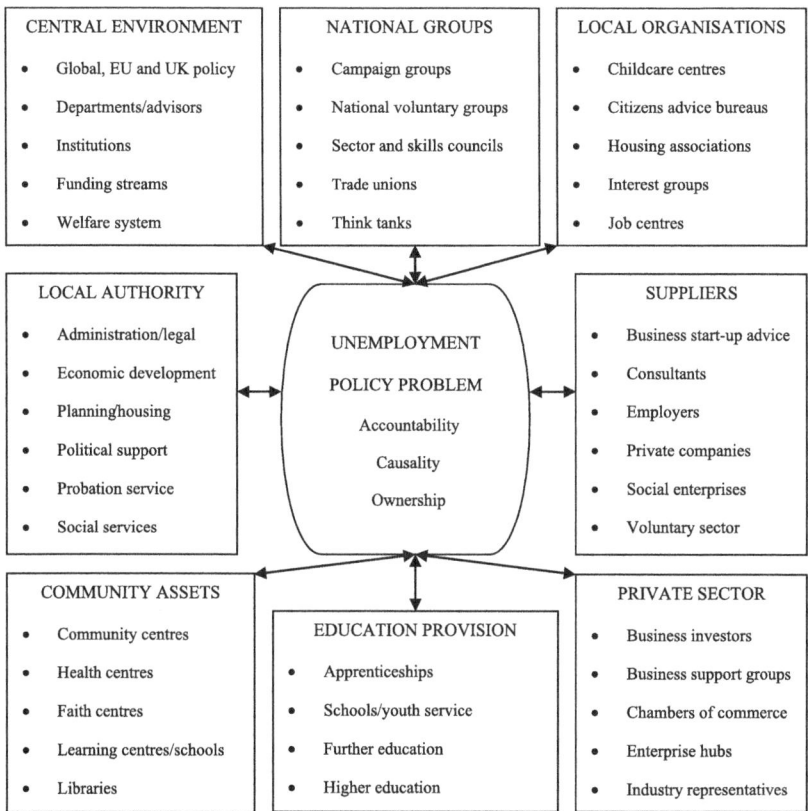

Fig. 3.1 Stakeholders in the unemployment policy field

'Government agencies, bureaus, divisions, and offices are becoming less important as direct service providers, but more important as generators of public value within the web of multiorganisational, multigovernmental, and multisectoral relationships that increasingly characterise modern government'. Politics communicates public values, administrators and managers defend policy aspects in the governance domain, but the roles, accountabilities and responsibilities for dealing with unemployment causality is not always clear. Despite governance rhetoric, policymakers distance themselves from the voices of unemployed people who have no right to speak or to a common settlement (Cinalli and Füglister 2010: 86–7).

Ironically, suppressing right-to-work activism maintains the status quo, poor-quality jobs and welfare dependency. Departmental change across the welfare, skills and business domains strain policy governance and reduce local government authority.

For example, in 1992 Local Education Authorities lost control of further education college management; skills policymaking powers were transferred to the Further Education Funding Council for England, which was subsequently replaced by the Learning and Skills Council (LSC) in April 2001. The LSC replaced 72 TECs, created a quasi-market in education, colonised the skills governance domain, funded adult learning and post-16 training, including sixth forms and further education (not higher education) and commissioned work-based learning and apprenticeships. The skills governance industry is expected to spend fast and hit targets; hence, governance, reputation and trust-building can be profitable for training providers inside networks. However, network accountabilities are blurred, and expenditure blunders, cover-ups and cases of mismanagement have surfaced. For example, a corrupt tendering ring was uncovered at the LSC, and former employees were sentenced for training contracts bribery and money laundering (see SFO 2010).

In 2001, the Department of Health and Social Security merged with parts of the Department for Education and Employment to create the DWP. In 2002, the DWP created the executive agency Jobcentre Plus (JCP) by merging the Employment Service and the Benefits Agency to administer Jobcentre services, the New Deal programmes and deliver labour market policies through partnership working (DWP 2004: iii).[18] The Home Office also wanted community safety partnerships (joint working between police, local authorities, fire and rescue authorities, probation service and health providers to reduce crime), so as to support unemployment reduction in urban areas and monitor unemployment data alongside crime incidence and drug misuse. Formal links between the Home Office and JCP were lacking during this book's study phase and, in 2007, the Probation

---

[18] In 2011, JCP lost its 'agency' status and became a brand name under the DWP. The business culture of JCP is at odds with network governance, and beset with operational difficulties—staff cuts, massive unrealistic targets, a poor reputation for inefficient customer service management, poor quality support services and outreach provision. Advisers are target-driven with little autonomy and quickly demoralised, they have poor training and the role is not respected, staff turnover is high and 'hard to reach groups' are not supported (Fletcher 2008a, 568, 571, 574). McNeil (2009: 7) suggests the lack of industry-wide qualifications for the low-paid personal advisers hinders effective service provision for clients.

Service's remit of supporting ex-offenders into work or training in England and Wales passed from the Home Office to the Ministry of Justice. In the same year, JCP established local employment partnerships that provided incentives for employers to appoint those who had been in receipt of JSA for six months or longer. These included recruitment subsidies, pre-employment training, work trials and work placements. Although JCP staff supported 'employer-responsive' partnerships, in effect, 're-branding and marketing of the existing Jobcentre Plus services', they disbanded in 2010, having had little strategic influence over job growth or sustainability (see Bellis et al. 2011). In June 2007, the Labour Prime Minister at that time, Gordon Brown, split the Department for Education and Skills (DfES) into the Department for Innovation, Universities and Skills (DIUS), and The Department for Children, Schools and Families. DIUS would oversee the National Skills Strategy assisted by the LSC. The NAO recommended the setting-up of networks between the DWP, DIUS, JCP, and LSC to assist pre-employment skills and in-employment support (NAO 2007).

In 2008, the DWP's Deprived Area Fund merged with the Neighbourhood Renewal Fund (NRF), managed by the Social Exclusion Unit (SEU) in the Department for Communities and Local Government (DCLG) (previously the Office for the Deputy Prime Minister (ODPM)), to create the 'Working Neighbourhood Fund' (WNF).[19] The DCLG has overall responsibility for regional policy. The £1.5 billion WNF fund aimed to tackle worklessness and low economic growth, and build on the Working Neighbourhood Pilot (WNP) managed by JCP or private companies in employment zones between 2004 and 2006 in 12 of the UK's most deprived areas.[20] Following evaluations of the WNP, Action Team for Jobs and New Deal for Communities, the Houghton Report on tackling worklessness through local authorities and partnerships (see Houghton et al. 2009) recommended a three-way approach: i) worklessness assessments, 2) devising a work and skills plan, and 3) integrating work and skills budgets. This would require local authorities working through Multi-Area Agreements, targeting worklessness, skills, housing and transport needs, to join up the following strategies: a Sustainable Community Strategy, a Child Poverty Needs Assessment, a City Strategy Plan, an Employment

---

[19] The ODPM's (previously the Department of the Environment, Transport and the Regions) responsibilities transferred to the Department for Communities and Local Government on 5 May 2006.

[20] Two Working Neighbourhood Pilots are case studies in this book; see Chaps. 5 and 6.

and Skills Plan and a Regional Strategy. Moreover, local authorities, LSPs and sub-regional partnerships should join up responses to explore co-commissioning services that meet DWP standards, align their budgets and use employer-led Employment and Skills Boards to guide governance arrangements. The most productive recommendation for citizens was a 'challenge fund' overseen locally to coordinate long-term unemployed people in temporary community jobs. Labour's Employment White Paper (DWP 2009) amplified the Houghton Report and recommended that partnerships coordinate a Work and Skills Plan based on 'Total Place'; a pilot for decision-making 'closer to the front-line, with agencies brought together locally to develop cost effective solutions based on a shared goal' (2009: 120). This would need to occur alongside the implementation of the following: 'The Flexible New Deal'; 'Pathways to Work'; 'The Future Jobs Fund' (to place unemployed youth aged 18–24 in temporary work, including social enterprises); the WNF; the 'Education Maintenance Allowance' (to enable poor students to continue in education) and the National Apprenticeship Scheme.

In 2008, the LSC announced that it would close and transfer its strategic planning role to the local authorities, but not before 2013. The DIUS merged again in 2009 with the Department for Business Enterprise and Regulatory Reform (created in 2007, having already taken part of its remit from the former DTI, whose research, science and innovation remit merged with DIUS) to form the Department for Business, Innovation and Skills (DBIS). DBIS funds post-19 learning in England (higher and further education, and community learning), skills and training, research and development, and regional and local economic development. Its former duties included Regional Development Agencies (RDAs) in English regions, and Business Link's (BL) local services, business start-up support, including personal business advisers, and workshops disseminating enterprise information. BL's regional advisory service disbanded in 2011 and its online business resource website, run by a private sector company on behalf of HM Revenue and Customs, cost over £100 million to build and operate from 2004 until its closure in 2012. The LSC dissolved in 2010 and its responsibilities were divided. First, oversight of 16–19-year-old learners was passed to local authorities, and four months later handed to the Young People's Learning Agency for England (YPLA) with nine regional branches administrating 16–19 funding to colleges and training providers. Second, post-19 skills and education passed to DBIS and the Skills Funding Agency (SFA), with a remit to handle the ESF, and work with the JCP, YPLA, the RDA, the DBIS Skills Investment Strategy and the Regional Employability Framework. In 2010, the Department for Education (DfE) replaced the Department for Children, Schools and Families. One DfE minister, responsible for FE, 16–19 year olds, the national apprenticeship service, the national careers service and data service, is also attached to the DBIS. The SFA leads in implementing apprenticeships and employer policy on behalf of the DBIS and DfE.

In 2012, the Education Funding Agency replaced the YPLA and Partnerships for Schools, and became responsible for funding the education of 3–19-year-olds, and schools and college estates, while local education authorities retain statutory responsibilities for 16–18-year-olds in accordance with the Apprenticeships Skills, Children and Learning Act 2009. The Connexions service (which provides information, careers advice and guidance for young people aged 13 to 19, and up to age 25 for people with learning difficulties and/or disabilities) devolved to schools without additional funding support. Consequently, young people's services have fragmented to such an extent that skills governance appears to be ungovernable, as the streamlining of the policy map 'is at the level of central government funders, rather than for the user' according to Gardiner and Wilson (2012b: 6). Lanning and Lawton (2012: 22–3) suggest the 'principal focus' is on 'individuals, rather than firms and sectors, with decisions about training and workforce development largely left to the market'. Heseltine (2012) calls for devolution of skills funding to LEPs; hence, government has tasked LEPs with producing local skills strategies that are appropriate to employers' needs.

Local government, meanwhile, lacks influence over careers advice provision, the Universal Credit system, the Youth Contract, and the £60 million youth funding cuts made between 2012 and 2014, which led to the axeing of 2000 youth workers' jobs (see a Freedom of Information response to a local government trade union, UNISON 2014). Then again, the DWP's £30 million Youth Unemployment Innovation Fund Pilot (2012–14) contracted 10 social investment partnerships for social impact and financial returns in selected areas to deliver the 'pay on results' model, targeting the employability of disadvantaged youths aged 14 years and over, which achieved 11,200 outcomes by the end of March 2014. In 2013, the Cabinet Office took charge of youth policy, although the DfE retains responsibility for local authorities' youth funding, including 14–16 work placements, and £7.2 billion has been ring-fenced to engage young people under 18 in education and training. However, the LGA perceives the shift of careers advice to schools as failing young people. Hence, the government's latest intervention, a £20 million employer-led self-sustaining Careers Enterprise

Company, will establish local advisory boards, oversee careers advice networks, collect employers' voluntary contributions and fund school enterprise advisers to improve young people's employability.

Following Britain's EU referendum in June 2016, Theresa May replaced David Cameron as the UK Conservative Party leader and Prime Minister. In the reshuffle the Department for Business, Energy and Industrial Strategy (DBEIS) replaces the Department for Energy and Climate Change, the DfE absorbs DBIS skills remit and international responsibilities transfer to a new Department for International Trade. DBEIS will develop and deliver industrial policy and lead government's relationship with business.

A note on industrial strategy

Industrial strategy may be viewed as any political process, government plan, policy, intervention, incentive or subsidy to expand preferred industries (such as, advanced engineering/manufacturing, digital, renewable energy, life sciences, food, drink and agriculture), on the pretext that market mechanisms alone cannot achieve economic growth. Hence, industrial strategy supports some old industries to survive and not others, subsidises research and development, encourages inward investment in regional clusters, often near small pleasant cities supporting new technologies and knowledge diffusion, funds large infrastructure projects, from house building to transport, provides enterprise and business support services complemented by finance, tax incentives or regulation policy, and stimulates supply and demand through skills enhancement and employability. Currently, LEPs add a regional or sub-regional dimension to local economic decision making and investment; like a modern equivalent of magistrates before elected councils were introduced. Whitehall's devolution of power to LEPs is a £12 billion 'Local Growth Deal', but LEP's performance is overseen by central government, not local government (see Chap. 5). However, funding instruments are not a wholesale solution for managing labour surplus in demand-deficient localities or effective pro-poor growth policy. Industrial policy wants the private sector to flourish, but education for innovation is often delivered in a vacuum from Britain's local labour structure, or serves rich and powerful regulators. Although industrial strategy responds to imperfect markets, there has been little consensus on who benefits. Rather, the real limits of supply and demand and the permanent condition of unemployment are ignored. For centuries, global firms exploit niche markets and low wages in developing countries for competitive advantage, but markets for winners that fear workforce empowerment are not socially responsible. Economic migrants may undercut the labour rate in domestic industries, from agriculture to hospitality, and profit margins sustain the managerial status quo in the supply chain. Government's poured millions into well known private sector firms to safeguard existing jobs, but ignores funding a social or green economy to boost job creation in weak economies for common societal benefits. A work and wages council might highlight discriminatory issues and respond to realist policy making, but only legislation supporting job outcomes for citizens can make policy effective for all.

Conclusion

Britain has not innovated to overcome neighbourhood unemployment. Networks oversee policies that cannot create enough demand or job surpluses. Traditional theories informing policy interventions have not tackled the problems (see Hasluck 1987; Moore et al. 1989: 36–38). Marxist/radical theories underpin the importance of area-based interventions to redress gross unfairness in labour markets, such as class, race or gender inequality. Keynesian/structuralist theories legitimise the transfer of massive amounts of public funding to industry and enterprise in order to 'grow' the economy during periods of depression. Neo-classical ideas underpin skills and training interventions to boost individual competitiveness and self-reliance in the free market. None of these theories have led to plentiful jobs for people of all abilities. The government's recent call for the creation of an additional 3 million apprenticeship places is another indicator of market failure. The former Labour government stated: 'For those who are capable of working, there will be no right to a life on benefits' (DWP 2008: 12). Yet without positive 'rights to work', social policy has worked to mask the effects of unemployment rather than to address its structural and institutional weaknesses and economic causes. A punitive welfare industry focuses less on linking people to sustainable jobs, and more on monitoring unemployed people: applying work tests and benefit sanctions, training people for non-existent jobs, and supplying employers, mostly retailers, with unpaid labour on 'work experience' schemes. Neighbourhood problems fester while networks report upwards to governing policy. The next chapter views unemployment policy through a historical trajectory of urban regeneration networks.

Bibliography

Amin, S. (1997). *Capitalism in the age of globalization: The management of contemporary society*. London: Zed Books.

Appelby, Y. and Bathmaker, A.M. (2006). The new skills agenda: increased lifelong learning or sites of inequality. *British Educational Research Journal*, 32 (5), 703–17.

Atkinson, R., & Kintrea, K. (2001). Disentangling area effects: Evidence from deprived and non-deprived neighbourhoods. *Urban Studies*, 38(12), 2277–2298.

Barr, N. (1998). *The economics of the welfare state* (3rd ed.). Stanford, CA: Oxford University Press/Stanford University Press.

Beatty, C., Crisp, R., Foden, M., Lawless, P., & Wilson, I. (2009). *Understanding and tackling worklessness volume 2: Neighbourhood level problems, interventions, and outcomes evidence from the new deal for communities programme*. Sheffield: Sheffield Hallam University.

Beatty, C., Fothergill, S., Gore, T., & Powell, R. (2011). *Tackling worklessness in Britain's weaker local economies*. A report to the National Worklessness Learning Forum. CRESR, Sheffield Hallam University.

Bednarzik, R. W., & Sorrentino, C. (2012). *30 years of the OECD employment outlook: Its historical beginnings, evolution and future direction as a major policy engine*. Paris: OECD Publishing.

Bell, D. N. F., & Blanchflower, D. G. (2013). Underemployment in the UK Revisited. *National Institute Economic Review*, 224(1), F8–F22.

Bell, S., & Hindmoor, A. (2009). *Rethinking governance: The centrality of the state in modern society*. Cambridge: Cambridge University Press.

Bellis, A., Sigala, M., & Dewson, S. (2011). *Employer engagement and Jobcentre Plus*, Research Report No. 742. London: DWP.

BBC. Radio 5 (5 April 2009). Jobless Training Courses 'Demoralising'. Donal MacIntyre show. Retrieved from http://news.bbc.co.uk/1/hi/uk/7982550.stm.

BBC. Radio 4 (13 August 2009). 'Jobcentre Plus: Is it Working'? Face the Facts, Producer K. Johnstone.

BBC. News UK. (23 March 2012). 'Leaked document suggests 'systemic fraud' at A4E'. Retrieved from http://www.bbc.co.uk/news/uk-17476415

BBC. (2013). Jobs for the boys? Panorama, Television Broadcast 13th May 2013, BBC1. 8.00–30 p.m. Producer Asif Hasan.

Brewer, M., & Shephard, A. (2004). *Has labour made work pay?* York: Joseph Rowntree.

Bulpitt, J. (1987). Thatcherism as statecraft? In M. Burch & M. Moran (Eds.), *British politics: A reader* (pp. 167–186). Manchester: Manchester University Press.

Cabinet. (1929). *Inter-departmental committee on unemployment: Report on 'We Can Conquer Unemployment and the Melchett-Tillett Report'* CAB 24/203,

C.P. 104 (29) 10 April 1929. Retrieved from http://www.nationalarchives.gov.uk/cabinetpapers

Cabinet Office. (2003). *Ethnic minorities and the labour market*. London: Strategy Unit, Cabinet Office.

Castles, F., & McKinlay, R. D. (1979). Does politics matter: An analysis of the public welfare commitment in advanced democratic states. *European Journal of Political Research, 7*(2), 169–186.

Cinalli, M., & Füglister, K. (2010). Networks and political contention over unemployment. In M. Giugni (Ed.), *The contentious politics of unemployment in Europe: Welfare states and political opportunities* (pp. 70–93). Basingstoke: Palgrave Macmillan.

Citizens Advice Bureau (CAB). (2014). *Citizens advice sees 60 per cent increase in problems with JSA sanctions 15 April 2014.* Press release. Retrieved from https://www.citizensadvice.org.uk/about-us/how-citizens-advice-works/media/press-releases/citizens-advice-sees-60-per-cent-increase-in-problems-with-jsa-sanctions/

CoDE. (2013). *Addressing ethnic inequalities in social mobility: research findings from the CoDE and Cumberland Lodge Policy Workshop*, Residential conference at Cumberland Lodge in Windsor Great Park held on 13–15 November 2013, Centre on Dynamics of Ethnicity.

Commission of the European Communities (CEC). (2001). *European governance. A White Paper*. COM (2001) 428 final. Brussels: CEC.

Committee of Public Accounts (CoPA). (2007). *Sustainable employment: Supporting people to stay in work and advance, thirteenth report of session 2007–08*. HC 131. London: The Stationery Office.

Committee of Public Accounts (CoPA). (2013). *Department for work and pensions: Work programme outcome statistics*. HC 936 [Incorporating part of evidence HC 814, of Session, 2013-14]. London: The Stationery Office.

Committee of Public Accounts (CoPA). (2014a). *Contracting out public services to the private sector*. HC 777 [Incorporating HC 791, of Session, 2013-14]. London: The Stationery Office.

Committee of Public Accounts (CoPA). (2014b). *Promoting economic growth locally, sixtieth report of session 2013–14*. HC 1110. London: The Stationery Office.

Conservative Party. (2008). *Work for welfare: REAL welfare reform to make British poverty history*. Responsibility Agenda, Policy Green Paper No. 3. London: The Conservative Party.

Considine, M., & Lewis, J. M. (2003). Networks and interactivity: Making sense of front-line governance in the United Kingdom, the Netherlands and Australia. *Journal of European Public Policy, 10*(1), 46–58.

Copeland, P., & ter Haar, B. (2013). A toothless bite? The effectiveness of the European Employment Strategy as a governance tool. *Journal of European Social Policy, 23*(1), 21–36.

Crisis. (2010). *Crisis' response to the BIS skills for sustainable growth consultation*. London: Crisis. Retrieved from http://www.crisis.org.uk/data/files/publications/Skills%20for%20Growth%20consultation%20response.pdf

Crisp, R. (2008). Motivation, morals and justice: Discourses of worklessness in the welfare reform green paper. *People Place & Policy Online*, 2(3), 172–185.

Crisp, R., Gore, T., Pearson, S., & Tyler, P. with Clapham, D., Muir, J., & Robertson, D. (2014, July). *Regeneration and poverty. Evidence and policy review*. CRESR, Sheffield Hallam University.

Croall, J. (1997). *Lets act locally, the growth of local exchange trading systems*. London: Calouste Gulbenkian Foundation.

Crouch, C. (2008). What will follow the demise of privatised Keynesianism? *Political Quarterly*, 79(4), 476–487.

Crouch, C., Finegold, D., & Sako, M. (2004). *Are skills the answer: The political economy of skill creation in advanced industrial countries*. Oxford: Oxford University Press.

Crowley, L., & Cominetti, N. (2014). *The geography of youth unemployment: A route map for change*. Lancaster: Lancaster University/The Work Foundation.

Daly, H. E. (2008). Growth and development: critique of a credo. *Population and Development Review*, 34(3), 511–518.

Davies, J. S. (2005). 'The social exclusion debate: Strategies, controversies and dilemmas', *Policy Studies*, 26(1): 3–27.

Department for Business Innovation and Skills (DBIS) (2011). *Job creation and destruction in the UK: 1998–2010*. London: DBIS.

Department for Communities and Local Government (DCLG) and Department for Work and Pensions (DWP). (2007). *The working neighbourhoods fund*. London: DCLG.

Department for Education (DfE). (2014). *Thousands more school leavers choosing apprenticeships*. Press release [online] 4 July 2014. Retrieved from https://www.gov.uk/government/news/thousands-more-school-leavers-choosing-apprenticeships

Department for Work and Pensions (DWP). (2004). *Delivering labour market policies through local and regional partnerships*. Report by the Policy Research Institute, Leeds Metropolitan University.

Department for Work and Pensions (DWP). (2008). *No one written off: Reforming welfare to reward responsibility: Public consultation*. Norwich: The Stationery Office.

Department for Work and Pensions (DWP). (2009). The Employment White paper. *Building Britain's recovery: Achieving full employment*. Norwich: Stationery Office.

Department for Work and Pensions (DWP). (2012). The Work Programme, December 2012.

Department for Work and Pensions (DWP). (2013a). *Households Below Average Income: An Analysis of the Income Distribution 1994/5-2011/12*. London: DWP.

Department for Work and Pensions. (2013b). *Support for the very long term unemployed trailblazer - longer term analysis of benefit impacts*. December 2013.
Department for Work and Pensions (DWP). (2014). *Quarterly Statistical Summary*, 14th May 2014.
Department for Work and Pensions (DWP). (2015). *New Enterprise Allowance Quarterly Official Statistics, April 2011 to September 2015*. DWP. 16th December 2015.
Douglas, T. (1995). *Scapegoats: Transferring blame*. London: Routledge.
EC (European Commission). (1993). *Growth, competitiveness and employment*. Delor's White Paper. Brussels: Commission of the European Union.
EC (European Commission). (1999). European Commission adopts 1999 employment guidelines. *Bulletin of the commission of the European communities*. Brussels.
EC (European Commission). (2014). *EU employment and social situation quarterly review – March 2014*. Luxembourg: Publications Office of the European Union.
Ernst, C., & Berg, E. (2009). The role of employment and labour markets in the fight against poverty. In *Promoting pro-growth employment* (pp. 41–68). Paris: OECD.
Esmark, A., & Triantafillou, P. (2007). Document analysis of network typology and network programmes. In P. Bogason & M. Zølner (Eds.), *Methods in democratic network governance* (pp. 99–122). Basingstoke: Palgrave Macmillan.
Eurostat. (2015). *Eurostat news release indicators, January 2015, Euro area unemployment rate at 11.3%*. Retrieved from http://epp.eurostat.ec.europa.eu/
Fairclough, N. (2000). *New labour, new language?* London: Routledge.
Farmer, R. E. A. (2013). The natural rate hypothesis: An idea past its sell-by date. *Bank of England Quarterly Bulletin, Q3*, 244–256.
Field, J. (2009, July 6). Able-bodies: Work camps and the training of the unemployed in Britain Before 1939. *The Significance of the Historical Perspective in Adult Education Research Conference*. Institute of Continuing Education, Cambridge University, Cambridge.
Finegold, D., & Soskice, D. (1988). The failure of training in Britain: Analysis and prescription. *Oxford Review of Economic Policy, 4*(3), 21–53.
Finn, D. (2000). Welfare to work: The local dimension. *Journal of European Social Policy, 10*(1), 43–57.
Fletcher, D. (2008a). Tackling concentrations of worklessness: Highlighting the limits of work-focused organisational cultures in the UK. *Environment and Planning C: Government and Policy, 26*(3), 563–582.
Fletcher, D. (2008b). Employment and disconnection: Cultures of worklessness in neighbourhoods. In J. Flint & D. Robinson (Eds.), *Community cohesion in crisis: New dimensions of diversity and difference* (pp. 99–118). Bristol: The Policy Press.
Foreign Office and Commonwealth (FCO). (2013). *Good business: Implementing the UN guiding principles on business and human rights*. Cm 8695. London: FCO.

Friedman, M. (1953). *Essays in positive economics*. Chicago: University of Chicago Press.
Gamble, A. (1987). The politics of economic decline. In M. Burch & M. Moran (Eds.), *British politics: A reader* (pp. 7–26). Manchester: Manchester University Press.
Gardiner, J. (2000). Rethinking self sufficiency: Employment, families and welfare. *Cambridge Journal of Economics, 24*(6), 671–689.
Gardiner, L., & Wilson, T. (2012a, June). *Hidden talents: Skills mismatch analysis*. London: London Government Association.
Gardiner, L., & Wilson, T. (2012b, January). *Hidden talents: Analysis of fragmentation of services to young people*. London: London Government Association.
Garraty, J. A. (1978). *Unemployment in history: Economic thought and public policy*. New York: Harper and Row.
Garside, W. R. (2002). British unemployment: 1919–1939: a study in public policy. Cambridge: Cambridge University Press.
George, H. (1879). *Progress and poverty: An enquiry into the cause of industrial depressions, and of increase of want with increase of wealth. The Remedy*. K. Paul, Trench & Company.
Ginsburg, N. (1999). Putting the social into urban regeneration policy. *Local Economy, 14*(1), 55–71.
Girling, J. (1997). *Corruption, capitalism and democracy*. London: Routledge.
Goetschy, J. (1999). The European employment strategy: Genesis and development. *European Journal of Industrial Relations, 5*(2), 117–137.
Goldsmith, S., & Eggers, W. D. (2004). *Governing by network. The new shape of the public sector*. Washington, DC: The Brookings Institution Press.
Green, A. (1998). The changing geography of non-employment in Britain. In P. Lawless, R. Martin, & S. Hardy (Eds.), *Unemployment and social exclusion* (pp. 95–115). Oxon: Jessica Kingsley.
Gregg, P., & Wadsworth, J. (2011). *The labour market in winter: The state of working Britain*. Oxford: Oxford University Press.
Greive Smith, J. (1994). Policies to reduce European unemployment. In J. Michie & J. Grieve Smith (Eds.), *Unemployment in Europe* (pp. 259–275). London: Academic Press.
Grover, C., & Stewart, J. (2002). *The work connection: The role of social security in British Economic Regulation*. Basingstoke: Palgrave.
Hall, P., & Soskice, D. (2001). *Varieties of capitalism: The institutional foundations of comparative advantage*. Oxford: Oxford University Press.
Hanf, K. I., Hjern, B., & Porter, D. O. (1978). Local networks of manpower training in the Federal Republic of Germany and Sweden. In K. I. Hanf & F. W. Scharpf (Eds.), *Interorganizational policy making: Limits to coordination and central control* (pp. 303–341). Beverly Hills: Sage.

Hansard, H. C. Deb. (2013, October 14). *Work and pensions. Oral answers to questions*, col.416-34.
Hansard, H. C. Deb. (2014, April 3). *Sanctioning of benefit recipients. Oral answers to questions*, col.1056-82.
Harris, J. (2006). So the 1980s were one long orgy of champagne and shoulder pads? What about the riots, poverty and bombings? The Guardian. Retrieved from http://www.guardian.co.uk/
Harvey, P. (2013). Wage policies and funding strategies for job guarantee programs. In M. J. Murray & M. Forstater (Eds.), *The job guarantee: Toward true full employment* (pp. 39–58). New York: Palgrave Macmillan.
Harvey, P. (2014). The new deal's direct job-creation strategy: Providing employment assurance for American workers. In S. D. Collins & G. S. Goldberg (Eds.), *When government helped: Learning from the successes and failures of the new deal* (pp. 146–179). Oxford: Oxford University Press.
Hasluck, C. (1987). *Urban unemployment*. Harlow: Longman Group UK.
Hasluck, C. (2011, June). *Low skills and social disadvantage in a changing economy*. South Yorkshire: UK Commission for Employment and Skills.
Hassel, A. (2006). Wage setting, social pacts and the euro: A new role for the state. (Amsterdam: Amsterdam University Press).
Heseltine, M. (2012). *No stone unturned: in pursuit of growth*. London: Department for Business, Innovation and Skills.
HM Treasury (2014). *Europe 2020: UK national reform programme 2014*. London: Stationery Office.
HM Treasury, DBERR, & DCLG (2007). *Review of sub-national economic development and regeneration*. Norwich: HMSO.
Hogarth, T., Hasluck, C., Devins, D., Johnson, S., & Jacobs, C. (2003). *Exploring local areas, skills and unemployment, employer case studies*. Report to Government. Nottingham: DfES publications.
Houghton, S., Dove, C., & Wahhab, I. (2009). *Tackling worklessness: A review of the contribution and role of English local authorities and partnerships*. Final Report. Yorkshire: The Department for Communities and Local Government.
Hudson, R. (2009). Life on the edge: Navigating the competitive tensions between the 'social' and the 'economic' in the social economy and in its relations to the mainstream. *Journal of Economic Geography, 9*, 493–510.
Huggins, R., & Williams, N. (2009). Enterprise and public policy: A review of labour government intervention in the United Kingdom. *Environment and Planning C: Government and Policy, 27*(1), 19–41.
Hull, K. (2009). Understanding the relationship between economic growth, employment and poverty reduction. In *Promoting pro-growth employment* (pp. 69–94). Paris: OECD.

Janoski, T. (1990). *The political economy of unemployment: Active labor market policy in Western Germany and the United States*. California: University of California Press.

Judge, D., & Dickson, T. (1991). The British state, governments and manufacturing decline. In G. Esland (ed.), *Education, training and employment, volume 1: Educated labour – The changing basis of industrial demand*. Wokingham: Addison-Wesley in association with The Open University.

Keep, E. (2007). The multiple paradoxes of state power in the English education and training system. In L. Clarke & C. Winch (Eds.), *Vocational education: International approaches developments and systems* (pp. 161–175). Abingdon: Routledge.

Keep, E., & Mayhew, K. (2010). Moving beyond skills as a social and economic panacea. *Work Employment Society, 24*(3), 565–577.

Kitson, M., Martin, R., & Wilkinson, F. (2000). Labour markets, social justice and economic inefficiency. *Cambridge Journal of Economics, 24*(6), 631–641.

Lahusen, C., Giugni, M., & Berclaz, M. (2010). Globalization and the contentious politics of unemployment: Towards denationalization and convergence. In M. Giugni (Ed.), *The contentious politics of unemployment in Europe: Welfare states and political opportunities* (pp. 171–197). Basingstoke: Palgrave Macmillan.

Lanning, T., & Lawton, K. (2012). *No train, no gain: Beyond free-market and state-led skills policy*. London: Institute of Public Policy Research (IPPR).

Lawless, P. (1989). *Britain's inner cities* (2nd ed.). London: Paul Chapman.

Leaman, J. (2012). Weakening the Fiscal state in Europe: The European Union's failure to halt the erosion of progressivity in direct taxation and its consequences. In E. Chiti, A. J. Menéndez, & P. G. Teixeira (Eds.), *The European rescue of the European Union? The existential crisis of the European political project* (pp. 157–192). Olso: University of Oslo.

Leitch, S. (2006). *Prosperity for all in the global economy – World class skills*. Final Report. London: The Stationery Office.

Link, B. G., & Phelan, J. C. (2001). Conceptualizing stigma. *Annual Review of Sociology, 27*, 363–385.

Local Government Association. (LGA). (2014). *Councils could help 95 per cent of young people into work or learning*. LGA press release 11 June 2014. Retrieved from http://www.local.gov.uk/media-releases/-/journal_content/56/10180/6265766/NEWS.

MacInnes, T., Aldridge, H., Bushe, S., Kenway, P., & Tinson, A. (2013). *Monitoring poverty and social exclusion 2013*. York: Joseph Rowntree Foundation and The New Policy Institute.

MacKay, R. R. (1998). Unemployment as exclusion: Unemployment as choice. In P. Lawless, R. Martin, & S. Hardy (Eds.), *Unemployment and social exclusion* (pp. 49–68). Oxon: Jessica Kingsley.

Mann, M. (2012). *The sources of social power: Volume 2 the rise of classes and nation-states* (2nd ed.). Cambridge: Cambridge University Press.

Marchant, G., Angelica, Y., & Hennessy, J. (2014). Technology unemployment and policy options: Navigating the transition to a better world. *Journal of Evolution & Technology, 24*(1), 26–44.

Marsh, D. (1992). Youth employment policy 1970-1990: Towards the exclusion of the trade unions. In D. Marsh & R. A. W. Rhodes (Eds.), *Policy networks in British Government* (pp. 167–199). Oxford: Clarendon Press.

Marx, K. (1867). *Capital, a critique of political economy* (Vol. 1). Translated from the 2nd German edition by Ernst Untermann. Chicago: Charles. H. Kerr & Company 1906.

McLaughlin, E. (1992). Preface. In E. McLaughlin (Ed.), *Understanding unemployment new perspectives on active labour market policies*. London: Routledge.

McNeil, C. (2009). *Now it's personal: Personal advisers and the new public service workforce*. London: Institute for Public Policy Research.

Michie, J., & Wilkinson, F. (1994). The growth of unemployment in the 1980s. In J. Michie & J. Grieve Smith (Eds.), *Unemployment in Europe* (pp. 11–31). London: Academic Press.

Mitchell, W. F., & Muysken, J. (2008). *Full employment abandoned: Shifting sands and policy failures*. Cheltenham: Edward Elgar.

Moore, C., Richardson, J. J., & Moon, J. (1989). *Local partnership and the unemployment crisis in Britain*. London: Unwin Hyman.

Mourshed, M., Patel, J., & Suder, K. (2014). *Education to employment: Getting Europe's youth into employment*. McKinsey Centre for Government. New York: McKinsey & Company.

Murray, M. J., & Forstater, M. (2013). *The job guarantee: Toward true full employment*. New York: Palgrave Macmillan.

OECD (2009). *The OECD economic outlook: Tackling the jobs crisis*. Paris: OECD.

OECD. (2014). *OECD better life index*. Available from http://www.oecdbetterlifeindex.org

OECD-ILO. (2014, September 10–11). *Promoting better labour market outcomes for youth*. OECD-ILO background paper for the G20 Labour and Employment Ministerial meeting, Melbourne, Australia.

Office for National Statistics (ONS). (2013, June). *Full report: Graduates in the UK labour market 2013. Numbers of underemployed in the labour market*.

Office for National Statistics (ONS). (2014). *Trends in self-employment in the UK 2001 to 2015*. Available from https://www.ons.gov.uk/

Office for National Statistics (ONS). (2016a). *UK labour market statistics: August 2016*. Statistical Bulletin.

Office for National Statistics (ONS). (2016b, February). *Young people not in education, employment or training (NEET): February 2016*.

Office for National Statistics (ONS). (2016c, June). *Educational status and labour market status of people aged from 16 to 24*. Dataset A06 SA. Available from https://www.ons.gov.uk/

Osborne, G. (2014). *Chancellor speech on tax and benefits reform.* Tilbury Port, 31st March 2014. Keynote address. Transcript retrieved from https://www.gov.uk/government/speeches/chancellor-speaks-on-tax-and-benefits

Osborne, G. (2015). Budget speech to parliament 8 July 2015. Transcript available at https://www.gov.uk/government/speeches/chancellor-george-osbornes-summer-budget-2015-speech

Özler, S. (2012). Adam Smith and dependency. *Psychoanalytic Review, 99*(3), 333–358.

Painter, C., Isaac-Henry, K., & Rouse, J. (1997). Local authorities and non-elected agencies: Strategic responses and organizational networks. *Public Administration, 75*(2), 225–245.

Peck, J. (2004). Full employability/empty promises: Old labour regulation under new labour. In A. Wood & D. Valler (Eds.), *Governing local and regional economies: institutions, politics and economic development* (pp. 205–229). Aldershot: Ashgate.

Peck, J., & Theodore, N. (2000). Beyond employability. *Cambridge Journal of Economics, 24*(6), 729–749.

Percy-Smith, J. (2000) Introduction: The contours of social exclusion, in J. Percy-Smith, (ed.), *Policy responses to social exclusion. Towards inclusion?* (pp. 1–21) Buckingham: Open University Press.

Philpott, J. (1997). Looking forward to full employment: An overview. In J. Philpott (Ed.), *Working for full employment* (pp. 1–29). London: Routledge.

Pollin, R. (2008). *Is full employment possible under globalization?* PERI Working Paper 141. Amherst: University of Massachusetts-Amherst.

Ramsden, M., Bennett, R., & Fuller, C. (2007). Local economic development initiatives and the transition from training and enterprise councils to new institutional structures in England partnership, discretion and local flexibility. *Policy Studies, 28*(3), 225–245.

Richards, D., & Smith, M. J. (2002). *Governance and public policy in the UK.* Oxford: Oxford University Press.

Ritchie, H., Casebourne, J., & Rick, J. (2005). *Understanding workless people and communities: A literature review.* Research report No. 255, for the DWP. Leeds: Corporate Document Services.

Roller, E. (2005). *The performance of democracies: Political institutions and public policies.* Oxford: Oxford University Press.

Rowthorn, R. E. (1991). De-industrialization in Britain. In G. Esland (Ed.), *Education, training and employment, volume 1: Educated labour – The changing basis of industrial demand* (pp. 34–56). Wokingham: Addison-Wesley Publishing Company in Association with The Open University.

Scharpf, F. W. (2001). *European governance: Common concerns vs. the challenge of diversity.* MPIfG Working Paper, No. 01/6.

Serious Fraud Office (SFO). (2010). *Four sentenced in Shropshire training contracts bribery and money laundering case*. Press release 04 June 2010. Retrieved from http://www.sfo.gov.uk/press-room/press-release-archive/press-releases-2010/
Shaikh, A. (1978). An introduction to the history of crisis theories. In *U.S. capitalism in crisis*. URPE reader, pp. 219–241.
Silver, H. (1994). Social exclusion and social solidarity: Three paradigms. *International Labour Review, 133*(5-6), 531–578.
Simpson, L., Purdam, K., Tajar, A., Pritchard, J., & Dorling, D. (2009). Jobs deficits, neighbourhood effects, and ethnic penalties: The geography of ethnic-labour-market inequality. *Environment and Planning – Part A, 41*(4), 946–963.
Skills Funding Agency (SFA). (2014). Statistical first release: SFA/SFR25. 16 October 2014. London: DBIS.
Smith, A. (1776). *An inquiry into the nature and causes of the wealth of nations* (Edwin Cannan, ed. 1904). Library of Economics and Liberty. Retrieved May 17, 2013 from http://www.econlib.org/library/Smith/smWN0.html
Social Exclusion Unit (SEU) (2004). *Jobs and enterprise in deprived areas*. London: Office for the Deputy Prime Minister.
Social Security Advisory Committee (SSAC). (2007). *Patterns of employment, benefit eligibility, and the rights and responsibilities agenda*. Paper No. 4.
Somers, M., & Block, F. (2005). From poverty to perversity: Ideas, markets, and institutions over 200 years of welfare debate. *American Sociological Review, 70*(2), 260–287.
Spicker, P. (1988). *Principles of social welfare*. London: Routledge.
Storm, S., & Naastepad, C. W. M. (2012). *Macroeconomics beyond the NAIRU*. Harvard: Harvard University Press.
Sullivan, H., & Taylor, M. (2007). Theories of 'neighbourhood' in urban policy. In I. Smith, E. Lepine, & M. Taylor (Eds.), *Disadvantaged by where you live?: Neighbourhood governance in contemporary urban policy* (pp. 21–42). Bristol: Policy Press.
Tabb, W. K. (1999). *Reconstructing political economy: The great divide in political thought*. London: Routledge.
Taylor-Gooby, P., & Daguerre, A. (2002). *National report on welfare reform*. Welfare Reform and Management of Societal Change Project, University of Kent.
The apprenticeships, skills, children and learning Act 2009. (2010). (Consequential Amendments to Subordinate Legislation) (England) Order 2010.
Therborn, G. (1986). *Why some people are more unemployed than others*. London: Verso.
Triantafillou, P. (2011). The OECD's thinking on the governing of unemployment. *Policy and Politics, 39*(4), 567–582.
UNISON (2014). *The UK's youth services: How cuts are removing opportunities for young people and damaging their lives*. London: UNISON.

UKCES (United Kingdom Commission for Employment and Skills). (2010, June) *Tackling exclusion: A scoping study into the employment and skills outcomes for people in or at risk of social exclusion.* Report. Yorkshire: UKCES.

UKCES (United Kingdom Commission for Employment and Skills). (2011, December). *Employer ownership of skills: Securing a sustainable partnership for the long-term.* Report. Yorkshire: UKCES.

UKSA. (2014). *Correspondence: From Sir Andrew Dilnot to Dr Eoin Barry Clarke, Labour Market Statistics, 17th March 2014.* United Kingdom Statistics Authority.

United Nations. (UN) (2015). Transforming our world: The 2030 agenda for sustainable development. A/RES/70/1. (UN). http://sustainabledevelopment.un.org

Walsh, K. (2009). *The development of employment policy in the EU leading up to European Employment Strategy 1997.* Worcester: Training & Employment Research Network.

Warr, P. (1987). *Work, unemployment and mental health.* Oxford: Oxford University Press.

Wolf, A. (2011). *Review of vocational education: The Wolf report.* Department for Education.

Work and Pensions Committee (WPC). (2012). *Youth unemployment and the work contract.* House of Commons. Second Report of Session 2012/13, HC151. Norwich: Stationery Office.

Young, A.R. (2010) The European policy process in comparative perspective, in H. Wallace, M. A. Pollack, and A. R. Young (eds.), *Policy-making in the European Union* (6th edn.). (pp. 45–68) Oxford: Oxford University Press.

CHAPTER 4

Urban Regeneration Policy and Governing Networks

British governments have used governing networks in urban regeneration policy as a strategy for tackling persistent unemployment in disadvantaged neighbourhoods. Networks govern in the sense that network participants steer financial incentives to meet specific regeneration policy goals and influence perceptions of problems. Networks have been a constant theme in the policy landscape, yet fail to deliver the anticipated neighbourhood outcomes. For example, the Blair government's national strategy for 'neighbourhood renewal' envisioned that, 'within 10–20 years, no-one should be seriously disadvantaged by where they live.' It stated that local people should have access to jobs and good quality services and be able to influence regeneration processes (SEU 2001: 24). This vision required multiple stakeholders to coordinate policy through centrally designed governing networks supported by problem exposition, research and consultation. Regeneration processes in England, however, are susceptible to 'neo-liberal preferences that dictate where investment occurs', while social regeneration 'has tended to be shunted into a separate discourse of community "renewal"' (Jones and Evans 2013: 227, 230).

Commercial regeneration in the last decade, for example, elevated some urban cities to 'world status'; however, the 'trickle-down' effect of large-scale spatial renewal did not penetrate poor neighbourhoods, where local economies stagnate from one political manifesto to the next. Neither do networks provide safety nets when industries abandon areas, as has been observed in many northern and coastal towns. The broad lexicon of 'regeneration' might even distract from political and economic shortfalls. Blair's national plan for 'joined-up' public services and networks supporting neighbourhood governance, for example, transformed into Cameron's one nation 'Big Society', where people find solutions through economic growth, self-help and communities working together, rather than from 'officials, local authorities or central government' (Cameron 2010). Yet shifting the ideology from a focus on the community sector to one on the private sector obscures those sectors' interdependence (Moore et al. 1989: 76).

Regeneration programmes have already been extensively evaluated (see for example, DTLR 2002; HM Treasury et al. 2007; Tallon 2013). Impact, however, can be difficult to disentangle, as change can arise from the effects of multiple policy priorities or seep beyond the spatial boundaries of the affected area (Moore et al. 1989: 120–31; Tyler et al. 2013: 172–3). Crisp et al. (2014) claim regeneration funding has 'focused on activities more likely to improve experiences of living in poverty in particular places, rather than to reduce poverty itself' (2014: 66). Likewise, Bailey et al. (1995) claim partnership agencies can only achieve marginal improvements, as objectives such as increasing the number of available jobs for target groups, increasing household incomes and reducing welfare dependency depend on economic trends beyond local agency (1995: 220–21). Community regeneration discourse represents a counterforce to neo-liberal hegemony beyond growth factors, yet community consultation processes can still be carried out more for political effect than for balancing social and economic outcomes. In practice, inclusive community participation is limited and underscores the fact that deliberating who participates in regeneration networks can distract from the question of who will benefit. As Papadopoulos (2007: 481) states: '… even if the network is pluralistic, the diversity of perspectives does not guarantee that these perspectives are representative of the society at large.' Undoubtedly, citizens influence and benefit from regeneration outputs, ranging from street furniture, community assets or refurbishment and art galleries to health facilities and transport infrastructure, and private sector subsidisation assists to create or sustain jobs. Yet billions of regeneration aid over decades has left thousands of unemployed people still hidden away in low-quality housing or sink estates with few employment prospects. Hence, this book is interested in finding out more about network outcomes.

This chapter considers the puzzle in three sections. First, it explores the scope of urban regeneration, and considers why regeneration may not acquiesce with the unemployment policy. The second section links different types of governing networks supporting urban regeneration policy for political change since the 1930s. The third section focuses on centralising networks since 1997. No single theory is adequate for an empirical investigation of urban networks. Hence, the chapter draws on wide-ranging theoretical frameworks, such as power elites, pluralism, Marxism, New Right perspectives (see Atkinson and Moon 1994: 7–11, 257–263), growth coalitions, urban regimes, policy network analysis (see Bassett 1996: 547–52; Davies 1999) and neighbourhood governance (Lowndes and Sullivan 2008).

The Unemployment Problem in Urban Regeneration Policy

Urban regeneration can be defined as processes for strengthening run-down areas, which may have legacies of industrial decline and weak local economies. Priorities include (i) worklessness, skills and business development; (ii) industrial and commercial property and infrastructure; and (iii) homes, community and environment (see Crisp et al. 2014). Activities may include the destruction, renovation or gentrification of an area to attract property-led investment, business-driven growth, commercial retail, industrial warehouse development, business hubs, cultural industries, superstore facilities on vacant land and accessible transport. They may also serve social goals, for example employment training, improved housing, better leisure provision, investment in health amenities and education facilities, and crime interventions. However, regeneration investment is selective, place-dependent, and impact varies. Policymakers view regeneration as a cross-cutting policy tool to transform the environmental fabric of a place and reset its reputation. Investment confidence depends on an area's viability to attract wealth-creation industries; this is assisted by land values, business-friendly policies, convenient locations, a potential workforce and good transport logistics. However, the regeneration process is not a wholesale remedy for neighbourhood unemployment and fails to create sufficient jobs for local people. For example, Liverpool's regeneration has not achieved the same level of economic success as that of nearby Manchester, despite receiving greater funding (see Jones and Evans 2013: 82).

Indeed, the effectiveness of regeneration processes is questionable. The notion of 'community cohesion', whereby no group or class is excluded from regeneration approaches, has had little economic impact on workless households; and the socio-economic gap between the poorest neighbourhoods and the national average has not closed (Turok and Robson 2007). Davies and Pill's (2012: 2208–11) comparative study of governing rationales in Bristol and Baltimore networks finds regeneration disparities between city investment and neighbourhood self-help approaches, the latter being a second-rate offering in the local power structure. Residents fear the gentrification effect of regeneration, as market-biased economies compartmentalise living spaces according to high or low incomes, thereby resulting in uneven spatial outcomes (Rydin 2011: 53). A regeneration study at Salford Quays found redevelopment had depleted community facilities and local services (see Jones and Evans 2013: 93–4). Culture-led, event-led or design-led 'shock of the new' regeneration is another way of transforming a place's identity outside the political arena of consent, but quality varies and the impact on communities is unknown (Evans 2005: 10, 19–20). Regeneration for boosting commerce, tourism, culture and place-branding revolves round money, profits and entertainment: how to spend, and reflect the effects of that spending. Gentrification of rundown properties increases land values, property and rent, and prices out locals from the area, as low-income people carry little weight in regeneration decision-making processes. Hence, regeneration involves managing community perceptions to minimise community resistance, as the outcomes for local citizens are uncertain. Even social regeneration approaches, from sustainable living, provision of community facilities and landscape recycling to mixed housing communities, may only provide a small job economy for local people.

Definitions

Political ideological preferences lead to changes in the definition of regeneration, with policy consequences. The former Labour government described regeneration as:

> ... a set of activities that reverse economic, social and physical decline in areas where market forces will not do this without support from the government (HM Treasury, and DCLG 2007).

In this version, the state rows rather than steers, 'if not economically then politically' (Davies 2003: 265). The Conservative-Liberal Democrat coalition government focused regeneration on business partnerships in eight favoured cities to improve infrastructure (broadband, transport, and housing), business rate deals and funding bids to boost jobs and growth (Cabinet Office 2011). However, a House of Commons Communities and Local Government Select Committee (CLGC 2012) inquiry pressed government to define regeneration and 'the nature of the problem it is trying to solve', and 'develop and publish a strategy that recognises the deep-seated problems faced by the most disadvantaged communities' (2012: 2). The former government's response was dismissive:

> ... it is not for Government to define what regeneration is, what it should look like, or what measures should be used to drive it ... Government has no plans to publish a national regeneration strategy ... prescribing outcomes, targets and measures or roles—would be inconsistent with our localist approach to regeneration. Instead, we are providing the tools, flexibilities, options and powers that will allow *local* partners to develop their *own* regeneration strategy to address their *own* priorities. In this way, there will be different strategies in different places to address different problems. (Underlining in the original, from a Government Response to the House of Commons Communities and Local Government Select Committee's inquiry of regeneration in CLGC 2012: 2–5).

Yet in other parts of the report, the government expects to come up with 'strategy', 'measures', 'private sector investment', and grants to specifically target 'areas with high levels of deprivation.' Moreover, government '... will note the progress of each local partnership in improving outcomes for local people through Local Enterprise Partnerships, Enterprise Zones and Rural Growth Networks', and oversee a gamut of neighbourhood initiatives (2012a: 4, 5, 10, 22).

Hence the paradox; government's 'non-intervention' localist policy is, in itself, an intervention (Moore et al. 1989: 76). As occurred in the 1980s, 'freeing-up the economy and reducing the role of the state, required a considerable degree of centralisation for its objectives to be achieved' (Atkinson and Moon 1994: 88–89). This concurs with Richards' (2008) view that successive British governments maintain a Westminster model of elite strategy-making and a centralised system of power over local governance.

Governing Networks 1945–97

The following two sections present a historical chronology of governing networks in the English urban regeneration policy context, from 1945 to the present day. Both sections examine why the politicised relationship between urban regeneration policy and the differentiated governing networks experienced difficulty in alleviating neighbourhood unemployment. Fig. 4.1 links network change to political administrations and policy field infrastructure. This section traces a timeline of governing networks through six network types from 1945 to 1997: closed networks; central-local urban partnerships; inner-city partnerships; market-driven networks; inter-organisational contracting networks; and public-private partnerships.

Closed Networks: 1945–68

In the inter-war period, the efforts of closed British networks often stalled, and were unrealistic or limited in scope, as can be seen from inter-war Cabinet minutes and reports. Economic policies at this time were ineffective as, by November 1932, there were approximately 2,858,000 registered unemployed. In response to this, the Treasury suggested the Cabinet Unemployment Committee might consider a permanent solution. Instead, they made three recommendations: (i) coordinate voluntary organisations assisting unemployed people across the country and provide grants for expenses, tools and equipment, (ii) occupy unemployed men and women in assisted schemes for no cash payment beyond the Unemployed Insurance benefit, including occupational activities (building playing fields and recreational grounds), special classes for unemployed people and physical training and (iii) support 'the Society of Friends' and local authorities to acquire land for use in allotment schemes (CAB 58(32) Appendix II 413–14). The Treasury advised postponing such schemes, but the Cabinet agreed that the Minister for Labour should allocate funds to voluntary organisations. The Prime Minister (Ramsay MacDonald) volunteered to oversee a new organisation involving ministers and a panel of experts to deal with schemes, trade and industry (CAB/23/72 C.P. 60(32): 440–1). Local authorities would also receive housing grants to recondition slum housing. High rents had also led to rent strikes, and tenants' associations were forming to campaign for social housing and tenants' rights.

Major policy event	Prime Minister	Politically constructed networks
Industrial strategy	Theresa May 2016-	
Devolution in selected areas for transport, skills, employment and business	David Cameron 2015-2016	
RGF Prime Providers RDAs disbanded Connexions disbanded Work Programme New Enterprise Allowance	David Cameron 2010-2015	Local enterprise partnerships (2010 -)
LSC disbanded Regional Assemblies disbanded	Gordon Brown 2007-2010	
LEGI JCP (merges ES and BA) LSC replaces TECs New Deal Programme Connexions RDA's Business Link Employment service Benefits agency	Tony Blair 1997-2007	Local strategic partnerships (2000 –2010)
	John Major 1990-1997	Public-private partnerships - SRB (1994 – 2002) Inter-organizational contracting networks-City Challenge (1991 – 1997)
Next step agencies TECs replace MSC 3.3 million unemployed Enterprise allowance Youth Training Scheme Community Programme Training Opportunities Scheme Skill centres Welfare benefit reforms	Margaret Thatcher 1979-1990	Market-driven networks (1979 – 1990)
	James Callaghan 1976-1979	Inner-city partnerships (1977 – 1979)
Employment, food, rate and rent subsidies Manpower Services Commission	Harold Wilson 1974-1976	
	Edward Heath 1970-1974	Central-local partnerships – Urban Programme and Community Development Programme (1968 – 1977)
Industrial Training Boards	Harold Wilson 1964-1970	
Post-war housing schemes Regional policy and investment Special Areas Act 1934		Closed networks

☐ Conservative ▨ Conservative-Liberal Democrat Coalition ■ Labour

Fig. 4.1 A timeline of British political influence in central-local governing network structure and central unemployment/employment policy infrastructure from 1960

Post-war, parliamentary governance networks supported the drawn-out process of civil service reform. The Northcote-Trevelyan Report (1854) recommended open competition for entry to the service to 'eliminate patronage', and 'silence three particular sets of critics': Chartists wanted to end corrupt aristocratic influence, sought enhanced efficiency and entrepreneurs desired greater economy (Lowe 2011: 30). However, the recruitment process excluded non-graduates; hence, reform was 'deliberately designed to be socially exclusive not inclusive' (2011: 30). Class bias in the civil service garnered more attention than the suppression of female representation, as until the late 1960s, governing networks were the preserve of professional white male enclaves controlling a system of privilege and recruitment; largely the Oxford-educated elite. This establishment supported closed networks outside formal democratic processes, in policy areas ranging from agriculture to housing and industry to health (Marsh 1983: 5; Richards and Smith 2002: 177–8). Corporate networks, business elites and trade unions also had privileged access to central government (Marsh and Locksley 1983: 38; Smith 1993: 33). Local government expanded from the 1950s to provide Welfare State services and amenities, and demand for raw materials and economic growth was buoyant (20% GDP) (Stoker 1991: 6). The rise in working-class collectivism and defiance against self-appointed authority, and the Labour Party's return to government in 1964, pressurised Whitehall executives, the administrative class and local town halls to modernise and address historic attitudinal problems; for example, hostility towards outsiders, isolated amateurish decision-making and reluctance to reform (Richards 2008: 24–27). Closed decision-making networks protected power and influence reserved to politicians, local government officials, architects and planners in terms of area-based reconstruction programmes for new town planning, urban growth and industrial change (Atkinson and Moon 1994: 21, 258). Citizens excluded from planning networks were unable to tackle the problems associated with poor-quality terraced housing, and, as a result, communities and amenities were lost to housing clearance programmes implemented during the 1950s and 1960s. Hence, the Skeffington Report (1969) (cited in Atkinson and Moon 1994: 180; Gilchrist 2004: 16) urged greater public participation in planning.

During the 1960s, housing estates reliant on state investment fell into disrepair, and a shortage of local jobs and facilities compounded social, environmental and economic problems (Power 2001: 734). Ethnic-minority residents removed from slum areas were not eligible for rehousing, and many settled in neighbourhoods adjacent to clearance areas (Edwards and Batley 1978: 30).[1] Poverty theory shifted to social pathology, blaming social problems on the behaviour and culture of residents from council-run estates. Thereby, instead of focusing on citizens' rights, paternalistic efforts evolved to reform behaviour through the provision of more welfare, social services, community housing services, and behavioural psychologists studied the poor. United Nations also endorsed community development from the 1950s. Thereafter, community and voluntary groups grew more prominent, encouraging active citizenship and self-improvement, from the social care of older people to youth clubs (Gilchrist 2004: 14–15). However, community activism is a contentious subject, falling between autonomous responses and state control, (see Craig et al. 2011 for a review of the debates).

In 1965, the Child Poverty Action Group formed to lobby ministers about rising family poverty. The group later claimed 'the poor had become "poorer" under Labour'; nevertheless, the unions refused to support them over the 'poverty trap' in wage negotiations (McCarthy 1983: 216, 228). In the absence of a central-local public interface, interest groups, associations and charities strengthened to fill service gaps and voice injustices (Mandell and Keast 2008: 179). Britain's insular global position, overshadowed by US industrial strength, prompted the government to renegotiate joining the European Community (French President de Gaulle had vetoed Britain's applications in 1963 and 1967) to recover a position of economic power.

Central-Local Urban Partnerships: 1968–76

By the late 1960s, the welfare state, physical capital accumulation and economic policy had failed to tackle structural inequality and inner-city poverty (Marsh 1983: 6). Social movements were also demanding gender and racial equality. In 1968, the Labour Prime Minister, Harold Wilson (who served during the period 1964–1970), launched major initiatives for local authorities in deprived urban areas to compete for assisted area status under the Local Government Grants (Social Need) Act (1969) and coordinate the Urban Aid Programme (UP) for coherent problem solving (Bache and Catney 2008: 413). The UP funded single projects in urban settings including advice information centres, special education projects, provision for children and community activities. None of which tackled 'the funda-

---

[1] The practice of discrimination in ethnic minority housing by public and private bodies continued until the Race Relations Act (1976).

mental issues of structural economic change that underlay the urban crisis'. (Pacione 1997: 24). However, the UP was criticised under Labour for being poorly conceived (Edwards and Batley 1978: 74–75). Ministerial responsibilities were confused, departments competed and undermined coordination and cooperation, and the inconsistent voluntary sector involvement unsettled local politics.[2] UP information was distributed ad hoc to the voluntary sector, problems lacked analysis, projects were not being cross-referenced or linked to programme objectives, and activities that warranted funding were predetermined. In 1968, the launch of 12 Home Office-coordinated Community Development Projects (CDPs) to encourage resident participation and neighbourhood self-help in deprived areas, caused resentment amongst local authorities. Several governmental departments administered CDP schemes and local action teams and university researchers documented the community development work. However, the CDP workforce rejected the social pathological perception of poverty and area-based strategy (Pacione 1997: 24, 26). Instead, they blamed uneven capitalist development and de-industrialisation, and supported tenant groups and neighbourhood councils to represent community interests and challenge the state (see Craig et al. 2011). Marxist, Foucauldian, Gramscian, Freirean and feminist theories offer perspectives on community action, paternalism and social policy professionalisation.

Edward Heath's Conservative government (1970–4) intended to reduce public spending and state intervention, but economic growth stagnated and unemployment rose to 1 million. Richards and Smith (2002: 96) claim Heath slipped from New Right policy to a Keynesian type intervention, the failure of which would later set in motion the Conservative Party's combative market-based approach to managing employment levels. In 1971, income policy escalated conflict between the government and trade unions, and the Industrial Relations Act (1971) restricted strike action. The car and aero-engine manufacturer Rolls-Royce Ltd. was nationalised. In 1972, the Department of the Environment's 'Inner Area Studies Team' enlisted consultants to investigate urban deprivation causation. Findings (published in 1976) were similar to those of the CDP (which had been abolished in the same year) and stated that economic and structural factors, not behaviour, caused poverty. Unsurprisingly, governments withdraw support for community development programmes if policy opposition is an outcome.

In 1973, Britain joined the EU. In the same year, oil prices quadrupled, resulting in inflation, high unemployment and reduced living standards

---

[2] Administrators included the Department of the Environment, the Department of Health and Social Security and the Department of Education and Science.

across Europe. The UK miners' strikes prompted the Labour government's return to power (Wilson/Callaghan 1974–79) and repeal of the Industrial Relations Act through the Trade Union and Labour Relations Act (1974). Labour inherited an economic recession, industrial closures, a skills export to the suburbs, unemployment at 1 million, high inflation at 25% and a disaffected private sector. Attempts to stimulate demand led to a crisis in the balance of payment, while stagflation (rising unemployment and inflation and low growth) undermined Keynesian policies. In 1976, an IMF loan required a reduction on the civil service workforce, and urban politics intensified.

Inner-City Partnerships: 1977–9

The first White Paper on urban decline, *Policy for the Inner Cities* (DoE 1977), identified four themes requiring national and local coordination: economic factors, physical environment, social factors and jobs.[3] Alongside UP, the Inner Urban Areas Act of 1978 gave Industrial Improvement Areas the power to assist private firms with loans and grants (Bailey et al. 1995: 45). Seven cross-sector Inner City Partnerships established teams involving central and local government, Manpower Services Commission and health authorities (Edwards and Batley 1978: 248).[4] The managerial style of these teams, however, attached too much importance to 'a "total" corporate approach and [the] need to coordinate the multiple agencies involved' (Parkinson and Wilks 1986: 293). Partnerships with industry, training and the voluntary sector were neither well established nor able to contest arrangements, and funding complexities hampered progress (Rhodes 1988: 343–64; Sullivan and Skelcher 2002: 59–60). Additionally, the White Paper failed to lead on the 1976 Race Relations Act, and the Commission for Race Equality lacked the power to ensure central urban initiatives were responsive to the needs of ethnic minorities (Atkinson and Moon 1994: 236). The UP remained a formulaic area-based grants system; its poorly coordinated policy direction reduced its effectiveness, and

---

[3] Hereafter, partnerships proliferated; see the list of 'Government initiatives for regeneration partnerships 1977–2001' in Sullivan and Skelcher (2002: 59, see Fig. 4.1).
[4] Post-war industrial policy had declined. Industrial support was ad hoc, hindered by poor relations between trade unions and the government and the separation of the banking system and industry. Research and development expenditure prioritised military over technological innovation and weakened British industry (see a history of manufacturing decline in Judge and Dickson 1991:12–17).

outcomes were negligible (Edwards and Batley 1978: 189; Atkinson and Moon 1994: 60–5). Inner-city policy materialised, but failed to establish synergy between government departments, and vertical local-central government relations continued. Community development separated local government, the voluntary sector and private sector. The political-economic climate worsened, traditional industries declined, and the Conservative's return to power in 1979 sought to address long-term unemployment through interventionist policy to promote 'enterprise' culture.

Market-Driven Networks: 1979–90

Thatcher wanted 'business in the community' to define urban intervention and rekindle the Victorian system of local business leadership (Lawless 1989: 127–32, 161–2). Property-led regeneration networks, business activism, market strategies, entrepreneurialism and wealth creation became the prerequisites for central funding and tackling unemployment (Bassett 1996: 540). In this way, free-market thinkers depoliticised the loss of job security, perceiving unemployment as an individual choice (MacKay 1998: 50–51). Citizens would need to draw on the market for support, not the state. Yet public expenditure on unemployment increased, as employment training support and job creation for inner-city needs were lacking (Atkinson and Moon 1994: 136, 138).

Local government services began outsourcing to actors from civil society in 'rowing agencies' while the centre 'steered' through regulation and 'managerial surveillance' (Cope and Goodship 1999: 6). The power shift was contentious:

> The more the balance of power shifted towards central government in the 1980s, the more it was able to insert its own values, methods and language into the new management practices and the more difficult it became for local institutions to shape the new methods in their own image and for their own purposes. (Burns et al. 1994: 85)

The phrase 'new management practices', as cited, refers to the 'New Public Management' (NPM), which was adopted in several countries including Britain to make public services more efficient through competition, marketisation and contracting-out (see Stoker 2004: 13; Pollitt 2003: 27). It intended to: (i) divide the core functions of large public organisations into smaller units; (ii) overcome institutional inefficiency through private-sector styles of management, performance measures, evaluation and pay-incentives; and (iii) reduce public services expenditure by coordinating contracts managed through tender-competition (Moore 2000: 100–1; Lane 2000: 147; Burns et al. 1994: 84). Its critics suggest that NPM increased the need for coordination and bureaucracy, and impaired service delivery, democratic processes, social values and outcomes for citizens. Decentralised, fragmented institutions struggled to deliver horizontally coordinated integrated services at street level (Perri 6: 2004: 110). Public governance exposed voluntary and private sectors to market conditions, but networks between funder and contract holders reaffirmed power relations and strained cooperative transactions (Painter et al. 1997: 229, 236). This contradicts Rhodes' (1997) expectations of governance:

> ... *governance refers to self-organizing, interorganizational networks* characterized by interdependence, resource exchange, rules of the game and significant autonomy from the state. (Rhodes 1997: 15 original emphasis)

Yet Rhodes acknowledges that government departments steered a wider set of actors in networks to progress governmental policy and goals (Rhodes 1997: 45). Voluntary associations had already used networks to confront inequalities, fill service gaps, campaign for local authorities to do more in poor neighbourhoods and challenge central government policies (Mahoney et al. 2000: 804). But grant aid was reduced, and the 'third sector' stayed independent by fundraising and mimicking business strategies to gain contracts. From 1980, public-private alliances, urban development corporations (such as the London Docklands Development Corporation (1981–98) that regenerated Canary Wharf) and enterprise agencies, levered funding to stimulate property-led regeneration in poor areas, emphasising places rather than people (see Sullivan and Skelcher 2002: 60). Market interventions, however, did not reduce poverty, and central urban policy confused provision (Taylor 2000a: 8–9). Coordination dysfunction amongst the many organisations promoting employment is said to be 'one cause of the continuing problem of urban unemployment' (see Atkinson and Moon 1994: 137). The Department of the Environment's City Action Teams (1985) and the Department of Employment's (DoE) Inner City Task Forces (1986) urged better public management, but coordination between civil servants and the public and private sectors did not improve. Schemes' multiple funding was criticised (National Audit Commission 1989), and teams decentralised to the Regional DoE offices and local authorities and a new timetable was devised for submitting UP bids (Atkinson and Moon 1994: 119).

The second major civil service reform was outlined in Jenkins, Caines and Jackson's *Next Steps* Report (1988). This process would free Whitehall's core executives to define policy frameworks, 'hollow-out' the 'congested state' system and relocate large numbers of civil servants and ministerial functions to national and sub-national executive agencies (Taylor 2000a: 47–48). These quasi-autonomous non-governmental organisations (quangos) appointed non-elected chief executives and boards. They expected senior business managers to redress the overloaded bureaucracy in service delivery functions, and provide business acumen, better quality services that met 'customer' needs, and a performance management culture to set standards, achieve savings, measure outputs and outcomes against targets, whilst retaining a high degree of discretion over pay and recruitment (Richards and Smith 2002: 108–09). Some agencies continued to govern with the closed network culture of the old rule-bound civil service, and authoritative technical decision-making. The Child Support Agency and Benefits Agency reorganisations were beset with bureaucratic problems, overspending, a lack of accountability, poor client communication, and unrealistic targets, and were preoccupied with image-management.

Quangos also fractured local authorities' identity as a public housing provider, and paved the way for Housing Action Trusts, established under the Housing Act 1988, to regenerate many inner-city social housing estates. The politics of property and profits divided national from local government. Right-to-buy-schemes enabled tenants to buy council homes or transfer to housing associations. Networks were at an ideological crossroads. Right-wing ideologies promoted 'markets, competition and individual choice'; the urban left wanted 'democracy, participation and collective control'; and the Liberal party chased a local government capable of a 'radical form of neighbourhood decentralisation' (Burns et al. 1994: 19, 24, 85, 146). Furthermore, the European Community's structural funding, which was distributed from 1988 onwards to help countries with poverty issues (including Britain) to address social exclusion, wanted community economic development to improve social and economic cohesion, thereby challenging Thatcher's neo-liberal regeneration programmes (Haughton et al. 1999: 209). Hence, the next Conservative Prime Minister, John Major (1990–7) coordinated authority over urban problems through 'horizontal integration' (Rhodes et al. 2003: 1401; Perri 6 2004: 121).

Inter-Organisational Contracting Networks: 1991–7

By 1991, national incomes fell by 2.4% and negative growth followed in 1992 and 1993 (Kitson and Michie 1994: 95). During the recession, the Conservatives rationalised and introduced competitive partnerships, referencing the Victorian tradition of enterprise and liberalism. As Michael Heseltine, Minister at the DoE, explained in a speech to the Manchester Chamber of Commerce and Industry:

> ... when I speak of the need for a sense of partnership in our modern cities, it is today's equivalent of that Victorian sense of competitive drive linked with social obligation. (Heseltine 1991 cited in Atkinson and Moon 1994: 122)

On this premise, Heseltine launched City Challenge (1991–3) in 57 UP areas; this was the first programme to promote partnerships and governance, integrated strategies, and competition for short-term funding for jobs and enterprise projects targeting cohorts. (Skelcher 2004: 32–33, Mawson and Hall 2000: 68).[5] To bid, local authorities had to form multi-sector partnership boards with local stakeholders and submit a plan to lever inward investment in rundown areas and address local economic needs and quality of life (Smith and Beazley 2000: 860). Regional offices coordinated the programme as a 'controller and contractor rather than partner' (Stewart et al. 1999: 11). Unlike inter-agency collaboration that has no central command agent inter-organisational contracting networks issue contracts to those groups with the capacity to complete funding applications with preset targets and achievable outcomes in a limited time scale, often developed in isolation from other applicants.[6] Challenge partnerships were not models of local policy coordination; they inserted state managerial power, regulated community processes and quantified resource outputs, whereby 'What counts is what is done' (Taylor 2000a: 8).[7] The bureaucratic system of data collection was costly, seldom utilised, and abhorrent to some of the 'partners', who had a different understanding

---

[5] City Challenge areas received £37.5 m over five years and aimed to attract private sector investment.

[6] See inter-agency collaborative and inter-organisational contracting-network distinctions in Page (2004: 603 see 2. in notes).

[7] For example, staff managing City Challenge-funded projects recorded and monitored vast numbers of outputs, including descriptive accounts of telephone calls made during working hours. Thereafter, monitoring became standard project management practice.

of 'partnership' and 'collaboration' (Sullivan and Skelcher 2002: 189). 'Making people write things down and count them ... is itself a kind of government of them' (Rose and Miller 1992: 187). Partnerships distanced mainstream providers from voluntary groups , who in turn, maintained policy-setting independence but the managerial politics over planning and budgets were likened to an 'assault' on localism (Wilson 1992: 172; Ball 1995: 10). City Challenge introduced policymakers to the term 'regeneration,' and launched training companies, many of whom devoted time and effort to work with the local authority and steer bids, yet it could not overcome the unemployment problems. As Hastings (1996: 255) comments, an outcome in which partnerships contracted separate projects to turn neighbourhoods round, supported public sector reform, imbued the private sector with social obligation and transformed culture across sectors for mutual appreciation, seemed unlikely.

Public-Private Partnerships 1994–2002

By the mid-1990s, funding bodies had fractured network accountabilities in local authorities. EU directives had created community-led economic and social regeneration partnerships pushing community development themes (neighbourhood governance, sustainable development and social inclusion) (Geddes 2000). The EU URBAN community initiative (1994–9) established networks for grass-roots groups and organisations to capacity-build, shape funding bids aligned to EU themes, and tailor training and employment support to neighbourhood needs. However, complex funding criteria, cumbersome bureaucracy and area-targeting sometimes overlapped or changed mid-programme, throwing the scheme's administration into chaos; moreover, support for residents was a postcode lottery as some estates or areas were not eligible. Nevertheless, local governance enlarged. Local council departments intensified collaboration with the voluntary sector; housing services now had a duty to engage residents in representative democracy and find out their needs; and the economic development unit facilitated network infrastructure for socio-economic decision-making. The next governance inception, implemented in 1994, was the Single Regeneration Budget (SRB), which integrated 20 urban regeneration programmes in England into a single budget, and encouraged public-private partnerships in urban locations to bid for funds. The prize-competition model combined private sector elements similar to the mature UP, and to the enterprise culture and competitive bidding process promoted by the City Challenge. Six annual bidding rounds followed. The Conservative government delivered the first three rounds and Labour took over the middle of the fourth round in 1997. SRB applications were scored by the 10 Government Offices for the Regions of England (GOR), which had formed in 1994 to integrate the regional offices of the DoE, Department of the Environment, Department of Trade and Industry (DTI) and the Department of Transport. GOR supervised SRB packages for the first four rounds and the Regional Development Agencies (RDAs) in the last two. They monitored EU structural funds and community economic development programmes for training and employment, guided SRB match funding and consulted local authorities on priorities in specific neighbourhoods (Bache 2000: 588).

SRB bids were assessed against national and regional criteria, not local policy or capacity to manage local partnerships; nevertheless, partnerships had to demonstrate several competencies to the government (Mawson and Hall 2000: 69). For example, a governance structure and coalition of local interests, a coordinated, strategic response to regeneration, offering good value for money, bending mainstream services to public resources, increasing public, private and voluntary resources in an area, and improving the image of the area and its vicinity (Rhodes et al. 2003: 1414). Project bids could also include the topics of employment, economic growth, the environment, housing, ethnic minorities, crime, health, and cultural and sports opportunities. Local authorities led 'community-inclusive partnership boards' in 53% of all SRB schemes (DTLR 2002: 13).

The effectiveness of SRB partnerships received mixed reviews (see Rhodes et al. 2003: 1414–17). These researchers (see Rhodes et al. 2003) also completed an evaluation for government of ten SRB schemes in different areas, from Wolverhampton to Rochdale. Although strategic regeneration improved, partnership configurations in one place failed to materialise or abandoned midway through; one partnership had inadequate supervision of locally delivered services; another failed to deliver community involvement in the regeneration process (DTLR 2002: 104, 118, 127). On one estate, inadequate responses to worklessness was seen as typical of post war regeneration initiatives, such that sectors failed to overcome cultural differences and work together; specifically, the entrepreneurial private sector, the regimented culture of mainstream programmes and their unsuccessful policies in the area, and culture of multiple deprivation (DTLR 2002: 8, 122). Cross-sectorial SRB partnership boards consisted largely of senior public sector managers, as voluntary sector partners lacked the resources required to participate (Sullivan and Skelcher 2002: 29,134; Mawson and Hall 2000: 69). Some critics viewed the SRB process as biased towards physical and economic development; moreover, saw its piecemeal approach to local economic and social development as a waste of resources, which undermined the legitimacy of the local authority and community institutions (Jones and Ward 1997: 155; Davies 2001; Tofarides 2003 104). Support for ethnic minorities facing discrimination and disadvantage was also lacking. Recent analysis of the SRB competition process throws doubt on whether it brought about any substantial efficiency gains in bidding tournaments, since winning competitors offered what they already did well and the competition was limited to residual projects (Ward and John 2008: 61). Despite the funding façade, the SRB stimulated a process of interconnecting local projects and allocating resources. Yet the weakness of its partnership accountabilities, exit strategies, guidance and institutional linkage diminished its full impact (Hastings 1996: 254). Rhodes et al. (2003) suggest that the SRB did not pay sufficient attention to outcomes. Between SRB rounds, the Jobseekers Act 1995 made job training compulsory. In 1996, the JSA replaced unemployment benefit and income support; claimants were required to accept any job offers and training offered. The Employment Service used private sector training providers and business start-up programmes to target the long-term unemployed and stimulate an 'enterprise culture'. Nevertheless, beyond SRB networks, the Business Support Service, the TECs and small organisations supporting employment issues were weakly coordinated.

GOVERNING NETWORKS AFTER 1997

Under Prime Minister, Tony Blair (who served during the period 1997–2007), the next Labour government issued new SRB guidance emphasising the capacity-building of local partnerships to develop cross-cutting policy in deprived areas. SRB partnerships paved the way towards governing problem areas using scaled-up networks in order to drive central state planning, active community leadership and stakeholder-associational policies (Mawson and Hall 2000: 70). In effect, governments can hijack partnerships with preferred ideologies and ring fence funding interests and network participants. Such that, Bache and Catney (2008), contrast SRB approaches, between Conservative governments 'economic rationalism' and New Labour's 'embryonic associationalism'.

For example, the Cabinet Office launched the Social Exclusion Unit (SEU) to tackle social and economic inequalities and identify the cross-cutting issues and causes of neighbourhood deprivation. Similarly to other central units, the SEU intended to counter the perceived silo mentality and lack of policy coordination between departments, and facilitate departmental 'joined-up working' (Ling 2002: 623; McGregor et al. 2003: 1). In addition, the SEU managed the Neighbourhood Renewal Unit (NRU) and reported to the Minister for Local Government, Regeneration and the Regions, as well as to a Cabinet-level committee chaired by the Deputy Prime Minister. The NRU focused on network infrastructure to combine policy areas and manage services more efficiently in disadvantaged neighbourhoods. Sullivan and Skelcher (2002: 21–2) have suggested that partnership working post-1997 differed from previous arrangements as partners were expected to focus on joint outcomes, 'which are not necessarily organisationally specific', and granted a longer time span in which to implement cross-cutting outcomes. Network idealism was fast-tracked, not to devolve power for 'modern governance' but to govern by network and defend the strategic interests of the core executive and the Westminster model (Richards 2008: 138–40). The network governance process involved 'filling-in' central control to reorder policy objectives, in 'response to the transaction costs that arise from collaborative working in a hollowed-out environment' (Sullivan and Skelcher 2002: 18–21). However, as the joined-up goals faded, the centre increased the pressure to identify with it, convoluting network processes for decision-making and raising their profile still more, as outlined by the policies we are about to outline.

The SEU's consultation paper, 'Bringing Britain Together: A National Strategy for Neighbourhood Renewal' (SEU 1998), contrasts the policy failings of previous governments with policy solutions. For example, the 'economic ghettoisation' of neighbourhoods requires unemployed people to help themselves; the erosion of social capital requires more contact, trust and solidarity between residents. Core public services in deprived areas had used separate budgets, targets and outcomes to improve national averages, rather than being accountable to the communities they served. The role of private sector services had been ignored, and required a strategy for joint action. No one had responsibility at neighbourhood, local, regional or national level to ensure services work together to achieve common goals, and the role of communities and others who could offer help had been discounted (SEU 1998: 4). Thereafter, 18 thematic Policy

Action Teams (PATs) published separate reports on facets of neighbourhood renewal. The PAT 17 report, *Joining It Up Locally* (DETR 2000), explored the best method of delivering local coordination nationally, regionally and locally. In the foreword to the report, Hilary Armstrong, Minister for Local Government and the Regions advocates that localities focus on central objectives:

> Joining up policy and delivery at local level will be crucial to achieving the ambitious goals of the National Strategy for Neighbourhood Renewal. (DETR 2000: 5)

Yet the report's summary of its findings implies that local steering and a 'bottom-up' approach works best:

> ... much of the successful local joint working that we have seen results from local initiative, not central direction. (DETR 2000: 9)

PATs also noted an overlap of management systems for different area-based initiatives. Researchers (see Stewart et al. 1999: 2; Cabinet Office 2000; McGregor et al. 2003) had already found that the national and regional government machinery and the design of area-based initiatives were primary constraints for local government in handling cross-cutting issues and that this structure had led to initiative overload and coordination failure. Nevertheless, structural problems lacked critical attention as (i) partnership working, (ii) injecting money into deprived areas and (iii) rallying communities to help themselves, were the vanguard of joining up, and PAT 17 recommended 'Local Strategic Partnerships' (DETR 2000: 54–5, 60–8).

Local Strategic Partnerships 2000–10

Local Strategic Partnerships (LSP) intended to close the poverty gap in deprived local authorities, and marked a new development in governing by network (Goldsmith and Eggers 2004: 9). Blair recognised that social problems join up and wanted networks and mainstream services to tackle problem pathways together. From 2000, the NRU issued £800 million Neighbourhood Renewal Funding (NRF) (2001–04) to 88 local authorities ranked highly on deprivation indices to deliver community planning through LSPs (SEU 2001: 10, 48). These LSPs would supersede existing area-based partnerships and 'network the networks' as a single governance body and concentrate decision-making authorities. The latter would include local government officials, the business community, the voluntary sector, institutions, funding quangos, regional government and development agencies in a pyramid of thematic hierarchical networks for tackling work and enterprise, crime, education and skills, health, and housing and physical environment areas, accountable to central monitoring systems. Using NRF and network management approaches, LSPs intended to coordinate community plans and goals. NRF also influenced voluntary sector community networks to mirror LSP themes, but there is little evidence that they negotiated power in neighbourhoods beyond the state (Taylor 2007: 299). Indeed, sectorial democratisation requires a cultural shift, 'without which structural manipulations and constitution writing will produce little positive result' (Peters 1999: 88).

In principle, the LGA accepted the LSP goals of making public services work better in localities, given that past programmes had not succeeded in this, but locally they wanted the freedom to negotiate the national targets.[8] Yet the literature cautions against over-commitment:

> ... indiscriminate cloning of fashionable models into areas where they are deeply inappropriate is a cardinal recipe for the production of reverse effects through over-commitment. (Hood 1998: 220)

As Sullivan (2004: 188–90) observes, LSPs shaped 'community governance' structures more like a 'community government' model.

Regional Governance

In 1998, the Regional Development Agencies Act devolved power to Scotland, Wales and Northern Ireland, and in England eight Regional Assemblies or Chambers were established to preside over eight RDAs (Richards and Smith 2002: 250–7). The London Assembly was established through separate legislation as part of the Greater London Authority. The

---

[8] Various strategic partnerships were already established, from the LGA's 'New Commitment to Regeneration' initiative, Health Action Zones, to Crime and Disorder Partnerships. Subsequently, national and community bodies welcomed *A New Commitment to Neighbourhood Renewal: A National Strategy Action Plan* (SEU 2001; Catney 2009: 50–3).

RDAs' vision was to pursue economic renaissance to 'enhance the employment prospects, education and skills of local people' (DETR 1998: 7). This assumed that localities worked to regional objectives, instead of the latter working for the former. As the Deputy Prime Minister at that time, John Prescott, set out in his vision:

> Regeneration Partnerships will have to make clear the linkages between housing, regeneration and other initiatives and demonstrate that their local strategies fit within and complement the regional strategies that the Regional Development Agencies will develop. (DETR 1998: 6)

Regional bodies established prominence within the multi-level regional governance chain, but lacked the powers or policies to combat regional inequalities, and 'remained relatively weak compared to the enduring power of central government' (see Tomaney 2004: 168, 180). The RDAs' regional economic strategy, for example, required integrating regional strategies across national policy spheres and consolidating partnerships between Jobcentre Plus (JCP), the Learning and skills council (LSC), the Small Business Service and the Local Authority, but their goals were impractical and protected self-referencing targets at different spatial scales (see also Crisp et al. 2014). The RDAs' governance role administrating the ERDF and ESF in respective regions adopted a 'light touch' approach, but left people muddled by initiatives on the ground. Furthermore, local agencies competing for co-finance (LSC and JCP) contracts lost points on the applications scoring system if they failed to endorse regional objectives, even though many had little contact or awareness of RDAs. Regional and sub-regional closed networks are also the means by which quango boards influence and distribute discretionary funding. The EU consulted the LSC, for example, on skills and training and business and enterprise needs in a region, and the LSC in turn consulted the RDA, but blocked them from consulting their networks (business and enterprise representatives) in case any applied for LSC contracts later on.

RDAs had little control over quangos, and vice versa. Richards and Smith (2002) predicted that 'new structures of dependency are likely to make greater coordination and joined-up government difficult to achieve because the government will not have direct control over the devolved bodies' (2002: 258). Regional Assemblies also had a precarious existence outside the public sector jurisdiction, and the Conservative Party's paper *Control Shift* (2009) stated its intention to abolish them upon re-election

(Conservative Party 2009: 27–30). However, Gordon Brown, Labour Prime Minister in 2007–10, abolished Regional Assemblies in 2010 and replaced them with local authority leaders' boards.

Local Enterprise Partnerships 2010

In 2010, the Conservative-Liberal Democrat coalition government replaced Labour's network infrastructure of nine RDAs, 360 LSPs and associated neighbourhood forums, which had been established under the NRU remit, with 39 Local Enterprise Partnerships (LEPs) covering local parts of England. They also abolished the Working Neighbourhood Fund (WNF) and national performance management frameworks, including, sub-regional/regional targets and multi-area agreements, local-area agreements and neighbourhood plans. In June 2010, the Secretary of State for Business, Innovation and Skills and the Secretary for Communities and Local Government invited local authority leaders and business leaders to establish joint local-authority/business-led LEPs, 'to rebalance the economy toward the private sector' and stipulated partnerships parameters, including role, governance and size. Furthermore, they requested LEP proposals 'reflecting the Coalition Government's agenda' to be returned by September 2010 (DBIS 2010a). Community leaders were not addressed.

As Pugalis and McGuiness (2013: 352) observe; 'regeneration is no longer locked into repairing the destructive tendencies of capitalist uneven development; rather, it is locked into assisting the quest for continued economic growth.' The White Paper, *Local Growth: Realising Every Place's Potential* (DBIS 2010b), offered guidance and criteria for assessing LEP proposals. The LEPs were expected to assimilate some of the RDAs' activities and operate as strategic bodies on a voluntary basis with no start-up or running costs, training or outreach costs. The RDAs' duties ended on March 2012, but the time taken to establish new structures left a massive funding gap on the economic landscape between 2011 and 2013. In 2011, a one-off £5 million LEP start-up fund was announced with the potential of £45 million local match funding. LEPs were approved by December 2011, but setting them up took longer. 'By March 2012, nine had published their growth plans' (NAO 2013: 29). A recent analysis of 39 LEP attributes and functions finds the average board membership is 15, generally equally split between public/private sector representatives, and 15 % of whom are women (see House of Commons Library 2014).

The public sector representatives are usually council leaders, local councillors, or from schools, colleges and universities. The LEP chairs should be private sector representatives; appointed chairs represent large national companies, from consulting firms to energy providers.

The main LEP activities are to deliver strategic economic growth plans to central government; and manage funding streams for private sector welfare to meet the challenges of global competition and a low carbon future; job creation; and to 'encourage partnerships working in respect to transport, housing and planning as part of an integrated approach to growth and infrastructure delivery' (DBIS 2010b). However, their self-reported assessments are not audited against initial criteria, nor is it 'assessed whether reported jobs have been double counted', and funding streams overlap for job creation in some areas (NAO 2013: 31). The fact that several departments are overseeing LEP initiatives, whilst using different reporting procedures and with varied understandings about localism and devolution, has led to confusion. Thus, since summer 2012, a local growth cabinet committee was established, chaired by the Deputy Prime Minister to 'help design strategy and identify gaps' (NAO 2013: 35). However, DCLG leads on LEP's and other reforms, such as devolution deals.

Trade unions have concerns that LEP governance has failed to engage them and that self-selective board appointments are mainly of white male private sector representatives and local authority leaders. Moreover, LEPs are trying to deliver in a depressed economy. The main funds and loans available across England include the £3.2 billion Regional Growth Fund (RGF), running from 2011/2012 to 2017 for match funding with private investment in areas affected by public spending cuts; and the Growing Places Fund (£730 million) for infrastructure projects intended to unlock economic growth, job creation, commercial space, transport schemes in locations with investment constraints and house-building. Local councils are also eligible for New Homes Bonus payouts from a grant of £2.2 billion between 2011 and 2015 for increasing Council Tax revenue raised through new house-building and bringing empty homes back into use. From December 2011, cities can access a £100 million capital funding pot for broadband infrastructure, and new powers through 'City Deals' to invest in growth, in exchange for demonstrating strong accountable leadership, efficiency and innovation. The government's 'Do It Your Way' press release (DBIS 2011) recommends two innovations to coordinate skills and jobs: first, City Apprenticeship Hubs to place apprenticeships with employers and administrate the apprenticeship programme; and second, to physically integrate employment support services. As stated by the press release: 'Instead of being passed from one service to another from Jobcentre Plus to town hall to careers advisor—all of that can be done under one roof where it makes sense to do so' (2011). From 2015, the Local Growth Fund will replace the RGF, and £2 billion a year from 2015–16 to 2020–21 will fund Individual Growth Deals in each of the 39 LEPs for major transport projects, superfast broadband and business advice for young people. The funds expect to create 419,500 jobs and build 224,300 homes.

Funds are also available to create jobs and boost businesses for 24 LEPs hosting Enterprise Zones (EZs) (operational from April 2012), which are selected through competitive bids to attract commercial and industrial businesses in underdeveloped areas. EZs can access Local Infrastructure Funding with a pot of £59 million to encourage large-scale housing and support enabling works, and a Capital Grant Fund of £100 million to support commercially viable infrastructure projects. EZ incentives include 100 % business rate discount up to £275,000 over five years, simplified planning approaches, Enhanced Capital Allowances for machinery and equipment, and tax allowance for building costs. LEPs and local authorities in the relevant EZ can retain a percentage of business rate growth for reinvestment in local economic development for up to 25 years. Four EZs received a further £15 million to encourage high-tech firms to locate near to universities. The LEPs are also managing some EU Structural and Investment Funds (combining the ERDF and the ESF (£5.3 billion) and EU strands for rural areas, agriculture and fisheries) for 2014–20. Some LEPs will manage 'The Youth Initiative,' a £170-million Department for Business, Innovation and Skills (DBIS)- and Department for Work and Pensions (DWP)-funded youth employment programme to help young people into work, and match funded by ESF and project partners to nearly £490 million. LEPs face similar problems to those of the business-led partnerships introduced by the Conservative government during the 1980s. These include government pressure to engage the private sector, tensions in different cultural areas, a reliance on individual commitment, accountability problems and capital subsidies, and the fact that support to small firms may only achieve a relatively small number of job outcomes and that those might be taken by people already employed (Hasluck 1987: 180; Lawless 1989: 68–72; Moore et al. 1989: 59; Harrison 2011; Crisp et al. 2014: 21–7).

While 'The Government wishes to see partnerships which understand their economy and are directly accountable to local people and local businesses' (DBIS 2010b: 12), failure to induce a receptive employment structure joined up to unemployment representation will remain costly to society. Parliamentary inquiries have warned LEPs to learn from the RDAs' democratic deficit and lack of clear strategy (BISC 2011). The NAO notes that major outputs are off track, and that the estimate for job creation by 2015 in EZs 'has dropped from 54,000 to between 6,000 and 18,000' (NAO 2013: 10, 31). Indeed, only 4649 jobs were generated by December 2013. The Growing Places Fund expected to deliver 217,000 jobs, 5300 business and 77,000 houses, but outcomes in 2014 are 419 jobs, 3 businesses and 155 houses. The RGF, at the three-year point, had distributed £750 million to companies, delivering 90,000 jobs and increasing job expectations from 328,000 to 550,000 for outcomes to the mid-2020s. Yet, actual jobs monitored or safeguarded to December 2013 stand at 44,400, while costs per job have increased from £30,000 in round one to £52,300 in round four (CoPA 2014: 6, 16). Funding 'surges' can lead to unrealistic goal-setting if needs are not properly assessed. The DCLG estimate £85 million underspend for Local Growth Fund projects in 2015–16. The NAO's recent assessment of DCLG's lead on LEPs, raises various concerns about LEP capacity, from the quality of local assurance frameworks to the varying level of financial transparency, furthermore, 87 per cent of LEPs did not disclose senior staff pay (NAO 2016: 35, 42–45). The presumption that growth funding will counter neighbourhood unemployment lacks evidence to support it (Crisp et al. 2014: 24–31). Local employers within regeneration areas may still not appoint unemployed people. These uncertain outcomes call into question LEP's leadership and the government's genuine commitment to support unemployed people.

Conclusion

This chapter identifies eight governing network trends in urban regeneration policy since 1945. Successive governments have established governing networks to endorse political authority over local authorities and coerce stakeholders to manage three strategies. Political strategy reinforces the ideology underpinning economy, unemployment, and public sector reforms. Coordination strategy implements the government's policy priorities and multiple funding streams. Governance strategy establishes relational parameters, rules, rituals and information-sharing to ensure participants support the policy framework and funding opportunities. High-profile networks try to make urban regeneration more inventive, yet despite their intentions, the policies and the quality of joint working have been unable to boost employment sufficiently in disadvantaged neighbourhoods or solve problems at the scale required. The lack of unemployment representation in networks has allowed distributive justice to drift. Poor coordination between government departments, multiple programmes and complex funding structures have put a strain on policy implementation and delivery on the ground. Consequently, multiple stakeholders vie for attention and funding, and this is difficult to coordinate in governing networks. Part 2 (Chapters 5 to 8) applies the investigative framework (see Chap. 1) empirically, which permits a systematic review of the range of networks in local authorities with high unemployment and their network outcomes (see Chap. 5–7).

Bibliography

Atkinson, R., & Moon, G. (1994). *Urban policy in Britain*. Basingstoke: Macmillan.
Bache, I. (2000). Government within governance: Network steering in Yorkshire and the humber. *Public Administration, 78*(3), 575–592.
Bache, I., & Catney, P. (2008). Embryonic Associationalism: New labour and urban governance. *Public Administration, 86*(2), 411–428.
Bailey, N., Barker, A., & MacDonald, K. (1995). *Partnership agencies in British urban policy*. London: UCL Press.
Ball, R. M. (1995). *Local authorities and regional policy in the UK: Attitudes, representations and the local economy*. London: Paul Chapman.
Bassett, K. (1996). Partnerships, business elites and urban politics: New forms of governance in an English city? *Urban Studies, 33*(3), 539–555.
Burns, D., Hambleton, R., & Hoggett, P. (1994). *The politics of decentralisation*. Basingstoke: The Macmillan Press.
Business, Innovation and Skills Committee (BISC) (2011). *The new local enterprise partnerships: An initial assessment: Government response to the committee's first report of session 2010–11, House of commons*. London: The Stationery Office.
Cabinet Office (2000). *Reaching out: The role of central government at regional and local level*. London: Performance and Innovation Unit, Cabinet Office.
Cabinet Office (2011). *Unlocking growth in cities*. London: Cabinet Office.
Cameron, D. (2010). *Big society speech*. 19th July 2010 [online]. Transcript retrieved from http://www.number10.gov.uk/news/big-society-speech/
Catney, P. (2009). New labour and joined-up urban governance. *Public Policy and Administration, 24*(1), 47–66.
Committee of Public Accounts (CoPA). (2014). *Promoting economic growth locally, sixtieth report of session 2013–14*. HC 1110. London: The Stationery Office.

Communities and Local Government Committee (CLGC). (2012). *Government response to the house of commons communities and local government committee report of session 2010–12: Regeneration.* Cm8264. London: The Stationery Office.

Conservative Party. (2009). *Control shift: Returning power to local communities.* Responsibility Agenda, Policy Green Paper No. 9. London: The Conservative Party.

Cope, S., & Goodship, J. (1999). Regulating collaborative government: Towards joined-up government? *Public Policy and Administration, 14*(2), 3–16.

Craig, G., Mayo, M., Popple, K., Shaw, M., & Taylor, M. (Eds.). (2011). *The community development reader: History, themes and issues.* Bristol: The Policy Press.

Crisp, R., Gore, T., Pearson, S., & Tyler, P. with Clapham, D., Muir, J., & Robertson, D. (2014, July). *Regeneration and poverty: Evidence and policy review.* CRESR, Sheffield Hallam University.

Davies, J. S. (1999). *Urban regime theory in critical perspective: A comparative study of public-private partnerships in UK local governance.* Doctoral Thesis, York: York University.

Davies, J. S. (2001). *Partnerships and regimes: The politics of urban regeneration in the UK.* Aldershot: Ashgate.

Davies, J. S. (2003). Partnerships versus regimes: Why regime theory cannot explain urban coalitions in the UK. *Journal of Urban Affairs, 25*(3), 253–269.

Davies, J. S. (2005). The social exclusion debate: Strategies, controversies and dilemmas. *Policy Studies, 26*(1), 3–27.

Davies, J. S., & Pill, M. (2012). Hollowing out neighbourhood governance? Rescaling revitalisation in Baltimore and Bristol. *Urban Studies, 49*(10), 2199–2217.

DCLG, & DWP (2007). *The working neighbourhood funds.* London: DCLG.

Department for Business Innovation and Skills (DBIS). (2010a, June 29). *Letter re proposals for local enterprise partnerships.* From the Secretary of State for BIS and the Secretary of State for DCLG.

Department for Business Innovation and Skills (DBIS). (2010b). *Local growth: Realising every place's potential.* Cm 7961. Norwich: The Stationery Office.

Department for Business Innovation and Skills (DBIS). (2011). *Do it your way – Deputy Prime Minister launches new 'City Deals'.* Press release, 8 December 2011. Retrieved from http://www.gov.uk/government/news/do-it-your-way-deputy-prime-ministerlaunches-city-deals

Department of the Environment (DoE). (1977). *Policy for the inner cities.* Cmnd 6845. London: HMSO.

Department of the Environment, Transport and the Regions (DETR) (1998). *Housing and regeneration policy. A statement by the Deputy Prime Minister.* London: DETR.

Department of the Environment, Transport and the Regions (DETR). (2000). *Joining it up locally.* Policy Action Team 17 Report. London: DETR.

Department for Transport, Local Government and the Regions (DTLR). (2002). *Neighbourhood regeneration: Lessons and evaluation evidence from ten single regeneration budget case studies.* Urban Research Summary, No. 1. London: DTLR.

Edwards, J., & Batley, R. (1978). *The politics of positive discrimination: An evaluation of the urban programme 1967–77.* London: Tavistock.

Evans, G. (2005). Measure for measure: Evaluating the evidence of culture's contribution to regeneration. *Urban Studies,* 42(5/6), 1–25.

Geddes, M. (2000). Tackling social exclusion in the European Union? The limits to the new orthodoxy of local partnership. *International Journal of Urban and Regional Research,* 24(4), 782–800.

Gilchrist, A. (2004). *The well-connected community. A networking approach to community development.* Bristol: The Policy Press.

Goldsmith, S., & Eggers, W. D. (2004). *Governing by network. The new shape of the public sector.* Washington, DC: The Brookings Institution Press.

Harrison, J. (2011). *Local Enterprise Partnerships.* Loughborough: Loughborough University.

Hasluck, C. (1987). *Urban unemployment.* Harlow: Longman Group UK.

Hastings, A. (1996). Unravelling the process of 'Partnership' in Urban Regeneration Policy. *Urban Studies,* 33(2), 253–268.

Haughton, G., Lloyd, P., & Meegan, R. (1999). The re-emergence of community economic development in Britain: The European dimension to grassroots involvement in local regeneration. In G. Haughton (Ed.), *Community economic development* (pp. 209–224). London: The Stationery Office with the Regional Studies Association.

HM Treasury. (2007). Increasing employment opportunity for all. In *The budget* (pp. 57–73). London: The Stationery Office.

Hood, C. (1998). *The art of the state, culture, rhetoric and public management.* Oxford: Oxford University Press.

House of Commons Library. (2014, June 27). *Local Enterprise Partnerships,* Standard Note SN/EP/5651.

Jenkins, K., Caines, K., & Jackson, A. (1988). *Improving management in government: The next steps, report to the Prime Minister.* London: HMSO.

Jones, P., & Evans, J. (2013). *Urban regeneration in the UK* (2nd ed.). London: Sage.

Jones, M. R., & Ward, K. G. (1997). Crisis and disorder in British local economic governance: Business link and the single regeneration budget. *Journal of Contingencies and Crisis Management,* 5(3), 154–165.

Judge, D., & Dickson, T. (1991). The British state, governments and manufacturing decline. In G. Esland (ed.), *Education, training and employment, volume 1: Educated labour – The changing basis of industrial demand.* Wokingham: Addison-Wesley in association with The Open University.

Kitson, M., & Michie, J. (1994). Depression and recovery: Lessons from the interwar period. In J. Michie & J. Grieve Smith (Eds.), *Unemployment in Europe* (pp. 75–96). London: Academic Press.

Lane, J.-E. (2000). *New public management*. London: Routledge.
Lawless, P. (1989). *Britain's inner cities* (2nd ed.). London: Paul Chapman.
Ling, T. (2002). Delivering joined-up government in the UK: Dimensions, issues and problems. *Public Administration, 80*(4), 615–642.
Lowe, R. (2011). *The official history of the British civil service*. Abingdon: Routledge.
Lowndes, V., & Sullivan, H. (2008). How low can you go? Rationales and challenges for neighbourhood governance. *Public Administration, 86*(1), 53–74.
MacKay, R. R. (1998). Unemployment as exclusion: Unemployment as choice. In P. Lawless, R. Martin, & S. Hardy (Eds.), *Unemployment and social exclusion* (pp. 49–68). Oxon: Jessica Kingsley.
Mahoney, W., Smith, G., & Stoker, G. (2000). Social capital and urban governance: Adding a more contextualized 'top-down' perspective. *Political Studies, 48*(4), 802–820.
Mandell, M. P., & Keast, R. (2008). Voluntary and community sector partnerships, current inter-organizational relations and future challenges. In S. Cropper, M. Cropper, C. Huxham, & P. Ring (Eds.), *The Oxford book of inter-organizational relations* (pp. 175–202). Oxford: Oxford University Press.
Marsh, D. (1983). *Pressure politics*. London: Junction Books.
Marsh, D., & Locksley, G. (1983). Capital: The neglected face of power? In D. Marsh (Ed.), *Pressure politics* (pp. 21–52). London: Junction Books.
Mawson, J., & Hall, S. (2000). Joining it up locally? Area regeneration and holistic government in England. *Regional Studies, 34*(1), 67–79.
McCarthy, M. (1983). Child poverty action group: Poor and powerless? In D. Marsh (Ed.), *Pressure politics* (pp. 212–233). London: Junction Books.
McGregor, A., Glass, A., Higgins, K., Macdougall, L., & Sutherland, V. (2003). *Developing people – Regenerating place: Achieving greater integration for local area regeneration*. York: The Policy Press and the Joseph Rowntree Foundation.
Moore, M. (2000). Competition within and between organizations. In D. Robinson, T. Hewitt, & J. Harriss (Eds.), *Managing development: Understanding inter-organizational relationships* (pp. 89–113). London: Sage in Association with the Open University.
Moore, C., Richardson, J. J., & Moon, J. (1989). *Local partnership and the unemployment crisis in Britain*. London: Unwin Hyman.
National Audit Commission (NAO). (1989). *Urban regeneration and economic development: The local government dimension*. London: The Stationery Office.
National Audit Commission (NAO). (2013). *Funding and structures for local economic growth*. HC 542. London: The Stationery Office.
National Audit Commission (NAO). (2016). *Local enterprise partnerships*. HC 887. London: The Stationery Office.
Pacione, M. (1997). Urban restructuring and the reproduction of inequality in Britain's cities: An overview. In M. Pacione (Ed.), *Britain's cities: Geographies of division in Urban Britain* (pp. 7–60). London: Routledge.

Page, S. (2004). Measuring accountability for results in interagency collaboratives. *Public Administration Review*, 64(5), 591–606.
Painter, C., Isaac-Henry, K., & Rouse, J. (1997). Local authorities and non-elected agencies: Strategic responses and organizational networks. *Public Administration*, 75(2), 225–245.
Papadopoulos, Y. (2007). Problems of democratic accountability in network and multilevel governance. *European Law Journal*, 13(4), 469–486.
Parkinson, M., & Wilks, S. (1986). The politics of inner city partnerships. In M. Goldsmith (Ed.), *New research in central-local relations* (pp. 290–307). Aldershot: Gower.
Perri 6 (2004). Joined-up government in the western world in comparative perspective: A preliminary literature review and exploration. *Journal of Public Administration Research and Theory*, 14(1), 103–138.
Peters, B. G. (1999). *Institutional theory in political science: The 'new institutionalism'*. London: Continuum.
Pollitt, C. (2003). *The essential public manager*. Maidenhead and Philadelphia: Open University Press/McGraw Hill.
Power, A. (2001). Social exclusion and urban sprawl: Is the rescue of cities possible? *Regional Studies*, 35(8), 731–742.
Pugalis, L., & McGuiness, D. (2013). From a framework to a toolkit: Urban regeneration in an age of austerity. *Journal of Urban Regeneration and Renewal*, 6(4), 339–353.
Rhodes, R. A. W. (1988). *Beyond Westminster and Whitehall*. London: Unwin Hyman.
Rhodes, R. A. W. (1997). *Understanding governance: Policy networks, governance, reflexivity and accountability*. Buckingham: Open University Press.
Rhodes, J., Tyler, P., & Brennan, A. (2003). New developments in area-based initiatives in England: The experience of the single regeneration budget. *Urban Studies*, 40(8), 1399–1426.
Richards, D. (2008). *New labour and the civil service: Reconstituting the Westminster Model*. Basingstoke: Palgrave Macmillan.
Richards, D., & Smith, M. J. (2002). *Governance and public policy in the UK*. Oxford: Oxford University Press.
Rose, N., & Miller, P. (1992). Political power beyond the state: Problematics of government. *The British Journal of Sociology*, 43(2), 173–205.
Rydin, Y. (2011). *The purpose of planning: Creating sustainable towns and cities*. Bristol: The Policy Press.
Skelcher, C. (2004). The new governance of communities. In G. Stoker & D. Wilson (Eds.), *British local government into the 21st century* (pp. 25–42). Basingstoke: Palgrave Macmillan.
Smith, M. J. (1993). *Pressure, power and policy: State autonomy and policy networks in Britain and the United States*. Hemel Hempstead: Harvester Wheatsheaf.

Smith, M., & Beazley, M. (2000). Progressive regimes, partnerships and the involvement of local communities: A framework for evaluation. *Public Administration*, 78(4), 885–878.

Social Exclusion Unit (SEU). (1998). *Bringing Britain together: A national strategy for neighbourhood renewal*. CM 4045. London: Cabinet Office.

Social Exclusion Unit (SEU). (2001, January) *A new commitment to neighbourhood renewal: National strategy action plan* (London: Cabinet Office).

Stewart, M., Goss, S., Gillanders, G., Clarke, R., Rowe, J., & Shaftoe, H. (1999). *Cross-cutting issues affecting local government*. London: DETR.

Stoker, G. (1991). *The politics of local government* (2nd ed.). Basingstoke: Macmillan.

Stoker, G. (2004). *Transforming local governance*. Basingstoke: Palgrave Macmillan.

Sullivan, H. (2004). Community governance and local government: A shoe that fits or the emperor's new clothes. In G. Stoker & D. Wilson (Eds.), *British local government into the 21st century* (pp. 182–198). Hampshire: Palgrave Macmillan.

Sullivan, H., & Skelcher, C. (2002). *Working across boundaries: Collaboration in public services*. Basingstoke: Palgrave Macmillan.

Tallon, A. (2013) Urban regeneration (2nd ed.). Abingdon: Routledge.

Taylor, M. (2000a). *Top down meets bottom up: Neighbourhood management*. York: Joseph Rowntree Foundation.

Taylor, M. (2007). Community participation in the real world: Opportunities and pitfalls in new governance spaces. *Urban Studies*, 44(2), 297–317.

Tofarides, M. (2003). *Urban policy in the European Union: A multi-level gatekeeper system*. Aldershot: Ashgate.

Tomaney, J. (2004). Regionalism and the challenge for local authorities. In G. Stoker & D. Wilson (Eds.), *British local government into the 21st century* (pp. 167–181). Hampshire: Palgrave Macmillan.

Turok, I., & Robson, B. (2007). Linking neighbourhood regeneration to city-region growth; Why and how? *Journal of Urban Regeneration and Renewal*, 1(1), 44–54.

Tyler, P., Warnock, C., Provins, A., & Lanz, B. (2013). Valuing the benefits of urban regeneration. *Urban Studies*, 50(1), 169–190.

Ward, H., & John, P. (2008). A spatial model of competitive bidding for government grants: Why efficiency gains are limited. *Journal of Theoretical Politics*, 20(1), 47–66.

Wilson, D. (1992). Cooperation and competition in the voluntary sector: The strategic challenges of the 1990s and beyond. In J. Batsleer, C. Cornforth, & R. Paton (Eds.), *Issues in voluntary and non-profit management* (pp. 169–180). Wokingham: Addison-Wesley in Association with the Open University.

PART II

Investigating and Analysing Network Impact

CHAPTER 5

Inner-City Network Cases

Chapters 3 and 4 considered how unemployment policy and urban regeneration assist or hinder network impact. Since neighbourhood unemployment persists, what effects do national policy, institutional infrastructure, and local culture have on network interrelations? Hence, this chapter and the next will investigate two local authorities in different regions in England and create formal and informal case network profiles to investigate the factors that impact on different networks, and the outcomes achieved at three impact levels. The analysis does not intend to disparage networks tackling local problems. However, if we do not look more critically at networks and the barriers to full employment, then our efforts will continue running up a down moving escalator. Within each local authority, two ward clusters were selected with persistent unemployment, but which had different local cultures, socioeconomic histories and structural conditions.[1] Both local authorities are comparable, since they operated two similar, centrally funded network interventions with remits for tackling unemployment, in addition to the primary agency, Jobcentre Plus (JCP): first, a Local Strategic

---

[1] From nine case area neighbourhoods studied, six fell within the worst 2 % of Lower Super Output Areas (LSOAs) from 32,483 LSOAs in England (ODPM 2004). LSOAs represent the deprivation statistics of small areas (approximately 1000–3000 people) since ward-level data can mask issues and ward boundaries can change. Note: the Index of Multiple Deprivation (IMD) was published in 2000 and 2004, initially by the ODPM, thereafter by the DCLG from 2007. The third and fourth editions in March 2010 and September 2015 were based on 326 local authorities as opposed to 354, because of local authority reorganisation.

Partnership (LSP) supported by Neighbourhood Renewal Funding (NRF); and second, the 'Working Neighbourhood Pilot' (WNP) funded by the Department for Work and Pensions (DWP) from April 2004 to April 2006. The latter aimed to work with the LSP to reduce unemployment in specific ward clusters. The chapters draw on official data and anonymous interviews with professionals associated with case networks during a period of worsening unemployment, between November 2004 and November 2005 when the ward cluster data aligned to Neighbourhood Renewal priority areas. The aim is to understand what happened before, and during network interventions, and provide progress updates where appropriate within the limited space available. However, caution is advised, as while the remaining chapters of this book interpret professionals' testimonies, it cannot claim to fully represent the networks described.

Each chapter is similarly structured. First, it will introduce the local authority case areas, representing disadvantage, high unemployment central interventions and case networks. Then, a comparison is made of ward cluster dimensions. A profile of formal and informal case networks associated with tackling unemployment follows, including their attributes, activities and relations. The final section compares the factorial impact on these cases using empirical evidence. This chapter reports on five network cases in the Labour-controlled London Borough of Tower Hamlets (LBTH) and two ward clusters. Chapter 6 reports on five cases in Great Yarmouth Borough Council (GYBC) and two ward clusters, during a period of Conservative control.

Tower Hamlets: More Jobs, Less Work

Tower Hamlets is a small, densely populated East London borough (total land 19.77 km^2), bounded to the south by the Thames, to the east by the London borough of Newham, Hackney to the north and the City of London to the west. The borough experienced the fastest population growth for a local authority in England and Wales during the last decade, rising from 196,106 (Census 2001) to 254,096 (Census 2011). The largest ethnic group within the borough is Bangladeshi, although this particular cohort decreased slightly, from 33.4 % in 2001 to 32 % in 2011. The White British cohort has decreased from 43 % in 2001 to 31 % in 2011; meanwhile, other Asian and African people have also made their home in Tower Hamlets

(LBTH 2012).[2] The borough has a history of domestic and foreign immigration, low wages, poor work conditions, long-term illness, premature death and child poverty.[3] During the Second World War, aerial bombing raids destroyed industries and 24,000 homes in the area, and the percentage of replacement social housing is the highest in the country, a large proportion of which fell below the government's decent homes standard when it was inspected in 2005 (LBTH 2005a: 27–8). Social rented housing stock improved over the past decade, but more than 20,000 people are on the social housing waiting list and overcrowding is a major issue (THP 2015: 6–7). The borough is renowned for its 'bare knuckle' politics, party political defections, and accusations of corruption and vote mishandling. The local council has been a Labour stronghold for decades, with the exception of a Liberal/SDP Alliance in 1986–90 and Liberal Democrat control in 1990–4. In 2010, Tower Hamlets' first directly elected mayor, Lutfur Rahman was suspended from the Labour Party and won as an Independent for the 'Tower Hamlets First Party.' Labour was elected in May 2014. The Labour MP's for Tower Hamlets are Rushanara Ali and John Fitzpatrick. In 2014, the Secretary of State for Communities and Local Government instated commissioners to oversee key LBTH administrative activities until 2017 following mismanagement claims. The mayor was removed from office and a new mayor, John Biggs was elected in June 2015.

The Tower Hamlets Economy: Two-Sides of the Same Coin

Tower Hamlets tells the quintessential tale of rich and poor. The international financial districts of Canary Wharf and the City account for 81 % of all employment in the borough, yet these wealthy landmarks also mask the poorest wards in Britain. Rich banking enclaves and business services generate incomes for those earning above the London average. Employment

---

[2] Over centuries, settlers in Tower Hamlets have experienced hostility; consequently perceptions of the Bangladeshi population are highly charged, and its place identity polarised 'between City and East End, gent and cockney, nation and outcasts' (Dench et al. 2006: 189).

[3] Tower Hamlets has a long history of entrepreneurship. In the seventeenth century, French Huguenot refugees developed a silk industry there, later supported by Irish weavers. The port expanded international trade. From 1795, seamen's homes were established to aid Asian seamen left destitute after delivering goods to east London. In the twentieth century, Bangladeshi men stayed to work in the clothing industry and find menial work in the area. Eastern European Jewish settlers, at the end of the nineteenth century, established a tailoring industry or became cabinet-makers. Before the First World War, many Germans had settled to work in local breweries and sugar refineries.

has grown 60 % in the past decade and there are 240,000 jobs in the borough.[4] The 'Tower Hamlets Employment Strategy' states; 'With nearly three jobs for every two residents, and with its economy expected to grow by up to 50 % in the next 20 years, Tower Hamlets is a place of opportunity' (LBTH 2011a: 3). Unpublished research, however, claims that 'the Borough's residents are not benefiting from the concentration of new employment in the Borough'; moreover, 'there is evidence that the portfolio of welfare-to-work and employability measures have had only a limited impact' (SQW 2005: i). Tower Hamlets consistently had the highest unemployment rate in London for years; between 2003 and 2013, the rates were 12.3 and 13.0 % respectively. However, since 2010, the districts of Barking and Dagenham and Newham have experienced large increases in unemployment, and Tower Hamlets is currently in third position. The English IMD ranks Tower Hamlets the third most deprived local authority out of 326 districts in England for the extent of deprivation in 2010 and 2015. Forty-seven per cent % of residents are on benefits, 33 % of family incomes are less than £20,000, 43 % of families receive working tax credits in 2012, and the borough's child poverty rate, while it has reduced, is the highest in the UK at 47 % (HMRC 2011; Trust for London 2013). Of the 12,380 residents who claim Incapacity Benefit (IB) (7.2 %), 44 % report mental ill health (LBTH 2011a). A higher incidence of physical and mental health problems for ethnic minorities possibly links to socio-economic disadvantage, perceived racial discrimination, experiences of racial harassment and lengthy periods of unemployment (Saffron and Nazroo 2002).

Apart from the financial districts, the borough has no large industries. The closure of the East End Docks in 1980 led to high unemployment for the white working class. The London Docklands Development Corporation regenerated the derelict land and established banking and finance districts; however, residents gained minimal benefit from employment in these areas and became dependent on the survival of retail shops and menial service jobs that paid low wages. Furthermore, there exists spatial disadvantages such as environmental degradation, and a lack of quality town centres and affordable spaces for small or new businesses. The number of self-employed people (16–64 years of age) in the borough peaked at 9,500 in 2004 and 19,600 in 2014, falling to 13,400 in 2015 (ONS Labour Market Profile). The borough's enterprise strategy

---

[4]The main employers are the financial services sector, public services and Bart's and London Hospital Trust.

suggests business information and financial support services 'are of limited scale and are probably not sufficient to fully resource local needs' (LBTH 2011b).[5]

Jobs, But Not the Right Kind

Tower Hamlets's contextual employment prospects are not working for local residents. Besides having the second lowest employment rate in London at 59.8 %, it also has the highest youth unemployment in London amongst 16–24-year-olds and highest number of long-term unemployed (LBTH 2011a: 19).[6] In 2005, an LBTH official said, 'the lack of local jobs is not the cause of unemployment.' However, the job economy is weighted toward specialist financial and business service occupations in Canary Wharf and the City, attracting commuters and graduates (probably of prestigious universities) from both home and abroad; relatively few residents gain access to these, and menial service jobs supporting the high-end workforce are not finite (LBTH 2005a: 8, 20). Local residents fill less than a fifth of local jobs (LBTH 2011a: 8). Education attainment is also polarised. The 2011 Census identified 29,366 people aged 16–64 with no qualifications, 40 % of whom were unemployed. Meanwhile, the number of people qualified to NVQ level 4 is above the London average and GCSE results are above the national average (see Table 3.4 in LBTH 2011a: 8, 27). However, compared with other local authorities, 'Tower Hamlets had the lowest proportion of graduates in full-time employment (36.3 %)' (Mehta and Rutt 2012: iv). Moreover, 41.4 % of highly qualified residents to level 4 and above worked in retail and wholesale industries 'This was twice the English average of 18.8 per cent' (LBTH 2014a). On census day 2011, 2859 graduates were unemployed. Ethnic minorities are more likely to face racial discrimination during selection processes, especially for private-sector jobs, and are 'less likely to meet with a job adviser, to have access to formal methods of recruitment or be aware they can study or work part-time whilst still claiming benefits' (GOL 2000: 52). Granovetter (1973) demonstrated the importance of acquaintances in job-searching, as opposed to strong ties amongst family

---

[5] The TH Community Plan 2002–2006 (LBTH 2002: 14) highlights the pressure on local enterprise.
[6] In 2007 the youth unemployment rate is 19.4 % compared to London average 10.6 % (LBTH 2008b: 48).

and friends. A US study claims organisational ties are equally important, and used more by black men and women than whites or Latinos (see Small 2009: 192–3). Minority high-school students were also more likely to find their first job through teachers acting as informal intermediaries. Yet informal job brokers are highly selective, and formal job brokerage services protecting their competitive edge may overlook those clients perceived as having greater needs.

The 'benefit trap' also undermines employment solutions. Private sector property rental in Tower Hamlets is substantially higher than the London average. Only 29 % of residents are owner-occupiers, private properties are unaffordable and half of the borough's population live in social housing, 41,230 new homes are needed and 23,000 people were on the council housing waiting list in 2013. High rents impact tenants' disposable income, and means-tested housing benefit tops up the difference between affordable rent and market-inflated rent, thereby preventing mass destitution. The root of these residents' 'benefit dependency' lies within a deficient capitalist structure, and benefits property owners, rental agents and employers. The housing benefit cap detailed in the Welfare Reform Act of 2012 restricts landlord exploitation, but reduces the number of rental properties available to benefit-dependent claimants, thereby increasing the likelihood of debt and financial exclusion, for which the borough's debt and welfare advice services are unable to meet demand (LBTH 2013a: 25). In 2008, 2000 homeless people in the borough housed in expensive rental properties were unable to 'make work pay' and the waiting list for social housing at that time was four years (LBTH 2008a: 136). Around 2200 council properties become available each year, but only 674 additional homes were built in Tower Hamlets between 2011 and 2012, despite the London Plan target being set at 2885 (LBTH 2013b: 82). The 'First Steps' 'affordable housing' scheme promoted by the Mayor of London is eligible only to working people on annual salaries of between £66,000 and £80,000. Nine London councils challenged the Mayor of London in the High Court over proposals to stop councils from being able to negotiate lower rents. However, the 'bid to protect genuinely affordable rents for local people' was overturned and the ruling allows social housing rent for new properties to be set at 80 % of the London market rate (LBTH 2014b), Thereby, 'social cleansing' is legalised. Poor-quality, low-pay jobs subsidised through the working-tax credit system constitute another benefit to employers. Consequently, LBTH was the first council to adopt the London Living Wage for agency staff contracted in cleaning, grounds maintenance, leisure services, social services and catering. The former Mayor of Tower Hamlets also supported the Living Rents campaign, and proposed taking social homes back into council ownership to redress the damage of the 1980s social housing auction.

In 2004, the Tower Hamlets Partnership (THP) commissioned research into recurrent unemployment following 'evidence that a high percentage (76 %) of Jobseeker's Allowance (JSA) claimants have repeat unemployment between 4 months and 1 year, indicating that many people are only finding short term employment.' Furthermore, 50 % of unemployed people 're-claim within a 4 month period' (THP CSPG 2003: 8; LBTH 2005b: 11; 2008b: 9). Unpublished research, however, obtainable through the Freedom of Information Act in 2012, suggests only 730 claimants experienced recurrent unemployment (defined as claiming more than once in the past year), and 64 % claimed twice during the sampling period (SQW 2005: 26–38). However, almost half of all Tower Hamlets's JSA claimants had been unemployed for over a year (2005: 17). To discover the reasons for recurrent unemployment, a number of claimants, employers and providers were interviewed about their training experience and needs, and personal and labour market characteristics. Most affected claimants reside in the west of the borough; 56 % have no qualifications, particularly Bangladeshi men aged 35–49 years who had worked previously in elementary and skilled occupations, such as machinists and garment workers. Claimants mostly cited structural reasons and the temporary nature of local work as causes of their unemployment; hence, they actively searched for work beyond the boundaries of Tower Hamlets. JCP's standardised provision was largely unsatisfactory; late intervention, its lack of tailored support and its inability to match skills to jobs disengaged clients and undermined their confidence (2005: 36). Commercially focused providers were perceived as understanding better clients' training needs; this was a view shared by employers. The volume and complexity of initiatives in the area confused clients and required improved coordination between providers, the council and JCP (2005: 52).

Coordination problems continued and, in February 2010, LBTH's Scrutiny Review Working Group on Reducing Worklessness Amongst Young Adults 18–24 asked the Prosperous Community Plan Delivery Group to commission research to map out provider services in the ward cluster areas. The LBTH's 'employment strategy 2011' notes that the survey conducted in 2010 found that several hundred welfare-to-work initiatives had added employment outcomes to their portfolio to access funding (including the European Social Fund (ESF) and the Working Neighbourhood Fund (WNF)) (LBTH 2011a: 46). 'This has to some extent fuelled an uncoordinated and complex array of overlapping provision' (2011a: 47). Consequently, employers and jobseekers would need to influence the commissioning process on 'what suppliers could offer', and facilitate dialogue between the DWP/JCP and the two or three Prime Contractors delivering the 'Work Programme' (WP) in the East London Contract Package Areas (2011a: 42–59). Moreover, 9,440 of the 10,440 DWP WP referrals in Tower Hamlets could have achieved a job outcome, but only 1,610 gained employment up to December 2013 (WP 2013). Despite private sector firms' 17.1 % success rate, local authorities lack influence over the central commissioning agenda. Nevertheless, the estimated 22,200 Tower Hamlets inhabitants seeking work (of which 14,800 are unemployed but economically active and 7,400 are economically inactive but want a job, based on the International Labour Organisation (ILO) measure) should be given the right to realistic representation (2011a: 19, 57).

Neighbourhood Profiles: East and West

This section compares the characteristics of two ward clusters: (i) Spitalfields/Banglatown and Bethnal Green South (SB); and (ii) East India and Lansbury (EIL).[7]

Demographics and Deprivation

SB comprises two wards near the City in the west of the borough. Ethnicity is largely Bangladeshi (58.1 %). The population in Spitalfields and Banglatown is 8382 (Census 2001) rising to 12,578 (Census 2011). The population in Bethnal Green South is 13,675 (2001), rising to 19,308 (renamed 'Bethnal Green', Census 2011). In 2000, Spitalfields and Banglatown ranked forty-sixth out of 8414 wards on the Index of Multiple Deprivation (IMD) (ODPM 2000).

---

[7] Note: ward data reporting is restricted to the ward boundary changes; in place as at December 2002. Bethnal Green South was formerly Holy Trinity, St Peters and Spitalfields. Spitalfields/Banglatown was formerly Spitalfields. Ward boundary changes occurred again in May 2014 at the council elections under the auspices of the Local Government Boundary Commission for England. Bethnal Green was formerly Bethnal Green South. Lansbury was formerly East India and Lansbury. Consequently, data availability is sometimes inconsistent.

EIL comprises two wards close to Canary Wharf in the east of the borough. Ethnicity is just over 50 % white. The population is 11,497 (Census 2001) rising to 14,859 (renamed 'Lansbury', Census 2011). In 2000, East India ranked one hundred and fifty-second and Lansbury ranked forty-seventh out of 8414 wards on the IMD (ODPM 2000). In subsequent years, EIL and SB had the same highest proportions of deprived LSOAs in the borough, and ranked in the bottom 20 % nationally (DCLG 2010; 2015). Child poverty in EIL is 51 %; in Bethnal Green South, 50 %; and in Spitalfields and Banglatown, 42 % (HMRC 2011). Both wards ranked in the top 10 % in the country for poor health, with rising numbers on sickness benefits (see the health deprivation and disability domain in ODPM 2004). Spitalfields and Banglatown had twice the expected mental health admission rate for adults aged 35–64 in England during the research period (Enum 2007: 24).

Unemployment

During the period of research fieldwork, both ward clusters had the highest number of claimants in the borough. Table 5.1 presents the categories of benefit claimants between 2002 and 2014. Youth unemployment reached 25 % in Spitalfields and Banglatown, and 26 % in parts of Bethnal Green South (see employment deprivation in ODPM 2004). On census day 2011, Lansbury had the highest unemployment rate in the borough at 18.8 % compared to the Tower Hamlets average of 12 % (LBTH: 2014: 9). More than 50 % of unemployed people in EIL had no qualifications, compared with 38 % nationwide (Census 2001). A decade on, 25.6 % of Lansbury residents aged 16–64 had no formal qualification (Census 2011). Recurrent unemployment was lower in EIL, but higher in SB, particularly amongst male Bangladeshi claimants with no qualifications, although areas west of the borough may have had more insecure job openings (SQW 2005: ii, 34). Unemployment continued to increase prior to the recession and then escalated. Arguably, the drop in unemployment rates, the first in over a decade (see recent figures to November 2014 in Table 5.1) should be treated with caution. Between December 2012 and December 2013, Tower Hamlets's local job centres sanctioned 6873 claimants; of these, 2891 cases were overturned, 3709 were cancelled and 538 had reserved decisions. Therefore, the '… benefit off-flow rates do not necessarily reflect positive outcomes' (WPC 2014: 29). The recent

Table 5.1 Working-age claimants in the LBTH ward clusters—November 2004 to November 2014

	SB							EIL						
	Nov-2002	Nov-2004	Nov-2006	Nov-2008	Nov-2010	Nov-2012	Nov-2014	Nov-2002	Nov-2004	Nov-2006	Nov-2008	Nov-2010	Nov-2012	Nov-2014
Total claimants	3460	3355	3365	3400	3440	3310	2855	2285	2320	2500	2510	2560	2650	2190
Job-seekers	1030	910	1000	1025	1140	1165	610[b]	560	525	625	640	785	880	445[b]
ESA and IB	1335	1380	1325	1385	1385	1350	1470	855	870	905	945	970	985	1010
Lone parents	485	445	430	365	295	190	170	545	580	595	550	420	335	280
Carers	205	210	235	255	290	335	375	110	130	155	170	195	235	265
[a]Income related	290	265	240	220	170	95	75	120	110	115	110	90	80	60
Disabled	90	100	100	105	120	135	125	75	85	90	80	85	125	120
Bereaved	25	45	35	45	40	40	30	20	20	15	15	15	10	10
Male	1970	1935	1900	1880	1890	1830	1510	1120	1105	1210	1200	1275	1325	1010
Female	1490	1420	1465	1520	1550	1480	1345	1165	1215	1290	1310	1285	1325	1180
16–24	455	530	500	505	525	465	310	340	370	405	430	420	455	330
25–49	2190	2040	2050	2000	1970	1865	1575	1395	1425	1530	1475	1515	1485	1210
50 and over	795	785	815	890	945	975	970	545	525	565	605	625	710	650
Out-of-work benefits	3140	3000	2,95	2995	2990	2800	2325	2080	2085	2240	2245	2265	2280	1795

Data source: http://www.nomisweb.co.uk

Figures relate to the 2003 ward boundaries. Data rounded to nearest 5. Claimant counts do not yet include Universal Credit (Data available from 2 March 2015).

ESA and IB Employment Support Allowance, replacing Incapacity Benefit, *SB* Spitalfields/Banglatown and Bethnal Green South statistics combined, *EIL* East India and Lansbury

[a]Others on income related benefit

[b]Decrease coincides with increase in sanctions and ESA

JSA claimant count increased in EIL and slightly decreased in SB between November 2015 and February 2016 (ONS Neighbourhood Statistics).

SB has the highest youth unemployment rate in the borough. Family migration history would also have impacted on SB's Bangladeshi youth, and local history. Despite the intense racial harassment which took place and racial murders committed during the 1970s to the late 1990s, the Muslim community appear 'comfortable to demonstrate cultural belonging in public spaces', suggests Begum (2008: 5.1). However, cultural experiences varied, beyond socially constructed identities associated with religion, nationalism and feminine/masculine boundaries, and some young Bangladeshi women yearned to modernise society through a globalised Islamic space, rather than a nationalistic Bengali one (2008). But the reasons for their unemployment were less clear. Thus, LBTH commissioned research into worklessness amongst Bangladeshi and Somali women in Tower Hamlets, given both cohorts have higher rates of economic inactivity and income poverty compared with other ethnic groups.

First, Mayhew and Harper's (2011) report finds that Bangladeshi women are more likely to have different barriers to work at different stages of their lives, from caring responsibilities to starting families at a young age, possibly due to cultural factors more than income (2011: 15, 17). Additional barriers to work include gender inequalities, low English proficiency and having few qualifications. Notably, in 2006, the Learning and Skills Council for England (LSC) depleted educational provision for English for Speakers of Other Languages (ESOL), leaving hundreds of LBTH residents on waiting lists for courses, according to Salman (2006). In 2007, providers were unable to contest LSC's decision to introduce tuition fees for ESOL. Funding for apprenticeships combined with ESOL was also abolished in the workplace (DBIS 2010). Perhaps cohorts were paying the price for the LSC's financial incompetence and lavish spending in other areas of its business, including staff meetings in top hotels and conference suites. Recent job-seekers with poor English on active benefits (JSA/Employment and Support Allowance (ESA) work-related activity group) are required to attend free language training or risk having their benefits sanctioned. However, Foster and Lane (2012) suggest that poor communication between ESOL providers, JCP and the WP and referral processes in London hampered outcomes, as ESOL providers running courses for JCP carry the financial risk if courses are undersubscribed (JCP may not make timely referrals) or, perversely, the client 'fails' if they transition to employment midway through a course, or are withdrawn and moved to the WP.

Second, Kabeer and Ainsworth's (2011) interview-based study of work patterns and worklessness amongst Bangladeshi and Somali women in Tower Hamlets yields testimonies of women's work experience, empowerment and family support, but also accounts of their struggles escaping domestic confinement, low aspirations and a marital ethos that is anti-work and anti-education. Practical work barriers include expensive childcare, unhelpful job advisers, the fear of losing benefits, and having skills but lacking job experience. However, the nuances are lost on LBTH's Employment Strategy (LBTH 2011a), which expects the (largely Muslim) population to develop skills and compete in the business arena. As Kabeer and Ainsworth's study (2011: 7, 37–8) notes, Sikh culture has integrated with city work culture, and is compatible in areas such as drinking alcohol and entertaining; but signs of an Islamic identity (headdress or beard) reduce job prospects, as Bangladeshis abstain from alcohol, and late-night working is difficult for women; hence, the council is perceived as a more tolerant employer. Thus, the private sector would need to play a greater part in addressing the discriminatory attitudes that contribute to social exclusion.

Environment

Both ward clusters are situated close to vibrant economies. SB is nearer to central London; therefore, it has economic and transport advantages over EIL. Commercial regeneration connecting the City of London, Liverpool Street Station and Spitalfields Market has pushed closer to the indigenous Bangladeshi population, which is segregated in run-down, overcrowded council estates and flats.[8] These areas have not seen housing investment programmes since a Task Force in the 1980s and are visited by few strangers. Tourists keep within the market areas, and Brick Lane's Indian restaurants and nearby pubs and clubs attract a vibrant night-time economy. Art and design industries have gentrified former warehouses, and the renovated Georgian and Victorian houses in the streets around Spitalfields Market have massively inflated house prices; a three-bedroom property can fetch over £3 million. Fainstein identifies the double irony of Spitalfields's urban development:

---

[8] Segregation also refers to a generational and social control system over women (Aftab 2007: 141).

On the one hand, the development coalitions that thought they had to bulldoze communities in order to transform them have been proved wrong; on the other, the defenders of community and diversity, while getting their own way in the short-run, seem only to produce a situation where they lose their dominance in the longer term. (Fainstein 2001: 146)

During fieldwork visits for this study, 'state-of-the-art' community libraries', called 'Ideas Stores' (IS) in each ward cluster, were evidently under- resourced. The booking system for accessing free computers (used by people on low incomes for internet job searches, for example), was overstretched due to demand. Recently, the 'Whitechapel Vision', a regeneration schema for luxury high-rise housing, some affordable homes and retail expansion, linked to a major CrossRail station for better access to Heathrow, will further envelop the affordable markets and small businesses at the south end of SB. Local officials view gentrification as a means of managing capitalism, while remaining oblivious to the limits of growth for local jobs. Potential for job creation is limited because the developers are unlikely to appoint multitudes of local long-term unemployed people and the demand for local training and construction programmes will outstrip provision in the timescale required. The investment will leave shiny buildings, and supposedly 5000 jobs (quoted in the scheme's 'Masterplan'), but there is no information about the types of jobs or training routes, and menial, low-pay, part-time jobs subsidised through tax credits will not provide a meaningful growth boost for poorer residents. In theory, the 'Masterplan', under the 'localism' remit, serves the Local Plan, and the London Legacy Development Corporation's 'Equality Impact Assessment' (LLDC 2013) should specify equitable growth; however, growth does not tackle the problem of unemployment sufficiently.

On the eastern side of the borough, EIL provides a stark contrast to the nearby financial district of Canary Wharf and the Docklands complex. As one official admits, 'We have created such a huge job opportunity but for local people to access those jobs is quite difficult.' EIL is largely a social housing economy following the transfer of seven isolated council housing estates covering five neighbourhoods to a housing association. This was led by 'Leaside Regeneration' (a registered social housing company) and Poplar HARCA (Housing and Regeneration Community Association) in 1998. Several more estates resisted, but eventually transferred between 2005 and 2007. Physical access across EIL was also difficult. One estate resembles an island, cut off by a dual carriageway on one side, the Docklands Light Railway on another and the Grand Union Canal on another. 'It has its own geographic issues in terms of its physical isolation from the rest of the world. Most of it is bound by roads and they have very few access points in and out, which does not help', commented one interviewee. Residents who fear using subways to reach public transport are unlikely to pursue evening jobs or study (Dench et al. 2006: 64–7). During the fieldwork for this book, the author observed that the housing disrepair on estates further afield from the main housing offices was staggering.[9] More recently, a pedestrian crossing has been installed on the A13 to enable residents to reach the Docklands Light Railway. When institutional and environmental failure is so entrenched, residents may find professional terminology difficult to contemplate; especially ideas such as localism, participation, governance and joined-up working.

Culture and Politics

SB and EIL have unique cultures and politics shaped by historical injustices. Housing politics and tenant group activism have long championed residential needs in EIL. Lansbury ward is named after the politician George Lansbury who was imprisoned in 1921 for his role in the council tax protest and Poplar's 'rates revolt'; subsequently, he became the leader of the Labour Party. During the 1980s, democratic politics and neighbourhood solidarity in the area grew as a form of counter-politics (see Burns et al. 1994: 202–13). Communities had no real control over the closure of the industrial docklands in 1980, and subsequent generational unemployment for white working-class families, disillusionment and economic deprivation fuelled race issues and contentious politics. More recently, Poplar HARCA manages its housing stock, consisting of 8500 homes in sprawling estates (many are built in the 1960s Brutalist style of architecture) like a planned community, with neighbourhood centres and committees for 'neighbourhood democracy', social support, and staff in place to help residents improve their employability. Nevertheless, while HARCA keeps up

---

[9] Community support services avoid the bleakest estates, as the author observed when carrying out the fieldwork.

with its social housing competitors, its capacity for economic regeneration is limited.

Conversely, migrant histories, global economic pressures, racial prejudice and local politics shape SB. In the 1970s, ethnic minorities' manufacturing businesses in the leather and garment trades fell victim to cheap imports and became bankrupt. Manufacturing continues to decline in London, falling 2.3 % between 2009 and 2012 (see Figure A6 in LEPN 2014: 69). Industrious people who lost jobs are now older, many are in poor health, and those with limited English have little chance of finding new employment. Only a culture of small-scale family enterprise remains, in food and local trades, and in independent arts and media industries which have converted warehouses, attracting new businesses to the area. In the 1980s, the Commission for Racial Equality criticised Tower Hamlets council for its housing allocation policies towards Bangladeshi families (Burns et al. 1994: 138). The white working classes were perceived to dominate consultative forums in Tower Hamlets and the only way the Bangladeshi community could express itself was at a political level; however, traditionally, Bangladeshi politics is divided between secular and fundamentalist forms of Islam (1994: 208). The rise of the 'Respect' political party during the 2005 general election also fuelled distrust in SB between long-standing Labour supporters, the community and the local authority (Dench et al. 2006: 219). Gender politics were also highlighted and anecdotal reports suggested the families of some young Asian women were restricting their travel on public transport to the further education college. One interviewee believed youth unemployment in SB lacked a gender dynamic and that 'no-one challenges existing practices, people are afraid it will affect their funding'. Alcohol- and drug-related crime rates are higher in this area because of large numbers of people visiting the local clubs and pubs, while racially motivated crimes increased following the 2005 London bombings. Several interviewees suggested mainstream providers avoided SB. One public sector manager associated the area with 'massive politics' and endless organisations receiving a lot of money but not getting people into jobs. A few interviewees mentioned a fraud case in 2003 when the council's regeneration funds were misused by a training organisation, as reported by Weaver in *The Guardian* (Weaver 2003). Consequently, both insiders and outsiders reproduce SB's reputation.

Overall Policies

Unemployment policy cascades through public bodies influencing national policy frameworks for getting people in London into work: such bodies include, for example, JCP, Business Link (BL), Connexions, the LSC and the London Development Agency (LDA) (responsible for the London Framework for Regional Employment and Skills, and Cityfringe Pathways to Employment programme).[10] Officials at a meeting to explore why Tower Hamlet's unemployment had not reduced were advised to relate partnerships and plans to regional institution's strategies:

> The trend towards co-financing programmes ensures that all employment and training related activities are intrinsically linked to the main strategies of the LSC and the LDA and any activities provided locally will have to identify how they relate to the priorities in these strategies. (THP CSPG 2003: 11)

One council official noted; 'I don't think the Regeneration Department here is good at joined up work ... our links are almost invisible although we are only on the next floor.' Later, the THP was advised to develop 'a strategic commissioning framework for regeneration funds in future ... to ensure specific interventions reinforce higher level strategic objectives' (LBTH 2008a: 134). Nevertheless, local regeneration and economic development plans lacked a detailed employment strategy. More recently, East London District JCP invited Tower Hamlets partnerships to designate a lead accountable body to manage funds and provide services to meet the needs of customers who cannot be supported by existing training provision such as the WP, Support Contract, Work Choice, SFA-funded college courses, and so on. The major problem, however, is finding sustainable jobs for the very long-term unemployed, which targeted programmes are still struggling to achieve (see Krasnowski and Vaid 2012).

Borough-wide providers and ward-based organisations offer generic employment support, or tailor services to specific needs. Evidently, local culture influences provision and policy preferences, such that SB promotes enterprise, advice and support for ethnic small and medium-sized

---

[10] The Localism Act 2011 replaced the LDA (established in July 2000 under the auspices of the Greater London Authority (GLA)) with a subsidiary corporation (the Greater London Authority Land and Property), and the Mayor of London is responsible for an 'Economic Development Strategy for London', with the power to create mayoral regeneration areas and mayoral development corporations.

enterprises (SMEs), and creative industries, whereas EIL provision was organised largely through housing policy. The borough's first employment strategy, developed in 2008, linked to an 'emerging Economic Strategy' and highlighted the uncoordinated funding streams and employment provision (see Appendix 5, and LBTH 2008b: 11).[11] It claimed that small-scale efforts will not tackle structural or generational unemployment and major change is necessary to overcome residents' poor perception of effort:

> Improving co-ordination of funded activity will not only increase value for money but will improve the opportunities for engagement with the local community. Despite the plethora of activity resident's perceptions ... are still of a borough with no jobs and not enough provision to help them secure jobs and training. (LBTH 2008b: 11)

Meadows's (2007) report *What Works for Tackling Worklessness*, for example, associates poor performance with the high turnover of personal advisers in London, the low quality of training for disadvantaged people and of job-search assistance for the work ready, and a need for subsidised work and on-the-job support.

Olympics Legacy

It was intended that the 2012 Olympic Games would create a jobs legacy for local people in the Games' 'Host Boroughs' (Barking and Dagenham, Greenwich, Hackney, Newham, Tower Hamlets, and Waltham Forest) (see LA 2007). Despite the rhetoric, the unemployment rate in Tower Hamlets leading up to and beyond the Olympics barely changed between 2010 and 2011 (13.0 %), 2011 and 2012 (12.5 %) and 2012 and 2013 (13.0 %) (ONS Labour Market Profile). The Greater London Assembly feared the London Skills and Employment Board and LDA would repeat the 1980s Docklands employment fiasco and marginalise residents. The Docklands programme had been overly reliant on property-led regeneration, with

---

[11] The Tower Hamlets Employment Strategy (LBTH 2008b) would ensure the delivery of national agendas and initiatives, seek 'flexibilities to allow national programmes to meet local need', and commission research to find out about the barriers to employment. This required the DWP, LSC and LDA to build on the City Strategy approach, and coordinate in the context of the emerging Multi-Area Agreements and delivery plan of the 'Prosperous Communities Community Plan Delivery Group' (2008b: 25).

little impact on the local employment market. Only 9 % of the construction workers employed in the Docklands project were from LBTH, new housing prices were unaffordable for young people, and education and training approaches to support unemployed people were inadequate, according to an unpublished report (see Atkinson and Moon 1994: 149). Host boroughs (known collectively as 'Growth Boroughs' post-Olympics), depended on the growth stimulus generated by the Games and on £20 million invested in three flagship legacy projects: the Six Host Borough project (6HB), the Employment Legacy project (EL) and the Construction Employer Accord (CEA), to assist long-term unemployed Londoners with finding lasting jobs (the measures included numbers of people who started on jobs, and numbers who sustained employment for one year or more). According to the 'Olympic Jobs Evaluation' final report (SQW 2013), only 735 people had achieved sustainable job outcomes under 6HB, and 63 through CEA; the EL was cancelled in early 2013, with no reported sustainable outcomes for Black Asian Minority Ethnic groups, gender equalities or disabled beneficiaries, although results can lag behind pre-employment support (2013: 10). The report frames the underperformance (in terms of outputs against targets) against London's depressed situation at the time. It claims the rise in the working-age population (which must have been foreseen) increased economic inactivity and made it difficult to support people into jobs; moreover, short-term' or 'zero-contract'-type jobs, if attained, are harder to identify as sustained outcomes, while some employees may require additional support and motivation on return to the competitive marketplace (see 2013: 7–8). The 'payment on results' funding model, however, led to an underspend of £12 million. This was due to the fact that subcontracted suppliers were unable to meet the targets and so scaled-back projects, reprofiled targets and shortened programmes and post-employment support, while one major private sector contractor 'did not have the in-depth knowledge of working with the economically inactive, the key target group for the project' (2013: 10). Hence, there was a double irony as the Olympic growth programme was defined by its limits: 'This was a time when the Programme was needed but which made it harder to secure its desired outcomes' (2013: 8).

Network Culture

SB's and EIL's different network cultures reflected local culture and place histories. Several umbrella organisations used SB as a base, and many smaller organisations served the local community, but lacked the capacity and cohesion to bid for big grants and were forced to adapt to the competitive stop-and-start funding rounds. One manager commented that a training provider network funded by ESF had delivered projects in 2000–3 but had then ceased operating as a result of tensions within the voluntary sector. One interviewee, a local ethnic minority worker, perceived a network culture of politics and cronyism. Media reports in subsequent years reiterated political and religious tensions. Two ethnic minority managers stated in interviews that SB lacked the coordination and leadership necessary for tackling unemployment.

EIL, in contrast, had a history of consensus-building networks, community development and tenant participation. The area had benefited from strong housing and regeneration institutions (Poplar HARCA and Leaside Regeneration) and pro-active leaders attracted inward investment for community regeneration, neighbourhood management and in-house unemployment support (ODPM 2006). This had been accomplished through the efforts of single-minded organisational self-interests and of more network brokers with strategic interests, job status and reputation. An interviewed official suggested Leaside Regeneration received funding to establish a community hub scheme because of their affiliation with Neighbourhood Renewal and their ability to contribute to policy discussion: 'I mean they won by public tender, but the fact is if you are in a network they know your worth.'

From 2001, the NRF sponsored the THP to manage the LSP and represent unemployment policy through Community Plan Action Groups (CPAGs) and Local Area Partnerships (LAPs). THP expected existing networks to report to these structures; for example, the Tower Hamlets Employment Consortium, established in 1999 (partners include the Council's 'Skillsmatch' jobs brokerage service, Tower Hamlets College, Education Business Partnership, East London Business Alliance, Poplar HARCA and JCP). While organisations and networks in SB and EIL had to comply with central policy for funding, they did not assimilate easily into the imposed network structures, and pockets of resistance were apparent.

Few interviewees mentioned the Tower Hamlets Infrastructure Network (supported by the Community Organisations Forum (COF)), despite its obtaining Home Office funding to produce a 'ChangeUp' plan (2004–14); this was a national programme to facilitate voluntary, community and social enterprise sector (VCS) change, provide quality services, engage in commissioning processes and improve communication and cooperation. The VCS were also experiencing multiple pressure, as the COF report outlines:

The VCS in Tower Hamlets is under pressure to act as a conduit of community opinion, whilst becoming more heavily involved in providing contracted services. In the process of changing to meet these challenges core services are being threatened. (THCOF 2005: 5)

In August 2008, the Cultural Industries Development Agency launched a ChangeUp Consortium to 'improve infrastructure support services' and address 'a breakdown of trust within the sector and the recent failure of Tower Hamlets Council for Voluntary Services' (THCUP 2008: 1). In October 2008, the THP website announced the WNF as the new governance structure for tackling worklessness (£32.6 million was allocated to Tower Hamlets for commissioning projects for delivery from 2008/2009 to 2010/2011). It gave organisations one month to submit bids despite the plea over decades for lead-in time to develop partnership bids that respect the governance requirements of local membership boards. In July 2009, the Ocean Estate New Deal for Communities (a ten-year £56.6 million local regeneration programme) launched 'Community Consortium Against Poverty' (CCAP), a community interest company involving 14 west of borough organisations. In 2009, WNF granted CCAP £1.875 million for an 'Access2work' service to secure job opportunities for residents, employment advice and training. CCAP wanted to establish a Third Sector Employment Network to engage the voluntary sector in strategic development for services relevant to local people. In 2009, the LSP created an employment subgroup to join up policy areas and link children's centres and parents groups to the City Strategy Pathfinder and Multi-Area Agreements. In January 2010, the Tower Hamlets Council Voluntary Service relaunched, with agreement for ChangeUp to maintain its role. Following recommendations made in the Houghton Report (2009), network configurations handling unemployment changed again in 2010, with the establishment of separate LBTH enterprise and employment task groups, yet their strategies lacked operational detail and network governance arrangements, with prime contractors delivering the DWP's Work Programme (LBTH 2011a; LBTH 2011b). Since prime contractors are not paid to work across policy domains, develop long-term goals, or deal with demand deficiencies, they cannot govern the problems. Neither can community development approaches as they stand.

The London Problem

The employment rate in London has lagged behind that of the rest of the UK for more than two decades, and high youth unemployment is a specific problem to the capital. Between 2001 and 2011, London's working-age population increased by 20 %, but workforce jobs increased by only 5 %; hence, 'the overall number of jobs has not kept pace with the working-age population' (see data criteria in Table 1, in Hughes and Crowley 2014: 15, 16). Across London, there are 'stark gaps between the employment rates of young people from White British and other ethnic backgrounds' (2014: 8–10). Many of London's jobseekers are already employed and seeking promotion, 87 % of part-time workers desire full-time work, and 930,000 were estimated underemployed in 2012 (Trust for London 2013: 51). There were five claimants for every JCP vacancy in London in November 2012, and '17 applications per vacancy' for apprenticeship places; thus, many people downgrade their occupations and seek lower-skilled employment, reports Hughes and Crowley (2014: 18–9). Krasnowski and Vaid's (2012: 33, 35) report *Right Skills, Right Jobs*, notes that London's ILO measure of unemployment peaked in early 2012 at 420,000, but also mention that only 72,000 vacancies are advertised per month; additionally about 368,000 economically inactive people would like to work. Since competition for first-level job entrants is acute in London, more people continue in education in order to delay unemployment or because of poor job prospects. Hence, more people in London are qualified to level 4 or above (23 %), compared with the rest of England (12 %) (Census 2011). Employers depress wages as the graduate surpluses are willing to work for low wages, and employment opportunities are unevenly distributed.

CASES: ENTHUSIASTS, PROFITEERS, REBELS AND OUTSIDERS

The title of this section caricatures five case networks (A–E) associated with unemployment in the case ward clusters. The 'enthusiasts' (Case A and Case D), had paid personnel to enthuse about the state's network goals. The 'profiteers' (Case B) refers to a private sector company contracted to create a network that would achieve targets and profits. The 'rebels' (Case C) resisted a network takeover bid. The 'outsiders' (Case E) were absorbed in the local ward culture. Table 5.2 compares case attributes (formal or informal type, size, actor role and geographic focus). Participants' job roles and responsibilities operated

Table 5.2 Numbers of contacts in the LBTH case networks

Cases[a]	A	B	C	D	E	Citation total
	CSPG[b]	WNP[b]	PAN[b]	EIL[c]	SB[c]	
Actors						
Local and strategic	3	2	11	2	5	23
Borough-wide strategic	10	3	2	3	4	22
Borough-wide private	1	–	–	–	–	1
London-wide strategic	4	–	–	–	–	4
Regional strategic	1	–	–	–	–	1
Case total	19	5	13	5	9	51
Interview total $N = 39$						

CSPG Creating and Sharing Prosperity Group (LSP), WNP Working Neighbourhood Pilot (Private sector provider managed), PAN Poplar Area Neighbourhood Partnership, EIL East India and Lansbury Wards, SB Spitalfields and Bethnal Green South Wards
[a] Includes members with a dual role or dual-network memberships
[b] Formal networks
[c] Informal networks

at different spatial scales: (a) *Local and strategic*—ward community workers or service managers; (b) *Borough-wide strategic*—borough-wide service managers or strategists; (c) *Borough-wide private*—private sector organisations; (d) '*London-wide strategic*'—London agencies; and (e) '*Regional strategic*'—South-East region executives. Cases A, B and C have formal remits for addressing unemployment. The number of interviewees totalled 39, with 51 citations; this is because two actors had a dual role, while seven were members of more than one network. Networks with more go-betweens, brokers (Burt 1980: 91) or mediators may trade more information (Kickert and Koppenjan 1997: 48). One actor with a dual role represented an employment service in Case A, and used Case D for proximity to LSP funds and a jobs brokerage service in Case C for referrals. Another actor used the go-between role in Case A to their advantage in Case C, in which they were involved. Case E had few go-betweens, and low knowledge transfer and resources. Figs. 5.1, 5.2, 5.3, 5.4, and 5.5 are identity-preserving sociograms of case networks (A–E) derived from social network analysis (SNA) and depict graphically the network structure and connectivity (see the method in Appendix 2).

Case A—Enthusiasts

Case A is a Community Plan Action Group (CPAG) called the 'Creating and Sharing Prosperity Group', representing unemployment matters, part of the Tower Hamlets LSP; it takes the form of a network hierarchy of five thematic CPAGs to strategically address cross-cutting policy issues, and eight Local Area Partnerships (LAPs 1–8) to facilitate citizens' engagement in NRF decision-making about expenditure in ward clusters.[12] NRF was also available to establish five voluntary-sector CPAGs with the purpose of influencing the community plan (THP 2002b: 30); however, they did not evolve. The remaining CPAGs are 'Living Safely', 'Living Well', 'Learning Achievement and Leisure' and 'Excellent Public Services.' A Community Empowerment Network elected a small number of members onto the five CPAGs. The THP operated from the LBTH town hall, supported by a team of staff, strong branding and marketing to self-advertise, a website and publications, including meeting minutes, reports, community plans and Local Area Agreements. They believed in seamless coordination:

> The THP will ... bring together local plans, partnerships and initiatives into a coherent framework ... capable of ensuring joined-up services. (LBTH 2002)

The ambitious vision required massive coordination and image-building to legitimise the LSPs' infrastructure and central investment.[13] CPAG protocol advised 'reducing the number of partnerships working to similar agendas' (THP 2002a: 6). One THP enthusiast described the process as 'managed autonomy'. CPAGs expected stakeholders to combine local services, improve the performance of the NHS, the local authority, the police and JCP, and bring about noticeable results from tackling health, education, employment, housing and crime. Three outcome levels (community, national and central government) were targeted, as stated:

> ... what is needed is a 'joining-up' of these resources and a marshalling of their totality in order to accelerate the rates of improvement in service delivery to the local community, and to sustain long-term improvements in relation to national priorities and government targets. (THP 2002a: 1)

---

[12] NRF ceased to exist on the 31 March 2008, replaced by a £1.5 billion WNF, announced on the 6 December 2007 with allocations from 2008 to 2011 (DCLG and DWP 2007: 10).
[13] NRF granted the THP £19.8 million in 2002–4 and £15.7 million in 2004–5. The allocation to CSPG for 2004–6 was £4.5 million, to be spent on tackling worklessness.

Case A set itself the ambitious targets of: (1) reducing unemployment and increasing employment rates for different groups; (2) tackling poverty by increasing benefit take-up; (3) providing training opportunities leading to jobs; (4) increasing the number of business start-ups; (5) improving the outcomes of services delivered through the third sector; and (6) increasing the number of day-care places and after-school clubs. CSPG also reported to the 'Excellent Public Services' CPAG; the latter being 'charged with underpinning the work of the other four CPAGs … to respond to emerging issues and themes challenging all of the CPAGs' (THP EPS 2005).

Case A's quarterly meetings, chaired by JCP, were formal and closed, with the minutes published on the THP website. On average, 50 % of the membership attended these meetings. Sectorial representation included local authority departments, sub-government quangos, business support organisations, a university and three voluntary-sector members. Members' interests ranged from youth, skills and employment support to health, regeneration and higher education; additionally, SB's small business interests and EIL's housing interests were represented. Participation was mandatory for 15 mainstream managers with borough-wide, London-wide or regional strategic roles. Meeting attendance was ad hoc and a major policy representative was not present. One interviewee, representing a key policy area with potential employment outcomes, expressed no interest in unemployment. Actors presented information updates at meetings, and many external specialists presented on specific subjects by invitation. In subsequent years, the Employment Task Group, an LSP subgroup, merged with the Employability Group of the 14–19 hub, supporting the transition from school to work, employment opportunities and career pathways.

Actual Contacts

Fig. 5.1 depicts the relational structure of Case A in response to SB issues; findings for EIL were similar. Only four ties were reciprocated (N4 and N6, and N8 and N19). Larger organisations (N2, N12 and N14) had resources and policy authority but they did not mobilise governance. This concurs with the hollow core theory discussed in Chap. 2. Central government reforms forced these organisations to take on governance roles, but their business culture, outreach work, and contractual relations with external subordinate organisations, restricted alliance-building. N12, for example, had the highest in-degree centrality score, meaning the largest number of actors were linked to it ($N = 10$); but they lacked contact with others having the lowest out-degree score, or had non-reciprocated ties. Another agency, N2, did not report contact with a business support organisation (N19). N4 and N6 had

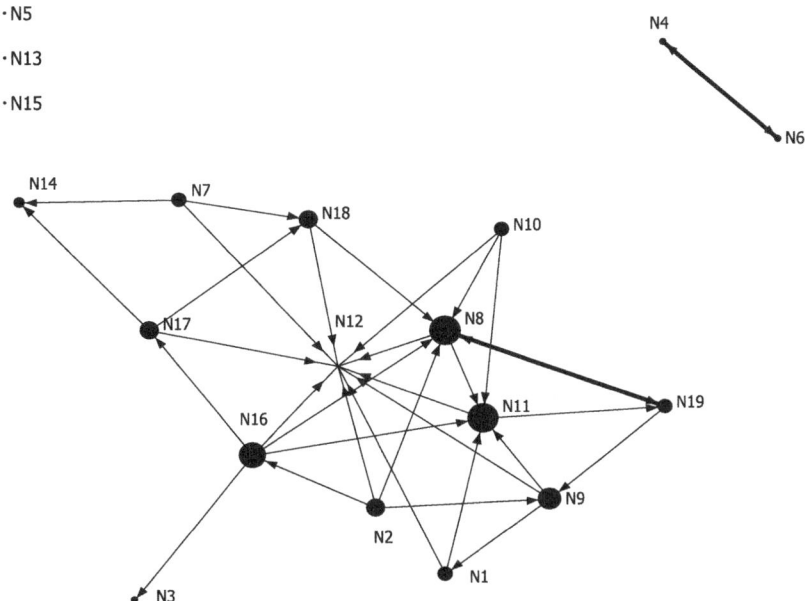

Fig. 5.1 Actual contacts in Case A (CSPG) relating to the unemployment issues in SB

'pendant ties'; connected by one contact only. A regeneration actor (N3) with funding interests had no contacts, but a powerful funder (N16) had made contacts. Some had 'indirect relations.' N16 had a two-step connection to N19 (N16 → N11 → N19). The relational gaps depended on 'bridging ties'. For example, N16 might have connected N3 with an employment support service N11.[14] The network literature claims network coordinators have greater centrality and power, yet in this case they had a low brokerage role

---

[14] N3 had networks supporting wards bordering the city in four local authorities including LBTH. They prioritised the development of a 3-year strategic plan, supposedly to join up the planning process and link with the London Development Agency's Economic Development Strategy and the London Skills Commission 'London's Framework for Regional Employment and Skills Action' (FRESA), sub-regional strategic plans for Business Link, and the Learning and Skills Council London East, the DTI's City Growth Strategy and the local authorities local strategic partnerships. Some of the projects benefited Tower Hamlets residents, institutions, including schools, and a pathway to work in the finance and business sector, Weavers Restaurant Trust to help the socially disadvantaged to work in the service industry and the college's raising of awareness amongst young people and structuring of progression routes to jobs in the public sector.

and the evidence suggests that priorities were administrative and concerned with monitoring, not joined-up working policies or concerned with relational aspects. Overall, this network resembled governance but not substance.

Case B—Profiteers

Case B represents the WNP that operated in EIL's employment zone in 2004–6.[15] WNPs received £3 million annually, with £1 million for a flexible discretionary fund 'to allow personal advisers, working with LSPs the flexibility to deliver services that best meet the needs of the local community … [and] … tackle the substantial and varied barriers that prevent residents in these neighbourhoods from finding or returning to work' (Dewson et al. 2005). The DWP assumed the LSP would 'bring together local partners and providers' and deliver effective partnership-working to 'combat engrained worklessness,' and that the WNP would 'access some harder-to reach customer groups', support the long-term unemployed through outreach and work-focused interviews, and break down the multiple barriers facing economically inactive residents (2005: 45). The DWP's decision to appoint a UK-wide private sector company (an accredited DWP provider) to deliver the WNP in EIL was contentious, as a housing association also provided employment support in the area. The firm managed the scheme from the top floor of a multistorey building in East India, adjacent to the social security buildings. The firm's promotional literature stated it was 'not interested in short-term fixes, but working with the right partners to ensure long term positive change in local communities'. To this end, the firm needed to understand the requirements of local unemployed people, allocate development time to work with community leaders, meet the complex funding criteria for the DWP's discretionary funds and handle administrative procedures (such as the four-stage payment for client outcomes), which meant increased contact with the DWP.

The MP for Bethnal Green South, however, raised parliamentary questions about payments to private contractors, since the performance statistics of the WNP were undisclosed. Meetings of a small steering group responsible for nominating WNP spending (£800 K) during the financial year were frequently cancelled due to personnel changes. The firm replaced three managers during the fieldwork period. Consequently, the

---

[15] Employment zones are areas where private-sector providers take over job-search provision for unemployed people who have been claiming JSA for 18 months or more, or who are below 25 years old and have already had time in the New Deal for Young People.

interview revealed that the managers' contacts were negligible. Instead of auditing the local needs or creating a joint plan, the firm relied on funding opportunists to generate ideas for bids, and the local network, PAN (Case C), a competitor. The DWP's mid-term evaluation found pilots had not progressed outreach or engaged with IB claimants, and that poor relations between partners affected outcomes (Dewson 2005: 49):

> ... one of the key objectives of the pilots was to assess how far the WNP produced area outcomes. There is certainly scant evidence of area outcomes at this stage in the evaluation. (Dewson 2005: 71)

The final report noted that LSPs 'rarely had a role in directing or steering pilot activity' (Dewson et al. 2007: 52). According to one official, the firm had tried to establish networks based on service level agreements, but small organisations felt inhibited from signing up. Guaranteed payments were earmarked for specific activities, rather than partnership-working outcomes. The firm's representative had a disparaging view of the LSP: 'These partnerships have been going on for the last five or six years and yet the unemployment levels in Tower Hamlets are extortionately high.' Moreover, the public sector took too long making decisions for small outcomes. However, an official complained that the WNP had not joined up with them, brought groups together for decision-making or provided guidance on the DWP's Flexible Discretionary Funding. In conclusion, it appears that the WNP had unrealistic goals, and lacked joint working, responsibility sharing and stakeholder ownership. The firm could not broker services to resolve the unemployment problems, and its values clashed with those of non-profit agencies. A local interviewee said that it had exploited voluntary sector resources to reduce its marketing expenditure and used local groups to reach clients for targets and profits. Another interviewee, a competitor, labelled the scheme a disaster.

Actual Contacts
Fig. 5.2 depicts an underdeveloped network. The WNP created a focus of interest but expectations were too high. It had personnel and bureaucratic pressures, conflicts over private sector values, and poor relations with its competitors (N2) and health services (N3). Many people in receipt of IB in EIL suffered with depression and back pain, yet the health representative was not fully involved. Connections sought by N5 were dependent on coercing the strongest institutions (N2 and N4) to support the commissioning process in the programme's short lifespan, rather than a needs-led approach. Relations may have developed for a short while beyond the

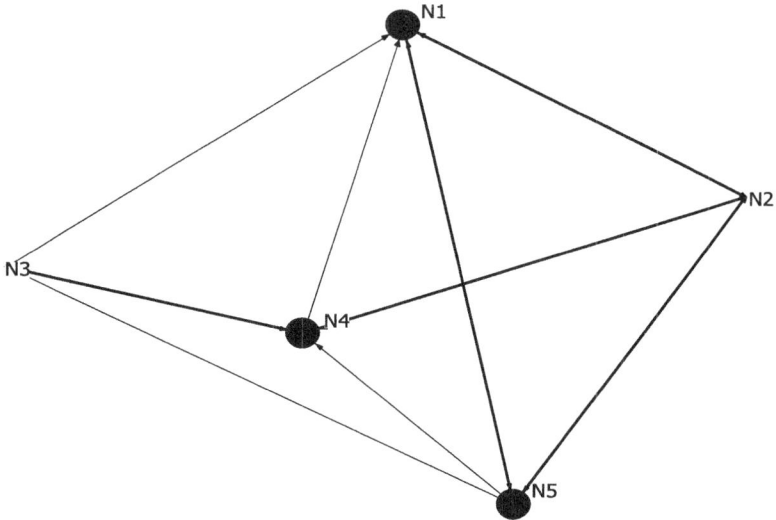

Fig. 5.2 Actual contacts in Case B (WNP) relating to the unemployment issues in EIL

fieldwork period, but the firm's priorities were job outcomes for payments, not joined-up working.

Case C—Rebels

Poplar HARCA established the Poplar Area Neighbourhood Partnership (PAN) as part of its social housing landlord duties to tackle poverty and low income. The PAN group explored here dealt with youth issues, education and the employment and training needs of residents. A third of the membership was comprised of Poplar HARCA employees associated with neighbourhood management, neighbourhood centres, youth support and employment support. A senior manager from Tower Hamlets College chaired quarterly meetings. Other representatives included Leaside Regeneration, Education Action Zone, Skillsmatch and the Ideas Store. Minutes contained detailed action points, development needs, local information and updates on local projects, programmes and funding opportunities, and local issues were aired (Poplar HARCA July 2004). An interviewee said the PAN was successful because people were enthusiastic, open-minded, visionary and pushing for employment, training initiatives and projects. Positive leadership and good relations between Poplar HARCA and Tower Hamlets College led to fertile contributions during meetings. Reportedly, this facilitated progression routes to college, information and support for unemployed people, and funding links between regeneration projects, trainers and business. One interviewee associated PAN with grass-roots politics and the struggle for autonomy: '...you've got a lot more different kinds of democracy going on here ... I suppose you could say anarchist not in the sense of guns and bombs but political where it comes out.' These comments refer to the LSP's interest in replacing PAN with LAP 7 (a small number of residents and groups, support workers and councillors). During the fieldwork period, EIL interviewees scarcely mentioned the LSP and the LAP 7, and they also received scant attention from community newsletters. One interviewee described the borough as 'the most difficult to get to the table'. Some referred to the lack of communication with LAP 7 or felt that the THP focused more on funding than on working together to achieve aims.[16] A government report, however, suggested that the neighbourhood management, THP and the LAP 'operations work in a relatively seamless fashion in practice' (ODPM 2006: 20).

Actual Contacts
Interviewees provided evidence of positive interrelations and helped combat uncertainty in the face of LAP competition. However, Poplar HARCA's in-house bonding may have led to complacency; in this respect, Fig. 5.3 suggests joint working was thin. For example, there was no basic needs audit or strategy to address unemployment issues, as one interviewee explained: '... what we don't have is either on a ward basis or on a Poplar basis a strategy to address the issues. I don't think there has been much mapping or analysis of where the issues are. I think it's quite a reactive process.' One worker believed work-based learning provision was inadequate and that JCP and the LSC hampered the network from supporting local people; however, they had no evidence to back up their claims. For example, regular reviews and needs-assessment could have identified gaps in provision and anticipated funding opportunities for a faster response. One interviewee suggested PAN, THP and specific local initiatives were

[16] A £150,000 LAP 7 underspend was spent across the borough after Tower Hamlets Community College withdrew from delivering an enterprise development centre. Some of the LAP 7 members were 'extremely concerned' that the money did not go to local organisations and community groups (THP LAP 7 May 2005b).

190 THE IMPACT OF NETWORKS ON UNEMPLOYMENT

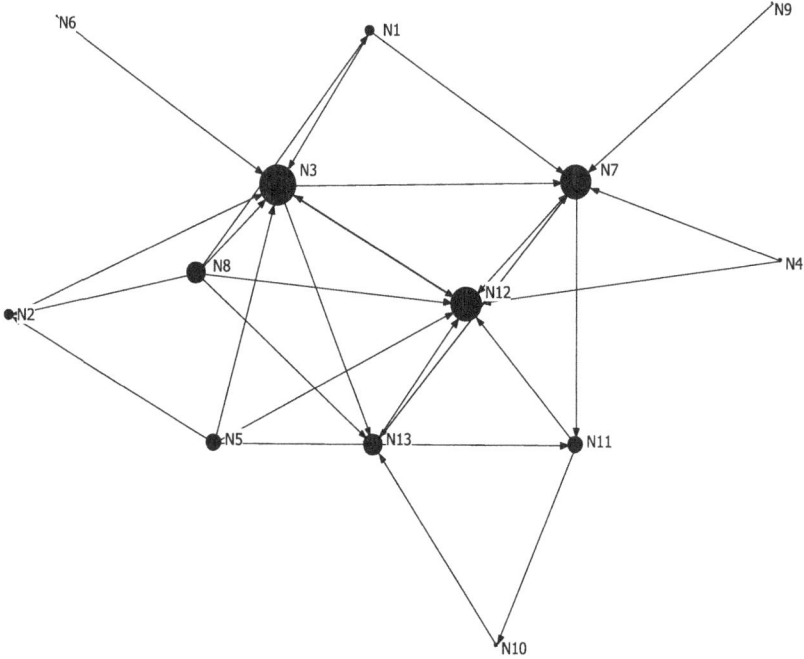

Fig. 5.3 Actual contacts in Case C (PAN) relating to the unemployment issues in EIL

not working in partnership. Moreover, outside the network, local information services were not coordinated well. During fieldwork, for example, different employment support organisations used a desk space at the Ideas Store on different days for jobs advice. Clients could pick up leaflets or speak to staff, some of whom had laptops and internet access to search for job opportunities. Over three days, the author had spontaneous conversations with different advisers at the desk and noted the findings (see a method for fieldwork activities in Appendix 2). Surprisingly, the advisers were unaware about the 'drop-in' services operating from the desk on different days, and the JCP outreach team said they had only just heard of the WNP. This suggests that Case C withheld information to subordinates, resulting in service duplication and confused provision. Although the internal network logic worked well, basic information was not being shared or developed through multi-agency protocol; as such its joined-up working was less effective externally.

Case D—Enthusiast

This small case network explores ego links between contacts reported by an official whose brokerage role was critical for achieving socially optimal outcomes for the neighbourhood. This individual stated that the nature of the role was : '… making sure that the services are working together, focusing on particular problems, actually adding value by working together and are actually not detracting from impact through not working together'. This requires a lobbying or influencing role. However, the official was unfamiliar with the details of key agencies relating to unemployment and reliant on the interface with JCP and the college. The interviewee continued: 'My needs will drive a lot of what the [neighbourhood] agenda will be and … feed in some of the reasons behind why we need to coordinate it, rather than doing the specific coordination of it.' Indeed, strategists often assert predetermined structure or planning tasks, to avoid negotiation and conflict. The contact expected local networks and residents to understand THP strategic purpose and LAP infrastructure, and overcome anxiety and rivalry that hampered progress. Another 'enthusiast' explained the process: '…what we are trying to do with the LAP is cut down the number of formal networks and make it all "the LAP"… there is to a certain extent a formal network called the PAN Partnership … but … we need the focus to be what the LAP is saying ..' Furthermore, 'I suppose they [PAN] have been quite effective but since the Local Strategic Partnership has come along you can't really live with two separate networks in one area … obviously they [PAN] are realigning their structures with the Local Area Partnership.' The assumption that PAN was aligning to the LAP was wrong. Central initiatives or departments inflate roles through power stimulus, structure and budgets.

Actual Contacts

Fig. 5.4 suggests the Case D contacts (N3) levered ad hoc strategic relations to implement the THP governance structure and did not join up policy or mainstream work. LAP meetings focused more on partnership charters and subgroup protocols than unemployment matters. One interviewee noted that education was a priority of the LAP, but no one from education was present. A neighbourhood organisation felt sidelined from meetings. A meeting organised by the contact to identify priorities for tackling unemployment decided to scope mentoring projects to improve aspirations, but proposals fell behind schedule and the LAP had to request details. Some members suggested expenditure distribution was

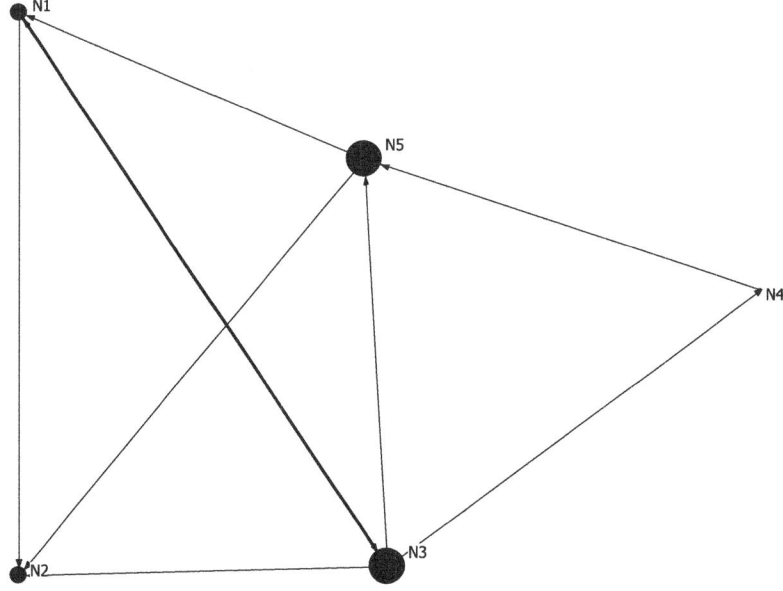

Fig. 5.4 Actual contacts in Case D (EGO)

undemocratic. Ill-defined goals, unrealistic planning timescales, weakly coordinated personnel and poor relations reduced the potential for the joint management of issues. One official takes a candid view: 'What we have got to address [is], why [did] so much activity happen and so little impact happen at the end of it.'

Case E—Outsiders

This network sampled SB-based organisations with a reputation for supporting unemployment, including voluntary-sector organisations specialising in ethnic minority support services which ranged from enterprise training, welfare advice and women's services to youth support. They operated with poor facilities and dwindling opportunities to share resources. Organisations can be territorial in order to attract loyal users, and the patronage of elites on management committees, power struggles, low connectivity, rivalry and uneven resource distributions are commonplace. Case E reflected some of these issues, and fragmented niche services developed. One interviewee said that no one worked jointly on unemployment. Another interviewee wanted to influence JCP to self-advertise. One contact had been approached by the THP to establish a formal network to explore unemployment and link with LAP 2, but progress was slow. Indeed, none of the interviewees appeared aware of the network initiative. Some participants believed the organisations lacked the capacity, time and resources to work jointly. Another explained how groups scrambled for funding before constructing a castle wall around them if successful, presumably to achieve targets.

One ethnic minority worker stated that groups were wary and suspicious of each other. Resource-rich organisations and mainstream services seemed to avoid SB, perhaps distrustful of cultural politics and cautious following a recent fraud case which had involved a local group and the Councils Regeneration Department. As one interviewee observed; '... the knock on effect ... is "you guys can't do it" and that is kind of the negativity that we feel we are getting ...'. Another worker found the local inertia and bureaucracy surrounding funding demoralising: '... there's an awful lot of this [LAP] and not much action and when you try to take action you tend to find that whether it is funding, indifference or jobs worth ... there is nothing really happening. I would suspect ... better networking goes on in Tower Hamlets between the commercial employment agencies ...'. Several actors were disappointed with the local networks. One felt the THP was ineffective and worn down: 'this is a tired, very tired set of people here, because they have been trying to reconfigure it, it feels like forever, and they are not getting where they need to get to'. Some criticised the lack of action in the local area partnerships, and one said: 'no one is doing networks'. Another suggested that organically grown networks worked better. Cross-partnership working was too difficult, according to one interviewee, and LSP staff lacked understanding of local needs. One interviewee claimed a proposal for a centre for parents to leave their children in a full-day crèche provision was inappropriate; half the Bangladeshi women did not want to work full-time and others did not want to work at all. Therefore, they needed to tailor projects to a 'women's economy', which would require more radical ideas that were not tied to outputs. The SB contacts had specific expertise, problem-solving and referral and resource-sharing potential, but the network was not utilised.

Actual Contacts
Fig. 5.5 depicts a star-shaped structure and demonstrates organisations' minimal investment in relationship-building. At the star's centre, N6 had the highest reputation across the borough for work in SB; they also

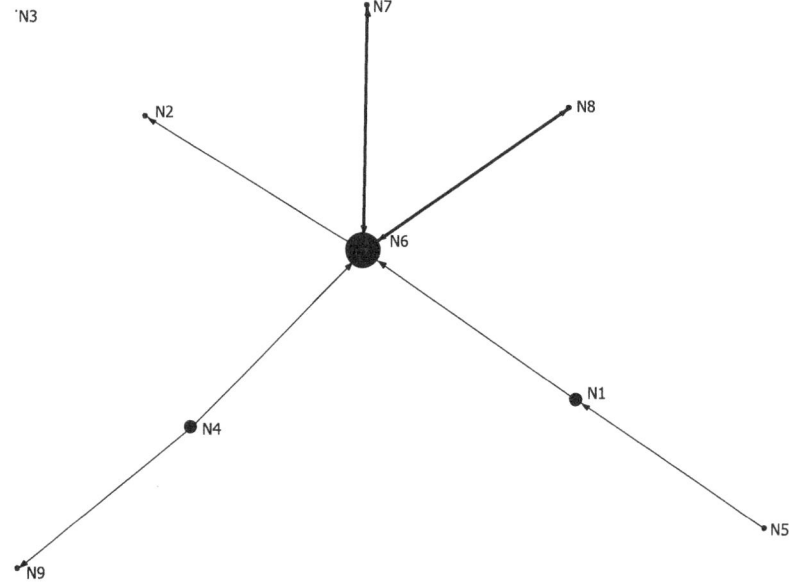

Fig. 5.5 Actual contacts in Case E (SB) relating to the unemployment issues in SB

provided snowball contacts (N3, N7 and N8). Although N6 had the highest centrality, it lacked reciprocal relations with N4, its close-by neighbour and competitor. One contact had a dual role—representing a Bangladeshi group in Case E, and community empowerment work in Case A. They had one contact in Case E but none in Case A; hence, there was a redundant bridging role. One project (N8) had the lowest out-degree and in-degree of all the results, yet its services were thriving. One interviewee claimed the area's isolation stemmed from cultural and gender issues: 'In terms of unemployment we deal with it ourselves ... that's because there is a gender issue ...'. Local people felt used by mainstream workers who wanted to tick the ethnicity box. Hence, another reason why organisations lacked interconnectedness was their wish to protect their values. Ties between different community facilities were absent (see N8, N9, N3 and N5). Providers had under-utilised facilities, overstretched services or underused expertise, rendering them vulnerable to external competitors capitalising on the next round of funding.

FACTOR COMPARISON

This section considers how five factors (see Fig. 1.1 in Chap. 1) impacted on Tower Hamlets's network culture.

Central Environment

The Labour-controlled council and the THP readily implemented the LSP infrastructure. They situated neighbourhood problems within a network management framework for joined-up governance. This ensured power arced back to the THP. The self-referencing LSP expected to be able to filter, defend, and legitimise all local activities. It captured job roles to provide continuous progress updates, staging information presentations and ticking central policy boxes, ranging from health data monitoring to disability. The allocation of short-term funding ensured groups supported THP objectives. However, members' ability to influence, plan and implement joined-up local mainstream services was limited.

CSPG (Case A) minutes, for example, considered the NRF allocation of £4.2 million was a substantial amount to spend on tackling worklessness over a two-year period (THP CSPG 2004). Elites become anxious when disseminating funding in case their careers are affected negatively by bad publicity, fraud or poor performance. Commissioning evidence-based research is one way to reduce expenditure risks and look active. Instead of acting on report recommendations to support local construction and training needs, for example, CSPG wanted to appoint another consultant to consider other best practice schemes (THP CSPG 2005). CSPG commissioned another consultancy firm to research recurrent unemployment in 2004–5 but failed to publish the findings or final expenditure of the £80,000 contract (LBTH 2008a). For several years, the consultancy firm would not release the report, and some LBTH officers had no memory of it. In 2011, a Freedom of Information (FOI) official stated they no longer had the report, and the consultancy firm did not reply to their request for a copy. In 2012, a further FOI request for evidence of expenditure, minutes and CSPG actions in response to the 'missing' report findings was ignored. In 2013, the Information Commissioner's Office (ICO) ordered LBTH to release the electronic report and minutes following an internal review, although final expenditure remains undisclosed. Ironically, network governance obstructs historians from tracking social welfare expenditure and 'public' records.

LSP participants also played tactical games to access funding, and counter-practices emerged. A prominent interviewee explained how they went with 'the grain' to draw down resources, but then 'kicking against it' to overcome double counting, or withholding client referrals in case they lost payments to other schemes. Some interviewees believed public-private WNP (Case B) relations delivered by a national firm under DWP central directives were unworkable. They perceived the WNP budget had disrupted local activities and fuelled tensions, and collaboration was not a two-way advantage. It was operationally distanced from the EIL community, lacked the capacity and legitimacy to steer local networks and collaborate with the LSP, and competed with Poplar HARCA's in-house employment services and regeneration agency partner, which were working to sustain funding for employment support projects. Case C, a self-governing network of local professionals, wanted independence, but the THP expected the LAP 7 to override it. Case D (the ego contacts of an official) had limited power to drive the LAPs and change institutional infrastructure to implement joined-up mainstream services. Moreover, LAPs (1–8) found NRF expenditure too bureaucratic. Central directives had the least impact on resource-poor Case E in SB.

Central Policy Rhetoric: 'Joined-Up Government'
LSPs played a role in supporting Whitehall policy for 'joined-up government.' Thus, interviewees were asked whether they believed 'joined-up government' was relevant to their work. Responses were analysed using phrase frequency and a coding system: 46 % stated it was relevant. Some strategic actors overstretched policy implementation realities. They expected joined-up government to reduce poverty, unemployment and assist organisations to achieve more sustainability instead of short-term initiatives. The LDA pursued 'joined-up' thinking on national targets, workforce policy and skills. One THP representative expected that the Government Office for London (GOL) would bring departments together, the NRU would track these and get others on board, and the LSP would involve the rest.

However, 39 % criticised the lack of joined-up government. One official emphasised poor communication between government departments: '… various fiefdoms at the Home Office, the DfES, the Learning and Skills Council and the various branches of the ODPM, like the RDA, the LDA and GOL are unbelievably not joined up'. Another interviewee commented that the Inland Revenue worked in isolation from the DWP, and neither joined-up with employers. One senior worker needed Welfare-to-Work policies to translate into local needs. Another officer believed the DWP did not align with local government interests. A manager said reducing the numbers of NEETS (young people not in education, employment or training) required the LSC and National Skills Agenda to be fed in to the Sector Skills Council, the DfES, the DTI and the DWP. Likewise, the Children's Bill required health, social services and education departments to work together to share information and identify priorities. Labour's Policy Action Team had also called for policy cohesion between Youth Services, Connexions and the Careers Advice Service. In practice, unemployment prevention depended on youth and community workers and teachers supporting young people by helping them to build their self-esteem and confidence, access good advice and information and engage in new activities in addition to academic work.

Departments' attention-seeking initiatives, short-term funding and lack of investment in existing networks generated local resentment. One official felt the mechanisms were too complex to benefit localities: 'London Government is horribly complex again and the institutional and political framework beneath that is incredibly complicated. All these are called partnerships but it is not at all joined up, so it's very difficult to build something or run a programme for unemployed people.' Another official believed target-setting, reporting systems and funding streams wasted the time of delivery agents and worked at cross-purposes on the ground. An executive provided several quarterly reports to one department. They believed services and LSP floor targets did not 'join-up', and that all the CPAGs needed to develop an unemployment response, not just CSPG.[17] A voluntary sector official noted: 'They have fractured nearly everything in sight, come up with little projects instead of investing in the networks on the ground. There's money following new ideas but they are not long term and embedded in the community'.[18] In other words, the means (reporting to the system, isomorphic activities, and

---

[17] Floor targets are government targets used to measure the success of LSPs and raise the quality of services to an acceptable minimum, or 'floor' level.
[18] Neighbourhood unemployment is likely to predict sustained area poverty (Kauppinen et al. 2011).

contractual games) became the end, and distracted from the lack of joined-up government.

Network Structure

LSP structure created power-dependences and 'rules of the game', but structural maintenance inhibited openness and locked out debate (Miles and Snow 1992: 69–70; Painter et al. 1997: 227). For Rhodes (2000:160), governing structures, including central initiatives and thematic 'zones,' use network management instrumentally; 'assumes that the centre can devise and impose tools to foster integration in and between networks and realise central government's objectives'. Thereby, a THP enthusiast believed 'semi-autonomous organisations working in little areas and doing their own things' had duplicated work; hence, they needed the LSP to bring organisations together for strategic rationalisation, and network structure provided the means. Cases A and D had image-managed mandated structure and network coordinators that linked to other LSP substructures to approve, monitor and evaluate NRF expenditure and initiatives. Case A structure maintained the status quo of differing professional culture and the disconnected central environment. Low connectivity amongst members seemed acceptable. Neither formal structure (coercive and centrist) nor informal structure (voluntary and independent) would organise direct coordination for handling neighbourhood unemployed. The Case E organisations focused on their own coping needs to survive, building bridges with small niche contacts to find funding, or adapting to the big institutions' controlling agenda.

Area-Based Factors

Networks without a neighbourhood focus are liable to distort unemployment realities or tackle problems superficially. Hence, the study asked interviewees what they understood about the unemployment problem in SB and EIL. Responses were analysed using a category and coding technique. Table 5.3 presents the percentage of local knowledge (services, history and demographics) in case networks directed towards the wards or borough-wide. For example, some interviewees could differentiate between the two ward clusters, noting that each area had different characteristics and community buildings, or emphasised specific problems, such as generational unemployment and barriers to work, including health, literacy and lan-

Table 5.3 Contacts in case networks aware of the unemployment issues in EIL, SB or borough-wide only (%)

Cases:	A	B	C	D	E
	CSPG	WNP	PAN	EIL	SB
	%	%	%	%	%
EIL	14	40	69	40	10
SB	28	NA	NA	NA	60
BW	57	60	31	60	30
Total[a]	100	100	100	100	100
Case total[b] Interview total $N = 39$ Citation total $N = 51$	$N = 19$	$N = 5$	$N = 13$	$N = 5$	$N = 9$

EIL East India and Lansbury; *SB* Spitalfields and Bethnal Green South; *BW* Borough wide; *NA* not applicable, the network did not operate in the locality

[a] Total citations are more or less than 100 because of rounding

[b] Includes members with a dual role or dual network memberships

guage. Conversely, one official in Case A explained why local perspectives did not matter: '… when y' know you have the highest unemployment in the country it doesn't really matter how the different adjustments between the wards are … I don't work on a ward basis at all or an area basis'. Case A participants had less local knowledge but more borough-wide strategic policy roles and resource influence (correlate Table 5.2 with Table 5.3). Case B and D remits could not target unemployment in an in-depth way. Cases C and E involved local experts, but they lacked a plan to oversee the provision for dealing with neighbourhood unemployment.

The second stage of analysis involved thematically categorising participants' unemployment knowledge and frequency statements totalled as percentages per case. Twenty-two unemployment factors were identified and organised into four categories: locality, structure, capacity to work, and behaviour. Higher scoring themes in Case A included government policy, such as skills and getting people off IB, and cultural barriers to employment, from English skills, literacy and numeracy, job perception, racism and religion, to job barriers for Bangladeshi women. SB workers in Case E noted local barriers, such as education, overcrowded housing and the detrimental effect on the community of losing its industrial work history. EIL workers in Case C emphasised historic and generational unemployment. High youth unemployment was scarcely mentioned. A senior council officer blamed the education system: '... the younger people coming through are not equipped to go into the work that is available'.

Perceptions of structural barriers to employment included job type, low pay and job decline. An Asian support worker suggested that conspiracy theories had arisen because ethnic minorities felt excluded from public sector appointments and the richer financial districts. A white manager believed the white working-class community and recent settlers like the Somali community could not access the 'word-of-mouth' work in SB amongst ethnic minority family-owned catering or trades businesses. Some felt the private sector should recruit more from the local population; hence, a subregional business alliance worked to help unemployed people without references gain agency work. A business strategist dismissed unemployment as an area in which to specialise, yet business start-up support is known to interest unemployed people. Again, actors' perceptions of the problem can weaken representation.

Notably, participants who lacked contact with clients overlooked the personal effects of unemployment, such as poor mental health or drug problems. This suggests certain types of actor had a limited understanding of the local situation. Conversely, an ethnic minority worker in Case E understood the effects of unemployment on individual behaviour, stating: '... they lacking of [*sic*] support, guidance, awareness, depression, motivation, building self-confidence ... these people need support, very close support, and motivation, and encouragement to develop the employability skills and increase employability in this area'. A local educationalist in EIL noted: 'There's a huge lack of confidence, so getting people onto that very first rung of training is a really major issue and then people move within those early stages but don't progress and that's another key issue.' Hence, the next section explores the organisations associated with assisting the unemployment issues in SB and EIL, and whether high- or low-reputational organisations participated in case networks (see methods, Appendix 2).

Network Agency: Organisational/Participants

In SB, 75 organisations were associated with tackling unemployment. Citation frequency ranged from low to high (1–16). Fifty-five per cent of organisations had one citation only. In EIL 74 organisations were associated with tackling unemployment. The citation range (1–22) was higher than SB, probably because a higher number of interviews took place in EIL. Fifty-three per cent of organisations had only one citation. Overall, more top scorers were based in EIL. Further analysis linked the high- or low-reputational organisations to case networks.

Contacts with a High Reputation
Reputational scores were highest for welfare and education institutions with resource power and authority. JCP attained the top overall score, $N = 16$ for SB, and $N = 22$ for EIL, including within cases. However, they were not reputational leaders on community matters and their input was limited in cross-cutting work and local economic development. JCP worked with co-finance partners and their outreach teams were expected to improve partnership working, yet they could not work flexibly with local groups beyond their contract regime with approved contractors and suppliers. One EIL-based interviewee considered joint working with JCP a futile idea: 'ineligible, regardless of how good it is, even if moving that individual towards employment'. While organisations contacted JCP for contracts or neighbourhood support, ties were rarely reciprocated. The second highest scorer, Poplar-based Tower Hamlets Further Education College, influenced Cases B, C and D and received $N = 12$ nominees for SB and $N = 13$ for EIL. The housing association Poplar HARCA yielded $N = 14$, the highest results for a case network organiser. Other top scorers were 'Employment Solutions', a long-running tripartite partnership involving borough-wide mainstream providers, JCP, the Tower Hamlets Further Education College and Skillsmatch, a job brokerage scheme coordinated by LBTH in Canary Wharf. One top scorer associated SB with negative local politics and possibly avoided the area to safeguard their reputation. Case B, the WNP, had also received $N = 13$ nominations. An EIL regeneration company actively seeking NRF funding received $N = 11$ nominations for EIL, whereas a city-side regeneration partnership closer to SB lacked nominations and was absent from Case A meetings. Borough-wide organisations with headquarters in ward clusters achieved higher scores in those areas. An SB-based ethnic minority enterprise project had $N = 8$ nominations but lacked the resources to make an impact elsewhere. A mainstream youth agency had $N = 7$ nominations for work in EIL where its headquarters were based, but none for SB, despite high youth unemployment there.

Contacts with a Low Reputation

Many organisations in positional network cases lacked an association with unemployment. Dead-weight or inappropriate organisations would have limited interest in effecting change or working together; moreover, network management lacked the necessary authority to coax passive actors or free-riders to tackle unemployment. Conversely, the low reputation organisations may act as needs arise or disseminate information outside the network. Case C (PAN) had more members associated with unemployment than Case A, but 38 % of Case C members had no reputation score for assisting with unemployment problems. The WNP (Case B) intended to tackle unemployment and health issues, yet a relevant actor lacked a purposeful role. One provider achieved zero in-degree scores for reputation, but the highest out-degree score for ties to others. The low scores attained by Case D suggest the coordinating agent had not fulfilled their role. Case E organisations in SB had low reputations and poor brokerage to other networks, although one organisation with the highest reputation score had a higher reputation outside the network. Overall, cases involved too many organisations with ambiguous reputations or limited agency to focus on neighbourhood unemployment problem-solving.

Network Processes

The government expected networks to operationalise its 'joined-up working' policy; namely through network governance and network management. Hence, interviewees were asked what they understood by the term, its relevance to their work, and who promoted it in the policy field. The following results were yielded.

Joined-Up Working: Associations and Meanings

Responses were organised into three categories: (i) general associations; (ii) value associations; and (iii) structural associations. Most respondents were familiar with the term 'joined-up working', although some SB ethnic minority groups were not, as a manager explained: 'It may be a new word they are trying to introduce but it is not well introduced yet anywhere in this area.' Groups that do not mimic funders' terminology are likely to be disadvantaged. A majority of SB actors thought joined-up working meant driving change through partnerships in neighbourhoods to improve services. Overall, the concept was not associated with improved outcomes for individuals. Only one provider suggested it referred to employers, schools, the DWP and advice centres pulling the policy chain together and working from a client perspective.

Particular values were associated with joined-up working, such as 'working together'. Even network participants with no connectivity to local organisations emphasised this; see for example this London-wide worker: '... my idea is various agencies whether they are local authority, civil servants, departments of state and voluntary sectors and private businesses actually should be working together for a common goal'. The importance of common interests and of sharing agendas, resources, expertise, cross referrals and goal direction were emphasised. A local worker added: '... agencies with a common purpose, common understanding, common goals, common principles and some sense of wanting to benefit each other and using people's expertise and skills[sic] being aware of people's weakness and strengths and being open about addressing the weaknesses'. In practice, critical reflection requires effective communication, risk-taking and team-building skills, but these are seldom implemented in networks. One support manager believed joined-up working requires trust and a change strategy of its own for structural reform to connect strategic planning systems with grass-roots delivery. An official emphasised coherent problem-solving: '... whether one system joins up with another shouldn't matter as much as that all the systems are focused on the same problem'. Ironically, this manager promoted one system: the LSP. Another manager associated joined-up working with power inequalities: '... it involves often sitting at the table with people who don't view you as an equal and to whom you are not an equal, it takes a lot of time that you are not funded for and ... a lot of political sophistication to really understand how to make it work for you'. A senior official suggested joined-up working required a competent coordinator, a role they believe they held: '... it's where you have a coordinated approach, where whoever is best ... are the ones who do the delivery to ensure there is no wastage of money'. However, this representative had perhaps overinflated their ability, and few could challenge their wasteful coordination as the agency had since been abolished. A local worker believed there was insufficient discussion or innovation surrounding joined-up working, or too much talking and little action.

Nevertheless, joined-up working was highly relevant to respondents' work. They repeatedly used phrases like; 'absolutely central to what we do', 'hugely important', 'crucial', 'it is important in everything we do' and 'everybody promotes it'. A manager reiterated its popularity and shortcomings: '... everybody is promoting it ... you've got the LDA promoting it you've got the LSC promoting it, everybody promotes it as a concept everybody loves the idea ... it all falls foul of the un-joined up funding'. Many interviewees, from the grass roots to strategic actors, believed they championed joined-up working and brought everyone together. For example, a local worker explained: 'I'm ... the broker between local residents ... helping them to identify what they think their real needs are ... then making public sector providers aware of the perspectives ... and trying to get them to work together not only between themselves but also with the other community and voluntary organisations providing a service.' The THP expected to lead joined-up working: '... its [*sic*] central to what we do ... bringing people together and getting them to identify their priorities and identify solutions together'. However, an ethnic minority representative suggested governance structures widened the gap between resource-rich and resource-poor voluntary organisations: '... the result is that most ... small community organisations operate in isolation ... [and] find it very difficult to be part of any meaningful network and benefit ... it creates a hierarchy within the voluntary sector'. A director noted that policies, from health to education and housing, stopped short of 'joint planning'. The joined-up working agenda gave the THP the power to join others up to central schema, rather than considering how best to join up in neighbourhoods.

Indeed, citation results for who promoted 'joined-up working' in the local unemployment policy sector, and frequencies, reveal low results across three categories: (i) organisations; (ii) networks/partnerships and initiatives; and (iii) strategic frameworks. JCP had the highest association for an organisation promoting joined-up working, but only scored 15 % ($N = 6$). Some believed JCP lacked the capacity to undertake joined-up working. One partner explains: 'JCP isn't powerful or doesn't employ staff with the depth of experience and intelligence to promote it or lead. I can't think of any other policy organisation that promotes it, the college promotes it and to some degree the Local Authority.' Tower Hamlets Further Education College gained a score ($N = 5$) of 13 %, followed by the LDA ($N = 3$) at 8 %. Low scores suggest no single organisation led on joined-up working policies for tackling unemployment. Only 10 % of informants cited two case networks (Case A (CSPG) and Case C (PAN)). The low promotion of joined-up working correlated with cases' low connectivity. Only $N = 4$ interviewees associated LSP strategic frameworks with being responsible for joined-up working. Perhaps the impetus of governance theories and structure for local joining up may not be entirely rational for organisations steering their self-interests.

Conclusion

Buoyant financial districts hosted by the inner city borough of Tower Hamlets provide a stark contrast to the case study neighbourhoods with their persistent unemployment, and demonstrate a type of growth that does not produce suitable jobs for the resident population and which increases living costs, particularly housing. Failure to regulate employers' and proprietors' behaviour (low pay versus high rental costs) has created the 'benefit trap' whereby work does not pay. Policymakers treat unemployment as a skill 'catch-up' problem, but fail to deal with the economic fault lines that disconnect cohorts. Across London, new job entrants face intense competition and businesses disengage with the apprenticeship system. The well-worn policy goal of supporting unemployed residents in improving their skills and competing for jobs in the City and Canary Wharf conflicts with the values of the Muslim population, since the wealth-creating industries are not a universally desirable work culture. Clearly, policies to address job design and the provision of meaningful work are lacking for all economically excluded cohorts.

Political legacies and historic labour injustices explain differences between ward cultures and shape local institutions, network culture and policy preferences. SB had enterprise confederations and trade alliances but no pro-local regeneration package, and community tensions sustained inadequate services. At a minimum, networks could have tackled ethnic minority advice and support for young people and people with mental health problems. EIL housing and regeneration bodies sought inward investment through competition rounds, but they had not reviewed why previous efforts had failed. Both localities depended on the local organisations' networking capacity to compete for central funds, which were not allocated on a needs assessment basis.

None of the five network cases profiled launched a sufficient attack on the social and economic needs of the localities. All the cases struggled to acquire or distribute resources. Well-resourced networks coordinated political preferences and the central infrastructure, and institutional leaders interpreted governance and crosscutting issues to maintain their power. The SNA techniques provided an empirical impression of 'joined-up working' for which the associated values lacked implementation in practice. Self-promoting networks had poor relational connectivity compounded by poor local knowledge in some cases; moreover, problem-solving was beyond their decision-making capacity or the issues were either undisclosed or not well defined. The first employment strategy, developed in November 2008, suggested that tackling unemployment required a major change (LBTH 2008b: 11). By 2016, austerity measures and changes in the employment structure assisted to reduce the claimant count, but unemployment remains above the national average, and more than 11,000 people in the borough receive ESA/IB. The next chapter explores the issues of high unemployment and network responses in Great Yarmouth Borough Council, where, at least superficially, the environmental and cultural aspects could not be more different.

Bibliography

Aftab, I. (2007). *The spatial form of Bangladeshi community in London's East end.* London: University College London.

Atkinson, R., & Moon, G. (1994). *Urban policy in Britain.* Basingstoke: Macmillan.

Begum, H. (2008). Geographies of inclusion/exclusion: British Muslim women in the east end of London. *Sociological Research Online, 13*(5), 1–10.

Burns, D., Hambleton, R., & Hoggett, P. (1994). *The politics of decentralisation.* Basingstoke: The Macmillan Press.

Burt, R. S. (1980). Models of network structure. *Annual Review of Sociology, 6*, 79–141.

DCLG, & DWP (2007). *The working neighbourhood funds.* London: DCLG.

Dench, G., Gavron, K., & Young, M. (2006). *The new east end: Kinship, race and conflict.* London: Profile Books.

Department for Business Innovation and Skills (DBIS). (2010, November). *Skills for sustainable growth.* Full Report.

Department for Communities and Local Government (DCLG). (2010). *Index of deprivation 2010.* Retrieved from https://www.gov.uk/government/statistics/english-indices-of-deprivation-2010

Department for Education (DfE). (2014). *Thousands more school leavers choosing apprenticeships.* Press release [online] 4 July 2014. Retrieved from https://www.gov.uk/government/news/thousands-more-school-leavers-choosing-apprenticeships

Dewson, S. (2005). *Evaluation of the working neighbourhoods pilot: Year one.* DWP research report No. 297. Leeds: Corporate Document Services.

Dewson, S., Casebourne, J., Darlow, A., Bickerstaffe, T., Fletcher, D., Gore, T. & Krishnan, S. (2007). *Evaluation of the Working Neighbourhoods Pilot: Final report.* DWP research report No. 411. Leeds: Corporate Document Services.

Enum, Y. (2007). *Tower Hamlets adult mental health needs assessment.* Tower Hamlets: NHS Primary Care Trust.

Fainstein, S. (2001). *City builders: Property development in New York and London, 1980–2000* (2nd ed.). Kansas: University Press of Kansas.

Foster, S., & Lane, P. (2012). *Analysis of English language employment support provision in London for JSA and ESA WRAG customers.* London: Centre for Economic and Social Inclusion.

Government Office for London (GOL). (2000, June). *ESF objective 3 draft regional development plan London 2000–2006*. London: GOL.
Granovetter, M. (1973). The strength of weak ties. *American Journal of Sociology*, 78(6), 1360–1380.
HM Revenue & Customs (2011) Snapshot as at 31 August 2011 - Local authorities. MSExcel worksheet. Published 20 Feb 2014.
Houghton, S., Dove, C., & Wahhab, I. (2009). *Tackling worklessness: A review of the contribution and role of English local authorities and partnerships*. Final Report. Yorkshire: The Department for Communities and Local Government.
Hughes, C., & Crowley, L. (2014). *London: A tale of two cities. Addressing the youth employment challenge*. London: The Work Foundation.
Kabeer, N., & Ainsworth, P. (2011). *Life chances, life choices: Exploring patterns of work and worklessness among Bangladeshi and Somali women in Tower Hamlets*. Report to the LBTH.
Kauppinen, T. M., Kortteinen, M., & Vaattovaara, M. (2011). Unemployment during a recession and later earnings: Does the neighbourhood unemployment rate modify the association? *Urban Studies*, 48(6), 1273–1290.
Kickert, W. J. M., & Koppenjan, J. F. M. (1997). Public management and network management: An overview. In W. J. M. Kickert, E.-H. Klijn, & J. F. M. Koppenjan (Eds.), *Managing complex networks: Strategies for the public sector* (pp. 35–61). London: Sage.
Krasnowski.K, and Vaid, L. (2012). *Right skills, right jobs*. London: London enterprise panel.
LEPN (Local Enterprise Partnerships Network). (2014). *Building local advantage: Review of Local Enterprise Partnerships area economies in 2014*. Athey Consulting Ltd, www.atheyconsulting .co.uk
London Assembly. (LA). (2007). *London olympic games and para-olympic games: The employment and skills legacy*. The Economic Development, Culture, Sport and Tourism Committee. London: Greater London Authority. March 2007.
London Borough of Tower Hamlets (LBTH). (2002). *Community economic development plan, 2002–06*. Retrieved from http://www.towerhamlets.gov.uk/
London Borough of Tower Hamlets (LBTH). (2005a). *Tower Hamlets regeneration strategy. Appendix 2: Evidence base*. London: Tower Hamlets.
London Borough of Tower Hamlets (LBTH). (2005b, January 26th). *Council meeting agenda item no. 7*.
London Borough of Tower Hamlets (LBTH). (2008a, May 6th). *Report of the scrutiny working group evaluating Neighbourhood Renewal Funding*. Overview and Scrutiny Committee Agenda Item 9.3., pp. 105–144.
London Borough of Tower Hamlets (LBTH). (2008b, November). *Tower Hamlets employment strategy 'getting neighbourhoods working'*.
London Borough of Tower Hamlets (LBTH). (2011a, April). *Tower Hamlets employment strategy*.

London Borough of Tower Hamlets (LBTH). (2011b, April). *Tower Hamlets enterprise strategy.*
London Borough of Tower Hamlets (LBTH). (2012). *2011 census: Second release – Headline analysis.*
London Borough of Tower Hamlets (LBTH). (2013a, July). *Tower Hamlets financial inclusion strategy 2013–2016.*
London Borough of Tower Hamlets (LBTH). (2013b, August). *Housing evidence base.*
London Borough of Tower Hamlets (LBTH). (2014a, June). Understanding skills and qualification levels in Tower Hamlets. Research briefing.
London Borough of Tower Hamlets (LBTH). (2014b). *Councils take Boris Johnson to court over 'affordable' rents.* News release March 14.
London Legacy Development Corporation (LLDC). (2013, October). *Local plan consultation document: Draft equality impact assessment.* London: LLDC.
Mayhew, L., & Harper, G. (2011). *Women and worklessness in Tower Hamlets: A multi-factor risk analysis.* Mayhew Harper Associates. London: NKM.
Meadows, P. (2007). *What works with tackling worklessness.* London: London Development Agency and GLA Economics.
Mehta, P., & Rutt, S. (2012). *Hidden talents: A statistical review of destinations of young graduates.* LGA Research Report. Slough: NFER.
Miles, R. E., & Snow, C. C. (1992). Causes of failure in network organizations. *California Management Review, 34*(4), 53–72.
Office for the Deputy Prime Minister (ODPM). (2000). *The English indices of multiple deprivation 2000.*
Office for the Deputy Prime Minister (ODPM). (2004). *The English indices of multiple deprivation 2004.*
Office for the Deputy Prime Minister (ODPM). (2006). *Alternative approaches to neighbourhood management.* Research Report 27. London: ODPM.
Office for National Statistics (ONS). *Labour market profile - Tower Hamlets.* See https://www.nomisweb.co.uk/reports/lmp/la/1946157257/report.aspx
Office for National Statistics (ONS). *Neighbourhood statistics.* See https://www.neighbourhood.statistics.gov.uk/
Painter, C., Isaac-Henry, K., & Rouse, J. (1997). Local authorities and non-elected agencies: Strategic responses and organizational networks. *Public Administration, 75*(2), 225–245.
Poplar Area Neighbourhood Partnership. (2004, July 9). *Education group minutes.* Poplar HARCA.
Rhodes, R.A.W. (2000). 'New Labour's civil service: Summing-up joining-up'. *Political Quarterly, 71*(2), 151-66.
Saffron, K., & Nazroo, J. Y. (2002). Agency and structure: The impact of ethnic identity and racism on the health of ethnic minority people. *Sociology of Health and Illness, 24*(1), 1–20.

Salman, S. (2006). Britain is teaching too few. *Guardian*, 20 October 2006. Retrieved from http://www.guardian.co.uk/
Small, M. L. (2009). *Unanticipated gains: Origins of network inequality in everyday Life*. Oxford: Oxford University Press.
SQW. (2005). *Targeting recurrent unemployment in Tower Hamlets (revised)*. Draft final report to the Tower Hamlets Partnership. London: SQW Ltd.
SQW. (2013, May). *Olympic jobs evaluation*. Final report. SQW Ltd.
Tower Hamlets ChangeUp Consortium (THCUP). (2008). *Infrastructure development fund*. Retrieved from http://www.towerhamletschangeup.org/
Tower Hamlets Community Organisations Forum (THCOF). (2005). *Tower Hamlets infrastructure development plan 2004–2014*. London: Tower Hamlets.
THP (Tower Hamlets Partnership). (2002a, August). *Community plan action group protocols*, pp. 1–12.
THP (Tower Hamlets Partnership). (2002b, May). *Developing a neighbourhood renewal strategy*.
THP CSPG (Creating and Sharing Prosperity Group). (2003, September 15). *Draft background research paper to the Employment and Community Cohesion Scrutiny Panel*.
THP (Tower Hamlets Partnership). (2015). Community Plan 2015.
THP CSPG (Creating and Sharing Prosperity Group). (2004, May 26). *Minutes*.
THP CSPG (Creating and Sharing Prosperity Group). (2005, July 15) *Minutes*.
THP EPS (Excellent Public Services). (2005, January 18). *Notes from the CPAG meeting*.
THP LAP 7 (Local Area Partnership 7). (2005a, February 16) *Minutes*.
THP LAP 7 (Local Area Partnership 7). (2005b, May 18) *Minutes*.
Trust for London. (2013). *London's poverty profile*. Website at http://www.londonspovertyprofile.org.uk
Weaver, M. (2003). Council staff suspended in fraud inquiry. *Guardian*, Friday 11 July 2003. Available at http://www.guardian.co.uk/
Work and Pensions Committee (WPC). (2014, January). *The role of Jobcentre Plus in the reformed welfare system*. Second Report of Session 2013–14, HC479. Norwich: Stationery Office.
WP (Work Programme). (2013). Work programme statistics website. Retrieved from www.gov.uk/government/collections/work-programme-statistics-2

CHAPTER 6

Seaside Town Network Cases

This chapter has four sections reporting on five network cases in Conservative-controlled Great Yarmouth Borough Council (GYBC), and two ward clusters.

GREAT YARMOUTH: FEWER JOBS, LESS WORK

Great Yarmouth covers a geographical area of 174 km^2 and is situated at the most easterly point of the Norfolk coast in East Anglia, about 20 miles east of the city of Norwich. The main coastal town, known locally as 'Yarmouth', is located on a spit between the North Sea and the River Yare. Two bridges link the town with the south of the borough. The town's constrained geography is problematic for industry and transport, and a third river crossing and a dual-carriage link road to Peterborough are long overdue. This has hindered the creation of jobs and growth, according to not only the county and local government, but also business leaders and the local media. Inland there are marshland areas for recreation and of national environmental significance.

The population has risen from 90,900 (Census 2001) to 97,277 (Census 2011), 94,215 inhabitants are White British. Other cohorts have made their home in the borough, including Asian, Chinese, Lithuanian, Kosovan and Western Europeans, particularly Greek Cypriot, Polish and Portuguese-speaking people who have formed communities, and have introduced, for example, Portuguese shops, cafes and bars to support migrant workers in the food processing and agricultural industries. The percentage of people aged 45 and under is lower than average for the eastern region and England as a whole, while the percentage of older people and retired people living alone is higher than average, and rising. This increases the pressure on carers and on voluntary and statutory services. For example, only 10.9 % of the population are aged 25–34, while 20 % of people are aged 50–64 (Census 2011). First and Second World War bombing raids destroyed most of the ancient town. Compared with other seaside towns, Yarmouth has a high proportion of overcrowded pre-1964 social housing, terraced streets and private housing in poor condition. More than 38 % of the social housing stock did not meet the government's decent homes standard in 2003. The town's administration changed from a County Borough to a Non-Metropolitan District Council in 1974, and its powers in education, libraries, police and social services transferred to Norfolk County Council (NCC). The town has been a stronghold for the Conservative Party since the 1970s, with the exception of Labour control between 1990 and 2000, and 2012 and 2014, and no single party had overall control between 1980–1983, 1986–1990 and 2014 to the present. In the May 2016 election, the Conservatives remain in charge of running the authority, and UKIP is now the second largest party in the borough. The Great Yarmouth Borough Council (GYBC) has adopted a political democratic governance model of leader, cabinet and scrutiny committees. Brandon Lewis, elected Conservative MP for Great Yarmouth in 2010, was appointed the Minister of State for Housing and Planning at the Department for Communities and Local Government (DCLG) in 2014.

THE GREAT YARMOUTH ECONOMY: THE GREAT DECLINE

Yarmouth evolved as a herring fishing settlement and its inhabitants battled to defend trade and stem tidal forces over centuries. Bailiffs, mayors and burgesses undertook fundraising campaigns to ensure the town's survival, and petitioned Norwich and Privy Councils for grant assistance and tax relief to establish a trading and environmental infrastructure. By the seventeenth century, Yarmouth was a wealthy town, and maritime and merchant oligarchies had developed trade links to Europe. Tourism developed from the eighteenth century onwards. By the mid-twentieth century, the herring fishing industry had collapsed due to over-fishing, and the town lost the economic contribution of Scottish seasonal migrant workers. Tourism waned from the 1960s onwards, as people sought cheap holidays abroad. The east coast rail link to London closed in the early

1960s as a result of road transport competition, although one rail line to Norwich remained. Nevertheless, Great Yarmouth remains a popular seaside resort and its fine beach, cultural heritage and marine and naval history attract holidaymakers and day trippers; tourism expenditure on local goods and services increased to £500 million in 2011 (GYBC 2012a). In 2003, the *Inte*great partnership facilitated £16.3 million environmental improvements to Great Yarmouth, linking the town centre, seafront and heritage sites.[1] The town's general decline is evident in its ageing property stock, vacant business premises and derelict under-utilised land; moreover, concerns over the risk of flooding and lack of investment in flood defences have stalled regeneration in some parts of the town.

Although large firms have moved production away from Yarmouth and neighbouring Lowestoft over the years, these towns' medium-sized port areas and logistics sustain the southern North Sea energy industry; offshore oil, gas and wind farms, marine-related businesses and industrial centres have been established, the latter ranging from advanced electronic engineering to food production and packaging. In Yarmouth in 2001, consent was obtained for a gas-fired power station. A large casino licence for development on the seafront was granted in 2007. A yachting marina has also been proposed. The loss of port worker jobs, redundancies and workforce casualisation resulted in trade union protest in 2009, prior to the opening in 2010 of a £32 million outer harbour, 'Eastport', which was a major economic asset equipped for cargo and grain export.[2] The planned ferry link to The Netherlands and plans for a container terminal failed to materialise; large cranes have since been removed and in 2015 the port was up for sale.

In 2002, Great Yarmouth was the most unemployment-deprived of 43 seaside towns studied, and then second place in 2008 behind Skegness (see Beatty and Fothergill 2003: 23; Beatty et al. 2008). The Index of Multiple Deprivation (IMD) 2007 ranked the borough fifty-eighth highest out of 354 local authorities for high levels of deprivation and fifty-seventh highest in 2010 (DCLG 2007, 2010a). Among its residents, 49 % live in

---

[1] The *Inte*great Partnership includes Great Yarmouth Tourist Authority, Great Yarmouth Heritage Partnership, Great Yarmouth Town Centre Partnership, Great Yarmouth Community Partnership, Seachange, Great Yarmouth Borough Council (GYBC) and Norfolk County Council (NCC). The funding partners are the East of England Development Agency (EEDA), European Regional Development Fund (ERDF) Objective 2, Heritage Lottery Fund, NCC and GYBC.

[2] Eastport took decades to raise capital and secure planning. In 1567, Yarmouth bailiffs attempted to fund a new harbour by entering the first lottery in England to raise money for public works, but the gamble with ratepayers' money was unsuccessful (Meeres 2007: 31).

the 20 % most deprived areas in the country, and the number of Lower Super Output Areas (LSOAs) in Great Yarmouth ranked amongst the most deprived 10 % in the country increased from 20.2 % since IMD 2007 to 22 % in 2010 (2010a). These problem domains rank highest for barriers to employment, education and income deprivation, while the general health of the populations falls below the average for England (2010a). There are 44,900 jobs in the borough (32,500 full-time and 12,400 part-time) (GYBC 2014a). Regarding household incomes, 44 % are below £20,000 per annum. Residents earn less than those who work in the borough, and weekly wages are lower than the county and national averages (GYBC 2012a: 21). At 15.7 %, Great Yarmouth has the lowest proportion of people engaged in higher education and reporting a degree qualification or above of any local or unitary authority in England and Wales; 13,083 people aged 16–64 have no qualification, and the number of people holding NVQ levels 1–4 is lower than national and regional averages, but improving (Census 2011; GYBC 2014a: 4). In 2013, 47.3 % of schoolchildren achieved five GCSE's graded A–C, compared with 62.2 % in England, and 53.9 % in Norfolk (2014a: 15). The majority of the town's 3000 businesses each employ fewer than five people. In 2010, 250 businesses started up, and 365 businesses deregistered; in 2012, 320 businesses started, but 335 deregistered (GYBC 2012a: 22; 2014a: 20). The number of self-employed people peaked at 4,100 in 2004 to 8,600 in 2013, falling to 4,500 in 2015 (ONS Labour Market Profile).

In 2005, the East of England Regional Development Agency (EEDA) supported the establishment by '1st East' of an Urban Regeneration Company (URC) for overseeing a 15-year 'employment-led regeneration' action plan for tackling the area's underperforming economy and declining town centres. This involved the following: coordinating local consultations; developing strategic planning; the creation of a Great Yarmouth Waterfront Area Action Plan to connect brown field riverside regeneration sites in the Southtown and Cobholm wards (SC) to the dock area and town centre improvements; and building 6000 new homes (GYURC 2007a: 12). The organisation 1st East disbanded in March 2011, and GYBC Economic Unit has been using its legacy to inform consultation stages for the Local Plan (under the Localism Act 2011), for example concerning core strategy, sustainability appraisal, development policies and site allocation by 2016.

In 2011, the 'New Anglia Local Enterprise Partnership' (NALEP) for Norfolk and Suffolk, took over from EEDA, Shaping Norfolk's Future (the economic arm of NCC's County Strategic Partnership) and local authorities

LSPs, including the Great Yarmouth LSP (GYLSP). The managing director of the NALEP is a former chief executive of Shaping Norfolk's Future, and previously a business editor of a family-owned media group producing regional newspapers and magazines, and four daily newspapers in eastern England. Primarily, the NALEP handles the government's regional aid for the following purposes: growing the subnational and local economy; tackling economic needs; encouraging private sector investment; patents and marketing; land development; building and equipment for offshore wind projects and so on; and supporting large and small businesses, from helping with employment and capital costs to training and low carbon initiatives.[3] However, it is not the sole agent of economic governance. For example, GYBC adopted an Economic Reference Group and an Employment and Skills Group, and NCC retains its strategic role in economic development.[4]

From 2011, the NALEP acquired Enterprise Zone (EZ) status for Great Yarmouth and Lowestoft, focusing on commercial activities that support renewable offshore energy, port sites and logistics in the energy park at South Denes and innovation at the Beacon Park business centre in Gorleston. Moreover, urban wards granted Assisted Area Status (2014–20) can access additional aid to enable business growth in areas with less prosperous local economies. Both Great Yarmouth and Lowestoft are among six Centres for Offshore Renewable Engineering in England, with partnership status between central and local government and LEPs. This ensures that businesses wanting to invest in off-shore wind manufacturing at port sites receive comprehensive support.

More generally, the NALEP was eligible for the Regional Growth Fund (RGF) until 2014/2015, including a £13 million 'Growing Places Fund'; this was a repayable loan scheme for investing in house-building, and creating jobs and employment spaces. A £12 million 'Growing Business

---

[3] The policies to incentivise regional growth over the next few decades are immensely challenging. The employment forecast for 2014–31 is not expected to grow; finance to develop new products is lacking; recent economic performance suggest productivity in terms of GVA is lower than the English average, business spending and investment has been low; the business stock is below the English average; employment participation is strong but part-time work is higher than the English average and wage growth is low; moreover, weak broadband and transport infrastructure makes global competition logistically inadequate and a self-sufficiency outlook is not even considered (NALEP 2013).

[4] Previously, NCC, Government Office for the East of England (GOE), EEDA and the East of England Regional Assembly had authority over the LSP for business and transport development for the subregion of Yarmouth and Lowestoft.

Fund' provides private sector investment grants for match funding to grow businesses, sustain and create jobs. An £11 million 'Grants4Growth' fund offers grants and technical advice to small and medium-sized enterprises (SMEs) in the East of England and Lincolnshire, on matters ranging from business growth, resource efficiency measures, energy-efficient buildings and transport and production processes to waste management, and small grants for the installation of energy-efficient appliances. Agri-tech and Carbon Innovation grants are also available. From 2015 to 2021, a £12 billion 'Local Growth Deal Fund' allocated the NALEP a further £173.3 million towards the creation of 6000 jobs and 350 apprenticeships, and the construction of 900 homes. The European Social Fund (ESF) 'Youth Employment Initiative', a £490-million fund supported by the Department for Business, Innovation and Skills (DBIS) and the Department for Work and Pensions (DWP), bypassed the eastern region. Nevertheless, the £86 million European Structural and Investment Funds (2014–20), will support implementation of the NALEP 'Strategic Economic Plan' (NALEP 2014a). Specifically, this will be used to create 95,000 new jobs and 10,000 new businesses by 2026 (2014a: 1, 4). LEPs will not officiate in funding strands; rather, the Department for Communities and Local Government (DCLG) will manage ERDF, the DWP will manage the ESF, and the Department for Environment and Rural Affairs will manage the EAFRD (European Agricultural Fund for Rural Development) and European Maritime and Fisheries Fund. However, the ESIF requires LEPs to establish stakeholder subcommittees to influence strategy, develop assessment procedure, identify potential co-funding with SFA (opt-in is £25,091,000) and Big Lottery Funding (opt-in is £7,443,803) and secure local support for projects. The NALEP's Skills Sub-Board, (established in February 2014 to support employers input in skills decision making and skills provision) was invited to self-nominate for one of two subcommittee places allocated to skills representatives. Prospects for joined-up governance is unlikely, given unelected strategists managing competitive funding processes maintain distance from providers; yet, community-led local development requires strategic coordination to join up provision and counter fragmentation created by competition. Moreover, subsidising the private sector to establish a green economy ignores the carbon footprint of crisis support caused by sanctions, from financial counselling to food parcel requests, and the impact of welfare cuts on Great Yarmouth economy; represents a £610 reduction per annum for every working age adult (see Appendix 1: 'Measuring the Impact of Welfare Reform, November 2013' in GYBC 2014b).

Job Losses and Insufficient Replacement Jobs

Similar to other UK seaside resorts, Great Yarmouth has battled seasonal and structural high unemployment for decades (Beatty and Fothergill 2003: 1). During the economic depression of the 1920s and 1930s, unemployed people were occupied in construction programmes to build a swimming pool, boating lakes and a racecourse. Public sites fell into disrepair after the heyday of tourism in the 1940s to the 1960s, and a funding campaign is underway for heritage skills and tourism projects to create training and local jobs: these include the restoration of the 1928 Venetian Waterways and historic waterways boats, and building boat storage and beach huts; the reintroduction of 1938 planting schemes; and improvements to North Beach, the Winter Gardens and boating lake (GYBC 2014c: 6–7). The subsequent emphasis of the programme of works is for volunteers, not paid employees. In 2016, the Heritage Lottery Fund and Big Lottery Fund awarded Venetian Waterways a joint grant of £1,020,800. DCLG awarded Pocket Parks funding (a USA idea for refurbishing derelict urban spaces) to 'Great Yarmouth and Waveney MIND' for a community allotment scheme and community wildlife walk. Traditional industries declined from the 1970s. Jobs today are 78 % service-based (health, retail, hotel accommodation, food and education) (GYBC 2014a: 5). However, there are too few jobs across East Anglia. Employment growth in 2009–12 was marginal at 0.1 % (LEPN 2014: 19). Job density in Great Yarmouth (2001–14) fluctuated between 0.72 and 0.75, resulting in insufficient jobs for each willing person in the workforce aged 16–64 (ONS 2014).

Opportunities are arising in sectors associated with intra-regional corridors of growth, interlinking logistics and a new superstore established in 2014. The EZ anticipates the creation of 9000 new jobs, 4500 indirect jobs and 150–200 businesses by 2025. A £3 million 'Building Foundations for Growth' fund accelerated the site's business-ready infrastructure, with £350,000 pinch-point funding to improve junctions at the EZ, and £6.8 million road improvements to facilitate growth in the south of the borough. In 2013/2014, public aid incentives saved globally connected EZ businesses £112,735 in business rate discounts, rising to £268,991 in 2014/2015. The NALEP retains additional EZ business rates, and GYBC receives 10 % of the funding 'to recompense for the business rates they would have been able to retain if the EZ was not in existence'. A further 35 % will be used to develop the EZ and the remaining 55 % will be absorbed by the LEP 'to support the delivery of its Strategic Economic Plan' (NALEP 2014b: 4–5). EZ firms are not obliged to employ or train local people. Appointments are often made via word-of-mouth, and recruitment needs are specific, with positions requiring applications from ex-military personnel to environmental science graduates, engineers to energy technology specialists. Minimum sector entry requirements are reportedly NVQ level 3 in Engineering (GYBC 2014d: 5–6). The Norfolk Chamber of Commerce developed a portal to broker supply chain links between local businesses and 'Sizewell C', a new nuclear power plant in Suffolk. In addition to existing wind turbines at Scroby Sands, Great Yarmouth has also forecast hundreds of construction jobs to build the £1.5 billion Dudgeon Offshore Wind Farm project. Recruitment, to date, is thirty specialist permanent staff, eight temporary staff for the initial phase, and tenders invited for offshore and base support (Lynch 2015). In 2016, EZ publicity cites £16 million private investment to the area and expects £50 billion investment in the energy sector over the next 20 years.

Major construction projects could increase demand for temporary and permanent housing, and private rental accommodation and house prices may rise. Housing affordability is already a problem for those on low wages and GYBC housing services reported an increase in rent arrears following welfare reforms (see GYBC 2014b); moreover, inappropriate benefit sanctions have added to financial insecurity (WPC 2014: 23–8).

Social enterprise and the voluntary sector attempt to boost small-scale demand. A vacant church donated to the GY Preservation Trust has been offered as a regional training ground in stonemasonry and traditional building skills (timber-framing, flint-knapping, thatching) for unemployed people. Meanwhile, the NALEP gambles on the 'gold-rush' effect of the green economy. Such approaches are only partial solutions. For example, 3340 of the 3720 DWP 'Work Programme' referrals made in Great Yarmouth to December 2013 could have achieved a job outcome, but only 570 did (WP 2013). At a 17.1 % success rate, private sector firms try to profit from the fact that the long-term unemployed are treading water.

NEIGHBOURHOOD PROFILES: NORTH AND SOUTH

This section compares the characteristics of two ward clusters: (i) the Central and Northgate wards (CN); and (ii) the Southtown and Cobholm wards (SC).[5]

---

[5] From December 2002, Regent Ward, the nineteenth most deprived ward in England, split into Central and Northgate, and part of Nelson Ward (ODPM 2000). Nelson Ward was the thirty-seventh most deprived ward. Southtown and Cobholm was formerly Cobholm

Demographics and Deprivation

Located in the north of the borough, CN includes the seaside town centre. The population is 7179 (Census 2001) and rising to 8298 (Census 2011). Ethnicity is 84 % White British, 11 % White Other, and 5 % Black and Minority Ethnicity (BME) (2011). Immigrant agricultural and seasonal migrant workers also use CN as a residential base (3.8 % of the working-age population).

Further south, across the Haven Bridge, SC is situated closer to industrial sites. The population is 4615 (2001) and rising to 5657 (2011). SC is 88 % White British, 8 % White Other, and 4 % BME (2011). In 2000, the IMD ranked CN four hundred and sixty-first and SC five hundred and sixty-eighth out of 8414 wards in England (ODPM 2000). A decade on, little has changed and CN and SC have LSOAs in the worst 10 %, out of 32,482 in England (DCLG 2010a). Across the ward cluster, 49 % of household incomes are below £20,000. CN has one LSOA ranked 513 (1 being the worst) for multiple deprivations, two for employment deprivation (ranked 241 and 2703) and one for crime (ranked 420). SC had one for employment deprivation (ranked 1697). Both had one for income deprivation and three for education, skills and training.

Unemployment

In both ward clusters, numbers of benefit claimants (Jobseeker's Allowance (JSA) and Incapacity Benefit/Employment and Support Allowance (IB/ESA)) increased before and after the 2008/9 recession. Table 6.1 presents the categories of benefit claimants between 2002 and 2014. Unemployment lowers in the summer, but the rate is still higher than the UK average. The Census 2011 identifies long-term unemployment at 7.7 % in CN and 8.2 % in SC. More women than men are affected, although the numbers of those working part-time were higher for women.[6] Unemployed people without qualifications accounted for 36.1 % in CN and 31 % in SC (2011). During fieldwork, the proportion of NEETs aged 16–18 reached 41 % in parts of CN, and 24 % in SC (Connexions 2006). NEET figures for Great Yarmouth have been falling; from 9.5 % in 2011, 7.5 % in 2012, and 6.0 % in 2013. However, data before 2011 is not comparable and relates to calendar

---

and Lichfield Ward. Data is sometimes inconsistent since the wards ranked in the ODPM 2000 refer to old wards.

[6] Wage inequalities were not national targets and financial well-being projects received the smallest amounts of Neighbourhood Renewal Funding.

Table 6.1 Working-age claimants in the GYBC ward clusters—November 2004 to November 2014

	CN							SC						
	Nov- 2002	Nov- 2004	Nov- 2006	Nov- 2008	Nov- 2010	Nov- 2012	Nov- 2014	Nov- 2002	Nov- 2004	Nov- 2006	Nov- 2008	Nov- 2010	Nov- 2012	Nov- 2014
Total claimants	1,380	1,350	1,445	1,460	1,680	1,795	1,635	745	730	785	830	925	945	840
Job seekers	395	380	440	425	550	675	390[a]	200	205	215	235	285	345	175[a]
ESA and IB	620	610	650	655	720	725	830	310	290	325	320	365	335	385
Lone parents	150	150	160	150	140	120	110	140	115	115	130	100	100	85
Carers	70	85	75	95	95	105	140	45	55	55	55	70	85	120
Income-related[b]	80	60	60	80	100	75	65	25	25	30	35	30	20	20
Disabled	45	50	45	50	60	80	90	20	35	35	45	65	55	50
Bereaved	20	15	15	5	15	15	10	5	5	10	10	10	5	5
Male	820	775	835	885	1,010	1,055	905	380	380	420	415	495	475	400
Female	560	575	610	575	670	740	730	365	350	365	415	430	470	440
16 to 24	225	205	230	265	320	365	310	100	110	145	145	190	190	140
25 to 49	710	685	745	780	875	905	815	430	410	400	445	480	495	440
50 and over	445	460	470	415	485	525	510	215	210	240	240	255	260	260
Out-of-work benefits	1,245	1,200	1,310	1,310	1,510	1,595	1,395	675	635	685	720	780	800	665

Data source: http://www.nomisweb.co.uk
ESA Employment Support Allowance, *IB* Incapacity Benefit, *SC* Southtown and Cobholm, *CN* Central and Northgate. Claimant counts do not yet include Universal Credit. (Data available from 7 March 2016).
[a]Decrease coincides with increase in sanctions and ESA
[b]Others on income-related benefit

age and school or college location, while recent data collection relates to residency and academic age at 1 September in a given year (see NCC 2014a: 1, and Figure 3). Nevertheless, NEET figures for Norwich, Great Yarmouth and Kings Lynn and West Norfolk are still worse than those of Norfolk and the national average. Likewise, numbers of JSA claimants decreased, following a peak in 2012 (see Table 6.1), but caution is required given that the DWP remove people from the claimant count if they do not sign on during a sanction period, and a parliamentary enquiry about 'off-benefit' flow outcomes is underway (WPC 2014: 25–6). In December 2012, Jobcentre Plus (JCP) reported to the Employment and Skills Group that harsher penalties were expected in future but that sanctions would not be undertaken lightly (GYBC 2012b: 5). Subsequent minutes raise concern over the high number of sanctions, but no further actions were recorded (see item 9, GYBC 2013a). Nevertheless, between December 2012 and December 2013, 2056 Great Yarmouth claimants were sanctioned; according to JCP data, 1251 sanctioned cases were overturned, 980 were cancelled and 288 received reserved decisions. Numbers of ESA claimants have also significantly increased since 2012. Claims on grounds of disability have doubled since 2002. Mental health is the most numerous category for ESA claimants, having risen in CN from 70 in 2010 to 440 in 2014; similarly in SC, numbers of mental health claimants have risen from 40 in 2010 to 170 in 2014 (ONS 2014). While ESA claimants in the Work-Related Activity Group are subject to conditionality (job search or work preparation), JCP appears more focused on sanctions than on helping people to return to (non-existent) work (WPC 2014: 21–2). Moreover, despite the national rise in employment, the JSA and ESA claimant count increased in CN and SC, between November 2015 and February 2016 (ONS Neighbourhood Statistics).

Environment

CN and SC environmental characteristics are very different. SC had three LSOAs in the worst 10 % in England for the living environment domain, while CN had two (DCLG 2010a). Parts of CN incorporate the town's main marketplace and shopping mall. During the study, many shops were empty despite retail relief schemes, and central amenities required renewal and transport planning to assist regeneration and overcome congestion problems (GYBC 2014c: 8–9).[7] Terraced streets and community facilities

---

[7] Development control and planning present major issues in Yarmouth; land for industrial development is limited due to poor ground conditions.

can be found further north. In the tourist area, amusement arcades, gaming rooms, fast-food outlets and gift shops provide summer jobs. The physical regeneration programme, *Inte*great, designed to benefit visitors, would not alleviate the crime and housing problems faced by residents. Hotel owners have been unable to refurbish properties to EU standards and many have sold them to unscrupulous agents for conversion to flats, but cheap private-sector rental accommodation has attracted homelessness problems and migrant workers; some of the latter have been exploited for cheap labour and are living hand-to-mouth in poor conditions (GYBC 2004). During fieldwork, the outdated main library had limited computer amenities and was only refurbished in 2009 following the receipt of National Lottery funding. A busy trunk road leading to the Haven Bridge separates Southtown (the riverfront has small marine businesses and rundown industrial buildings), and Cobholm, a densely populated working-class community with narrow terraced streets, few shops and a convenience store. A health and children's centre near a dilapidated hotel provides a focal point, and a food factory is located on the outskirts.[8] Recently, a neighbourhood management scheme organised a Community Tree Planting event and 420 trees were planted in Cobholm. Superstores and industrial parks two miles away are gradually expanding; some town centre shops relocated there. SC is near the river and has a high risk of flooding. Following a tidal surge in 2007, neighbourhood management teams established community resilience networks to better prepare Yarmouth communities to cope with emergencies. Since March 2013, the Office for Standards in Education, Children's Services and Skills (OfSTED) placed Cobholm primary school under special measures for serious weaknesses and for failing to meet the government's floor standards in 2011 and 2012. The school has more than twice the national average of children with a pupil premium (eligible for free school meals or under local authority care), and more than twice the national average number of children with a Statement of Special Educational Needs. Family Connectors operate in SC as part of the Government's Troubled Families and Early Help agenda.

Culture and Politics

Historical legacies of ward clusters were evident. In the seventeenth century, Cobholm was an island with a salt-making industry and was linked to

---

[8] Single Regeneration Budget 5 (SRB5) funded (£12 million) the health centre, a credit union, business advice and improvements to education and skills training, including an FE college engineering department, and community safety projects. Funding ceased in 2006.

Southtown by a bridge.[9] Southtown had developed a busy fish wharf and an extensive shipbuilding industry by the nineteenth century. Traditionally, boat owners lived in Southtown and boat workers in Cobholm. The Census (2011) identifies main occupations in both ward clusters as highest for wholesale and retail (21 %). CN has slightly more inhabitants employed in accommodation and food service (14 %), and in manufacturing (13.4 %), while SC has slightly more employed in human health and social care (13.2 %). In CN and SC respectively, 51 % and 56 % drive to work, while 25 % walk. A manager commented that people expect local work, as their great-grandparents did, but this tradition needed breaking as people should walk across the bridge to work. Yet jobs at the superstores were limited, and Norwich also had high unemployment. One interviewee said that people left Yarmouth, 'to make better their circumstances'. CN's east-side community was perceived as transient, like that of the neighbouring 'Nelson' ward, which has the highest level of deprivation and unemployment in Norfolk (43 % claimed unemployment benefit in 2013). The Nelson Ward was overlooked by the Working Neighbourhood Pilot (WNP) in 2004, presumably because there were few services to promote it. SC has more generational ties, young families and groups. Consequently, community narratives were shaped around migrants' cultural interests in CN, and family-orientated activities for people rooted in SC. The theme of unemployment barely featured in community governance until the implementation of the Working Neighbourhood Fund (WNF). Thereafter funding streams influenced organisations' behaviour in order to deliver pre-employment support and use community development to engage unemployed citizens; however, the lack of suitable jobs was not acted upon, consequently, the status quo was maintained. Notably, job competition extends across the New Anglia area, such that 101,200 workless people were classified involuntary unemployed (NALEP 2013: 4).

Overall Policies

Unemployment policy in Great Yarmouth cascaded through public bodies influencing national policy frameworks for getting local people into work, such as JCP, Business Link (BL), Connexions, and the Learning and Skills Council for England (LSC) and the Government Office for the East of England (GOE), EEDA and NCC. However, the local economy did not expand fast enough; the living environment was neglected, and

---

[9] For historical facts see Meeres (2007: 149–55).

the town's physical infrastructure and amenities declined. From 2005, 1st East intended to work in partnership and develop action plans for physical regeneration aligned to the GYBC Local Development Framework for ministerial approval; however, the GYLSP were scarcely mentioned in their plan, while the Economic Forum was not mentioned at all, despite similar objectives to: '… generate economic growth, create jobs and unlock economic potential' (GYURC 2007a: 19). The Economic Forum believed the URC action plan consultation lacked contact with the community and business (see GYURC 2007b: 17).[10] In recent years, the exit of retailers from the declining town centre prompted a full council meeting to review retail growth policies to stimulate business and encourage new entrepreneurs; these policy ideas included food and drink festivals, a market stall 'tables for a tenner' initiative, as well as free Wi-Fi facilities to increase dwell time (GYBC 2014e: 6). The same council meeting saw objections raised by some members over the appointment of an interim GYBC chief executive on a high salary (circa £200,000) whose role, ironically, was to steer funding cuts. The economic consequences of the borough's low-pay economy and monetary circulation was not mentioned. Conversely, Liverpool City Council is devising an employer charter with minimum employment standards, given half of Liverpool's job vacancies are in agency work. Flexible working policy (zero-hour contracts) and deregulation (an alternative to hiring permanent staff), lead to increases in inferior pay and working conditions, and particularly affect women. Researchers claim gender disparity in Norfolk and Suffolk reduces GDP and could be 'the starting point for a possible LEP manifesto around economic equality' (NALEP 2014c: 108). Policies that lacks equality perspectives will find inequitable practices harder to change later on.

With few large firms in Great Yarmouth, enterprise and business start-up support is seen to fill a gap. Hence, in 2006, the Local Economic Growth Initiative (LEGI) provided £8.7 million to stimulate local enterprise in the area.[11] A LEGI evaluation suggests that the initiative had a limited impact on unemployment, but that it contributed to a culture change by promoting enterprise to residents and schools, as well as advocating for local

---

[10] Two years previously, the Economic Forum had identified that more joined-up working was needed with the URC board and that this would require a mechanism through which to disseminate information (see minutes, GYLSP 2005).

[11] LEGI was a joint programme between the ODPM, HM Treasury and the DTI that distributed £419 million to 20 local authorities in two competitive funding rounds between 2006–07 and 2010–11 (DCLG 2010b: 28). GYBC received in total £12.64 million LEGI funding, managed under the auspices of 'enterpriseGY' Board from August 2006.

supply chains, partnership working and a continuum of dedicated business support, from start-up to growth (DCLG 2010b: 86–7). From 2007, 'enterpriseGY' was established in town-centre provision to provide free training for SMEs, ranging from hairdressers and gardeners to voluntary, community and social enterprises; moreover, a 'catalyst' centre provides office facilities to support new start-ups or business expansion. Post LEGI, GYBC and the Big Lottery Coastal Communities Fund contracted A4E to deliver a business start-up service. Moreover, it acts as a 'partner' for other enterprise schemes, such as a two year Skills Enterprise and Assets project offering comprehensive support for budding social enterprises, assisted by Business in the Community's 'one to one' volunteer business mentors and Great Yarmouth College's Alchemy Business Centre (GYC). The latter runs bespoke courses in establishing social enterprises in addition to free SME skills training solutions, from health and safety and customer service skills to bookkeeping. In 2013, Great Yarmouth College received £400,000 to support SMEs from a £2.4 million ESF fund shared between six colleges in the LEP area (GYBC 2013b: 5–6). In 2016, enterpriseGY secured £650,000 Coastal Communities funding over two years to enhance business start-up support (approximately 100 businesses per year).

Additionally, in 2011 an East region daily newspaper (owned by the media group, mentioned above) established 'Future50' to support and promote Norfolk and Suffolk SMEs. In February 2015, the media group and a regional enterprise service partner received RGF Round 6 to expand the programme and distribute discretionary grants for business growth. A Freedom of Information request to DBIS states the partnership received £1,500,000. The service is separate from a £3.9 million NALEP Growth Hub programme coordinating Norfolk and Suffolk business support in 2014/15, through Greater Ipswich and Greater Norwich City Deals via Lancaster University's £32 million DBIS 'Growth Hub' pot. Businesses can apply to both funding streams for grants. Norfolk Chambers of Commerce and Suffolk Chambers of Commerce also provide business support and training. The incoming ESIF, however, expects LEP 'Growth Hubs' to simplify the landscape of business support.

In 2005, the Home Office and DCLG 'Safer Stronger Communities Fund' provided £1.6 million via GOE for a neighbourhood management scheme, operating under the name 'ComeUnity', in parts of CN and south Yarmouth.[12] This aimed to meet the Local Area Agreement targets

---

[12] Neighbourhood Management expanded in response to the Local Government White Papers *Strong and Prosperous Communities* (DCLG 2006) and *Communities in Control, Real People, Real Power* (DCLG 2008).

managed by NCC, including community-focused activities, unemployment support and small grant distribution. In 2005/2006, the Community Fund, (now the Big Lottery Fund) provided £193,522 for Great Yarmouth Refugee Outreach and Support (GYROS) to assist newcomers with settling and integrating into the local community, which it did with the provision of language skills training, volunteering opportunities and job clubs. ComeUnity received a further £581,000 for 2007/08 to deliver neighbourhood projects via thematic working groups (see Appendix 5).

In December 2007, the Labour government announced that Great Yarmouth would receive £7.1 million from the WNF, successor to the WNP in CN, and expand learning, low skills, enterprise and worklessness assistance into other deprived wards. The neighbourhood management model also extended into three neighbourhoods. Subsequently, SC and the Halfway House Estate launched 'Make it Happen' in 2009 to engage local residents in community initiatives and resolve local problems, from those related to family needs to issues connected to youth and community resilience (CU 2010). NCC used the WNF in 2009 alongside the Future Jobs Fund for a work placement scheme. Written evidence presented to the Work and Pensions Committee praised the Future Jobs Fund for its six-month wage subsidy to employers and simplified bureaucracy; as a result, many voluntary sector organisations created entry-level jobs for young people. However, NCC believed the voluntary sector jobs lacked long-term sustainability, and the private sector was reluctant to engage; 52 % of participants returned to JSA (WPC 2010: Ev w62–65). Nevertheless, there was general opposition to the Future Jobs Fund's closure in 2011. In the same year, the Work Programme replaced 17 WNF providers in Great Yarmouth with one East Region prime provider (with a quarter of the £3.3 billion Work Programme contracts) who, in turn, subcontracted a county-wide training provider to deliver provision. Some unemployed people reportedly return to JCP on completion of the two-year programme with inadequate IT skills and basic skills support (see Appendix 1, GYBC 2014b: 12). Yet JCP's transition to the 'Universal Jobs Match' online job-search provision requires JSA claimants to upload CVs online and provide evidence of internet job searches. In Great Yarmouth, young people on work experience were reportedly being used by JCP to assist clients with IT, furthermore, provision was perceived as disparate and requiring coordination (see item 8, GYBC 2013a).

In 2012/2013, JCP, GYBC, GYC, Norfolk Community Foundation and ComeUnity, representing the six most deprived LSOAs in the town, pooled resources into a neighbourhood budget totalling £67,500 to push 'employability' support further into CN and SC, prepare long-term unemployed residents for anticipated EZ jobs, and raise community representation in economic growth policies. Funding provided a full-time 'Target Opportunities' neighbourhood-based support worker in partnership with Voluntary Norfolk, a volunteering organisation for brokering engagement between large and small charities and public sector. Initially, Target Opportunities was a neighbourhood pilot run by ComeUnity in 2007 to demonstrate social return on investment in community learning and pre-employment support. Between 2007 and 2010, Target Opportunities supported 90 people into full-time employment, and more than 500 people received employment, training and volunteer support (CU 2010: 4). Four organisations received the remaining budget to support unemployed people into current or future job opportunities in growth industries, and attract further investment. GYROS was commissioned to provide ten work placements at its community cafe located within the public library, while a social enterprise cycle hire and repair workshop in development received £37,500 capital investment from Norfolk's Community Construction Fund. In 2012, the Big Lottery Coastal Communities Fund awarded GYBC a further £600,000 funding to get 300 people into jobs, create 200 businesses and social enterprise support. The GYBC continues to fund a neighbourhood manager and the voluntary sector funds a community development worker. Ward budgets (£2000 per annum) were also reinstated, given the government's drive to strengthen ward governance. For example, the government's Office for Civil Society commissioned the Community Development Foundation to distribute £30 million match funding to wards with high levels of deprivation and benefit claimants, and establish Community First panels for deciding local priorities and facilitating projects to meet identified needs. CN's panel received £33,910, and SC received £50,865.

Rather than embed support, ad hoc funding disguises the job dearth. For example, the Coastal Communities Fund Round 1, delivered through the Big Lottery Fund for economic development activities promoting sustainable jobs and growth in coastal communities, provided Suffolk County Council £762,000 (from a pot of £24 million in 2012/2013) to create 200 employer-led apprenticeships in the low-carbon-energy sector across three counties (Essex, Norfolk, Suffolk), raise awareness about the opportunities in the field and close the skills gaps in engineering-related fields. As a direct result of the latter, 17 young people entered into engineering apprenticeships in the Great Yarmouth area (GYBC 2014f). Round 2 (a pot of £27 million in 2013/2014) provided a workboat business in Yarmouth with £165,000 to build an all-weather facility to undertake maintenance and construction work all year round. The borough's 'Economic Strategy 2011–2016' also cites 15 strategies at national, regional and local level that could have an effect on the local economy (GYBC 2011: 2–3).

Network Culture

Network governance evolved gradually in Great Yarmouth. Following the Environmental Protection Act 1990, an Environmental Forum was established. Then, in 1996, a Community Planning Partnership developed an additional Economic Forum for private and public sector actors alongside a social partnership. Business leaders had long been shaping local economic development in decision-making networks.[13] From 2000, the GYLSP established four forums and community network structures (see Appendix 5). This partnership was expected to contribute to the Norfolk County Area Agreement, revive the sustainable community strategy, promote a core strategy for a local development framework, establish a strategic approach for community cohesion, seek ways of expanding neighbourhood management and decide on approaches for area-based grants (Fox 2008: 3).

From 2002, the GYLSP aimed to tackle unemployment by: (i) expanding the local economy and reducing the reliance on seasonal unemployment; (ii) establishing community economic development schemes, including social enterprise and a credit union; and (iii) encouraging business growth and support for existing businesses (GYLSP 2002).[14] Unemployment responsibilities, however, were not well-defined. One interviewee had met with JCP, the LSC and the NCC to share information and discuss the dysfunctional elements of the labour market; however, little progress was made for clients, one of whom noted: 'I would say that thus far it hasn't got terribly far in terms of joining up the delivery end. Another interviewee mentioned that the Single Regeneration Budget (SRB) partnership in SC, where community relations were more stable, had provided training and a successful community asset, but that the funding had ceased in 2006. CN had no formal network history, apart from a residents' association.

---

[13] For an account of merchant leaders in the government of Great Yarmouth in the 17th century, see Gauci 1996: 78–88.

[14] Between 2001 and 2008, GYLSP received £12.3 million NRF to meet national floor targets.

According to one interviewee, community development workers preceding the SRB had 'axes to grind ... and hostile and negative to statutory organisations'. Subsequently, community development funding tamed adversarial relations, but the hierarchical LSP forums and community networks did not integrate. An interviewee reflects on the dichotomy:

> There is this ideal Nirvana end state where every individual resident is actively engaged in their community, they turn out to vote and they come along to community meetings and make their views felt and these views seamlessly pass into public policy which then delivers back to the community what it wants. You will never get there but we're moving in that direction and I think the level of real community engagement in both of these areas is improving.

The community engagement to which the interviewee refers is that of power-sharing; however, in reality, cross-reporting was limited between networks. Executives expected subordinate workers to deal remotely with community issues. A member of the LSP economic forum, for example, did not have the opportunity to read the minutes of the community partnership meetings. Another member had no idea which areas the community partnerships covered. In general, people only understood the groupings to which they belonged, rather than the overall LSP structure (Fox 2008: 8).[15]

County networks also distributed regional programme funding for separate projects, such as EEDA's Investing in Communities (2007–10). Kickstart, for example, received £825,000 for mopeds, rider training and moped hire or loans to assist people on low incomes in rural areas overcome public and private transport barriers to reaching work, apprenticeships, work placements or training. NCC's Norfolk Ambition representing its Sustainable Community Strategy for Norfolk (2008–2011) wanted to link economic development with overcoming employment barriers and community cohesion. To support the incoming WNF (2008–11) and reduce service duplication, GYBC and Great Yarmouth Community Trust worked in partnership to support a consortium of 17 voluntary sector

---

[15] A decade later, despite network evolution the NCVO states 65 per cent of LEPs have no voluntary sector representation, and 18 per cent 'have little or no engagement at all with the voluntary sector' (NCVO website, accessed March 2016: https://knowhownonprofit.org/how-to/how-to-get-your-head-around-local-enterprise-partnerships-leps).

organisations to deliver pre-employment support projects, outreach work and volunteering for those furthest away from the market, and more than 200 people moved from welfare to employment. Third-sector 'capacity building' programmes also aimed to strengthen partnerships and protocols in tendering processes and reporting to monitoring systems. For example, 'Enable and Engage' was a two-year EU-funded programme launched in 2009 to assist providers in the handling of co-finance 'learning and skills' contracts (O'Gorman 2011). Nevertheless, the Great Yarmouth 17-strong consortium failed to gain Work Programme contracts, despite GYBC backing. The Norfolk County Strategic Partnership also established a Norfolk Employment and Skills Board to join up strategy, and a countywide consortium of 40 organisations in 2009 delivered a £8.4 million Future Jobs Fund to help 1300 young people aged 18–24 to find work or progress to further education. Despite concern that up-skilled individuals may not find work, the County assumed an up-skilled workforce would attract quality jobs (see item. 5. NCSP 2009). Great Yarmouth Employment and Skills Group is now coordinated by Voluntary Norfolk with support provided by a GYBC-funded Employment and Skills Coordinator. Local leaders perceive its aims, objectives and agenda as sufficiently different from the GYBC Economic Reference Group; successor to the Economic Forum, and a Great Yarmouth Youth Advisory Group. However, NALEP officials preparing the ESIF Community-Led Local Development model failed to involve or update the group, despite the rhetoric which recommends involving civil society organisations (see item 3 and 4, GYBC 2013a). In reality, institutions secure positions; thereafter, the new policy brand enters the civic landscape ad hoc through governing networks. The group has since been informed that up to £1 million ESIF is earmarked for community-led development in Great Yarmouth to support skills, jobs and business–start-ups between 2014 and 2020 (see item 4, GYBC 2014f).

The Norfolk Health and Wellbeing Board offer another network angle, and recognise the link between mental and physical health and unemployment:

> ...the mental health of people who are not in work can deteriorate further and lead to significant financial and social problems... The health impacts of unemployment should be considered as there is a correlation between increased unemployment rates and increased early male death and disability... Reducing long term unemployment will improve health outcomes. (NCC 2013: 28, 38)

The Board's duty is to join up commissioning for health, social care, public health and other services provided by NCC, GYBC and Great Yarmouth Voluntary Sector Partnership, and Great Yarmouth and Waveney Primary Health Care Trust. However, despite 11,000 people in Norfolk claiming IB/Severe Disability Allowance in 2011 for mental health reasons, the Norfolk Joint Health and Wellbeing Strategy 2014–2017, whose campaign features the strapline 'Working together for a healthier, happier Norfolk', omits to represent worklessness (NCC 2013: 28; NCC 2014b). Treating mental health-related problems (addictions, obesity, muscular skeletal conditions or depression) without preventing the unemployment that contributes to negative health impacts appears a costly and perverse policy. Likewise, networks repeatedly promote the same policies of enterprise, skills, infrastructure and investment without achieving the job expectations.

CASES: THE OLD BOYS' CLUB, A CHERRY PICKER, ISOLATORS, THE INDEPENDENTS

The title of this section caricatures five case networks (A–E) associated with unemployment in the case ward clusters. Some interviewees used the term 'old boys' club' to describe a historically male-dominated governing network (Case A). The 'cherry picker' (Case B) selected its partners. Isolators (Case C) are separate self-regulating networked organisations that failed to join up. Finally, the 'independents' (Cases D and E) sometimes formed networks but mostly acted autonomously. Table 6.2 compares case attributes (formal or informal type, size, actor role and geographic focus). The job roles and responsibilities of participants are concentrated at different operational spatial scales: (i) *Local and strategic*—ward community workers or service managers; (ii) *Borough-wide strategic*—borough-wide service managers or strategists; (iii) *Borough-wide private*—private sector organisations; (iv) *County-wide strategic*—Norfolk agencies; (v) *County-wide private*—Norfolk private sector; and (vi) *Regional strategic*—eastern regional executives. Case A, the Economic Forum, is the only formal network associated with unemployment. The interview total is 47 and the number of citations is 50 because one actor had a dual role and two had dual membership. Cases had low network brokerage. Cases B (WNP) and C (CN) operated in CN with no go-betweens. One actor with a dual role represented a community empowerment group in Case A (Economic Forum), and a community facility in Case D (SC). The identity-preserving

Table 6.2 Numbers of contacts in the GYBC case networks

Cases[a]	A	B	C	D	E	Citation total
	EF[b]	WNP[c]	CN[c]	SC[c]	BW[c]	
Actors						
Local and strategic	1	2	2	4	1	10
Borough-wide strategic	11	2	2	3	7	25
Borough-wide private	4	–	–	–	–	4
County-wide strategic	3	–	–	1	2	6
County-wide private	1	–	–	–	–	1
Regional strategic	4	–	–	–	–	4
Case total	24	4	4	8	10	N = 50
Interview total N = 47						

EF Economic Forum (LSP), *WNP* Working Neighbourhood Pilot (Job Centre Plus), *CN* Central and Northgate wards, *SC* Southtown and Cobholm wards, *BW* Borough-wide only.
[a]Include members with a dual role or dual network memberships
[b]Formal network
[c]Informal network

sociograms (see Figs. 6.1–6.4) represent cases (A, B, D, E) connectivity (Case C sociogram is missing because the sampled organisations had no connectivity) and draw on a mapping exercise completed during interviews conducted in Great Yarmouth between May and November 2005, as discussed in the case profiles that follow.

Case A—The Old Boys' Club

The Economic Forum reported to the GYLSP. A lead officer supported the GYLSP executive board, and policymakers and managers considered cross-cutting issues in separate thematic forums: Economic, Social, Learning and Environmental. A website detailed the LSP governance structure, priority neighbourhoods, forum meeting updates, minutes, expenditure and reports. GYLSP marketing and promotions were modest compared to the Tower Hamlets LSP, although discourse for power-sharing and governance were similar:

> … brings people together to create a 'joined-up' approach to services and to allow the views of the wider community to be heard. (See http://www.gylsp.org.uk/lsp/index.php, accessed on 12 March 2006).

The Forum, however, was separate from the voluntary and community partnerships and presented a means of containing potential combatants. For example, a Forum member expected the voluntary sector to be silent at meetings, as they explained:

> ... a very well-meaning lady who was at one of the meetings from the voluntary agencies partnership wanted four seats on the Economic Forum. It's very difficult to explain to her that that's not the place she should want to be. If she wants to come along and listen and report back to the others, as I said to the [...], one comes and reports back, but it was the wrong place, do you know what I mean, so her own enthusiasm was her own wrong worst enemy. It was getting in the way.

By silencing the community representatives, the network safeguarded against groups that had misgivings about its authority or threatened its hegemony. Indeed, '... other partners (including those from the private sector) are rarely questioned about their legitimacy or representativeness' (see Note 15. p. 229; Taylor 2003: 134). One interviewee commented that the town's economic forum reputedly was an 'old boys' club', in that of the 24 members, only three were women and there were no minorities. Forum members included economic strategists and planners, business leaders, mainstream elites from areas ranging from business support to education and the GYBC chief executive, but no representative of JCP. A private sector representative chaired its quarterly meetings. Two members from Cambridge did not attend meetings. Four members were Norwich-based. Informants stated that the Forum would attract jobs and wealth creation but there was no statement of aims to that end or of aspirations around the numbers of jobs or industries they wanted to attract. Some members' professional interests interlinked and benefited the projects they supported. Meeting items frequently concerned the reporting needs of central and regional government. One member tried to stop these issues' domination of the meetings:

> ... you get the minutes thrust out and the statements from the council ... by the time that's all done you will be at the end of the...[expletive] meeting.

According to one interviewee, NCC would build consensus through Shaping the Future, and Norfolk elites would decide 'what should be happening in those particular areas'. One member attended county committees to try and stop the duplication of business representation and upstaging the Chamber of Commerce. A private sector interviewee said that public sector members darted from one initiative to the next, yet a business could not run like that. Another member thought the council chamber meetings venue stifled networking. One member believed attendance at the Forum had declined as it had turned into a talking shop, and business input at meetings had reduced: 'when we first kicked it off it had a much higher level of accountability and personal responsibility from business leaders'. In March 2006, a JCP representative was invited to attend a Forum meeting due to poor performance meeting NRF targets for jobs. The minutes note job vacancies were dwindling, data to compare wards was lacking and WNP data would not be available until the end of the year. It was questioned how performance achievement might reflect national decisions, such that reducing gas industry funding had employment implications beyond local control. Members thus resolved to look at monitoring of targets at the next meeting (GYLSP 2006b: 65-6). Thereby, passing responsibilities to other agencies in the interim. The learning and enterprise funding priorities of the incoming WNF in late 2007 to support unemployed people in deprived areas in Great Yarmouth, overlapped with the LEGI. The Forum wanted to lead and commission projects (not bids), but their role was in job creation and working with employers; hence the Social Forum objected. Henceforward, the Economic Forum would share commissioning within a new Employment and Skills group, link with LEGI, and join JCP to the LSP executive board. As a result of a fall in attendance, the Economic Forum intended to split into four thematic sub-networks: (i) image; (ii) enterprise; and growth, (iii) employment and skills (supply); and (iv) infrastructure/land and availability/plans, and report to the Forum twice per year (GYLSP 2008).[16] Members were expected to set targets, monitor performance, and provide an economic response for the Great Yarmouth Sustainable Community Strategy, which feeds into the Norfolk County Strategic Partnership and NCC's Local Area Agreement. One senior official raised the concern that skills, employment and business were not separate packages and that they required brokers to bridge supply and demand and group interests, similar to a 'clearing house', which the LSC might facilitate and fund (GYLSP 2008: 3). While brokerage can be helpful, it can also

---

[16] Notably the proposed sub-network arrangements were still being discussed a year later (GYLSP 2009).

erode transparency. Indeed, a non-Forum member notes that the training needs viewpoint always focused on Great Yarmouth College or the LSC, both of which represent further education or regional training issues, at the expense of the voluntary sector providers delivering training on the ground.

Several Economic Forum members shared concerns about its performance and that of the LSP networks in general. One interviewee said that there was little joined-up working or consensus between the LSP Forums and GYBC's 2020 vision, and that representatives pushed their own agendas over the distribution of NRF. Consequently, networks lacked objectivity as people promoted their own interests or wanted funding channelled to them, as hinted by another interviewee. One official believed cross-cutting work posed a challenge because NRF was perceived as a social response; moreover, directing funding towards the economic pot led to tension. The location of an enterprise centre with business units highlighted the tension between the purist business outlooks of the Chamber of Commerce and the community-focused LSPs. A local worker said networks break up because members with short-term funding 'revert to their own targets'. Several comments referred to a duplication of effort; for example, seeing the same people around the table at different meetings, or, too many doing the same thing in multiple networks. Private sector interviewees remarked that this was confusing from a business perspective and that the approach was limited in its efficiency. Another said the Forum lacked a sense of purpose as a result of its longevity, and had 'not really achieved an awful lot except for hot air ...'. Although the Economic Forum brought the issues to the forefront, a local representative doubted its effectiveness.

Actual Contacts
Interviewees were asked who they linked with in relation to unemployment in CN and SC. The sociogram (Fig. 6.1) depicts the actual contacts of the Economic Forum in relation to CN. The network had weaker ties in SC but the differences were minimal. A potential broker (N9) had the top in-degree ($N = 12$), and out-degree ($N = 8$) centrality score, meaning more Case A participants contacted them and they themselves contacted more participants. However, their role was semi-bureaucratic rather than related to brokerage, as the others they contacted were not well connected. Only seven ties were reciprocated in the network and five were between N9. Ties

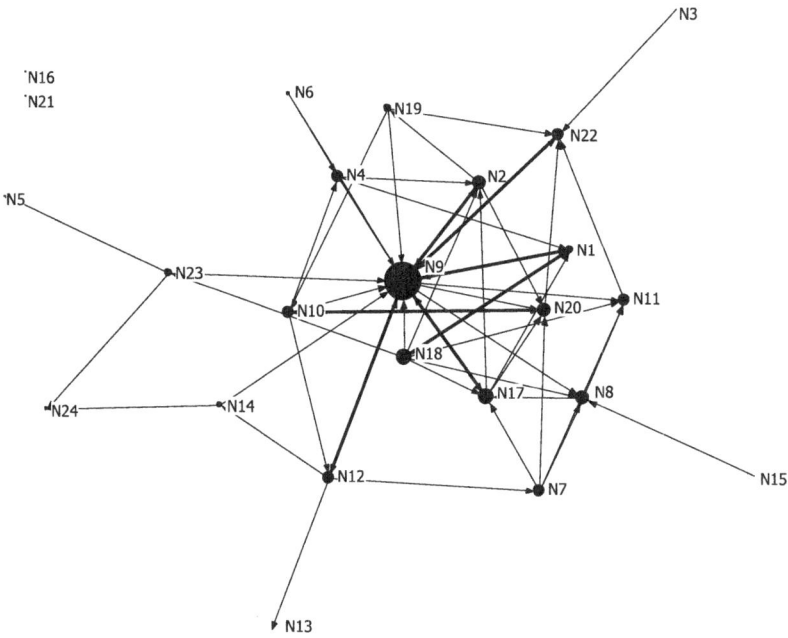

Fig. 6.1 Actual contacts in Case A (EF) relating to the unemployment issues in CN[17]

reflected the external needs of government programmes or silo working. The next highest in-degree scores ($N = 5$) for N20 and N22; both had low out-degree scores of $N = 2$ and $N = 1$ respectively. N4, N5 and N15 had 'pendant ties', meaning they connected by one contact only. N13 had no contacts, though N12 contacted it. Several actors had 'indirect relations', N6 had a two-step connection to N1 via N4 (N6 → N4 → N1). N17, a powerful agent might have bridged ties between a local provider, N8, and a related policy funder, N1. The two isolates were a major employer, N16, and a local authority service, N21. Arguably, the separate actors' issues around this topic were incompatible, with many actors' roles or responsibilities undefined, or interest was minimal.

---

[17] Case A findings were similar across CN and SC ward clusters.

Case B—Cherry Picker

The DWP selected the CN wards for a WNP (2004–6) in response to the area's high worklessness, sickness and crime rates. JCP managed the WNP annual budget of £1 million. The WNP and two community projects worked in close proximity in shared shop-front premises adjacent to the town centre.[18] The projects were competing for clients to achieve targets, and both offering careers and learning advice. One interviewee suggested that this was a conflict of interests and that the arrangement did not offer any benefit to the customer. As a result of its delayed start, the pilot had limited networking activities and partners and potential providers were cherry-picked for commissioning. The DWP documentation stated that the WNP would: (i) work closely with the LSPs; and (ii) bring local partners and providers together to overcome long-standing barriers to employment (Dewson 2005: 45-47). Despite the vision, the scheme did not develop a network because it perceived that networks already existed. One contact doubted the Economic Forum's usefulness to the WNP: 'There's a lot of information changing hands and I don't want to be critical, but I don't think it is much of a goer really'. One interviewee said that the LSP decision-making process was cumbersome as individual groups decided priorities which went up to the LSP board for decision making, and these decisions were then passed back to the groups. The Economic Forum had neither involved the WNP, nor had the WNP given a presentation to the Forum until halfway through the pilot (GYLSP 2005). An evaluation report said that WNPs had focused on operational delivery and commissioning and used LSPs to sign off funding, not for strategic partnership working (Dewson et al. 2007: 52). Another study could not find evidence of Great Yarmouth's WNP having been effective in reducing benefit rates (Selby 2008: 28, 49).

Actual Contacts

The pilot's short timescale meant it prioritised project outcomes, rather than governance outcomes. Consequently, relations were sparse during the early stages of this project (see Fig. 6.2). Success depended on commissioning organisations that could meet targets and handle WNP bureaucracy (in the form of countless forms and checks). Two contacts

---

[18] In 2007, the author returned to these premises; however, they were vacant with no sign of a forwarding address or contact telephone number, and the neighbouring shops had no information.

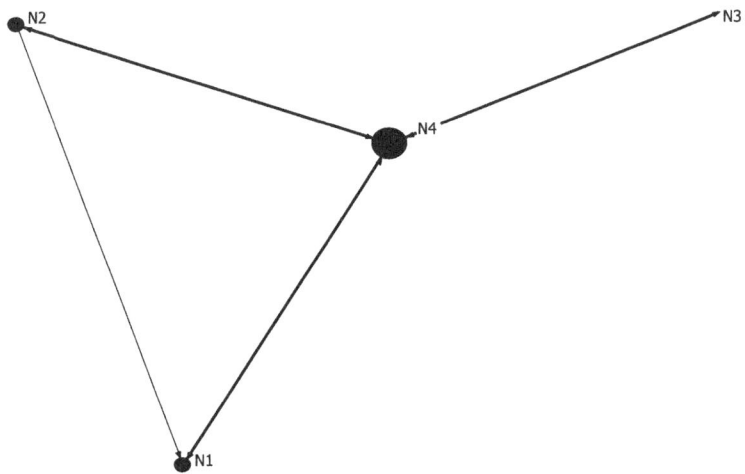

Fig. 6.2 Actual contacts in Case B (WNP) relating to the unemployment issues in CN

offered ideas for expenditure. One of the contact's own connections later received WNP funding; moreover, this contact wanted to influence JCP staffing issues. Another contact offered support and a community resource to distribute information to residents. Among the other contacts cited, one appeared inactive, and another organisation had ceased operating. One worker perceived the WNP as desperate to spend money; ideas came from the grass roots or were imposed from the top, but the important things in the middle became confused and lost. JCP could have instigated a local stakeholder group to lead in the pilot, identify potential projects, mentor the manager and legitimise expenditure; however, the Norwich Jobcentre had staffing problems to address. It left the manager to sort out the details of the ill-planned pilot. Moreover, in 2006, all JCP services were affected by a large national staffing reduction.

Case C—Isolators

This case investigated whether four organisations with a reputation for supporting unemployment issues based in CN formed a network to address neighbourhood unemployment or work separately. Two organisations with different learning programmes and funding streams supported families and children. One had received the government's Sure Start funding to refurbish a centre for children and families in a prominent location. JCP provided a free phone in the foyer and job cards on a community noticeboard so that parents could perform a job search. The other organisation, located out of town in a shabby ex-office complex, supported young teenage mothers and female offenders. The third organisation, located in a residential area, supported people with mental health issues, including a small craft workshop, advice and counselling sessions, and job support. The fourth organisation was in the process of establishing a community partnership for the LSP. Social network analysis (SNA) revealed that Case C did not network in relation to unemployment issues. Instead of building local connections, the four organisations contacted 15 other organisations with a borough-wide focus on unemployment issues. None had contact with Case A, nor had they been contacted by Case B. Institutions undertaking a strategic role could have acted as go-betweens, but organisational self-determination appeared to hold more importance than place-shaping. This is common in neighbourhoods as organisations strive to develop niche services and build a loyal clientele; keeping success in-house avoids the hazards of unreliable partners or competitors. One interviewee stated that no one had promoted joined-up working on unemployment matters. One commented that local organisations' lack of contact with each other about unemployment was detrimental to service users: '... if nobody is talking to each other about what they are trying to develop within their own organisations or link their service users to, then we are not going to get a proper coordinated approach and a consistent message about what's out there'. One organisation had assumed the LSP and community empowerment network were dealing with unemployment problems in Nelson Ward. Another interviewee had ambitions to involve unemployed people, but the partnership work was still in its infancy. The 'network organisations' had not identified a self-interest or place-based need to represent unemployment through a local network.

Case D—Independents

This network involved eight contacts associated with unemployment in SC. Two of these contacts were private-sector training organisations, four were voluntary sector-managed including two community resources projects, an information project and a community development scheme, and the remaining two were public sector-led. None of the contacts had led discussions about unemployment issues; they had dealt with issues informally as needs arose around interests, contracts or job roles. The two mainstream contacts were JCP and the SRB 5 programme, which was ending (seemingly without an exit strategy for partnership working). A health resource centre hosted a community partnership, and a community learning information project supported BME groups. The neighbouring children's centre, a successful social enterprise, employed local people. A training provider in Southtown managed a JCP contract for people with health problems. If a client completed their course and required additional support, they were referred to another JCP training provider in Cobholm with a contract guaranteeing them provision of access to a job club, CV support and training in interview techniques. Relations had been reciprocal before they moved from Cobholm, but JCP had reduced the training provider's contract and staff redundancies and enmity had followed.

The organisations prioritised different unemployment issues. One contact desired face-to-face information for parents regarding family tax credits and support in getting back to work. Another worker needed more support from JCP for clients with limited English skills and booking job interviews on the phone. One interviewee had suggested resident groups tackled dog mess, litter and the drugs problem and they invited professionals to meetings who could make a difference, but no one discussed unemployment at these meetings. The interviewee believed these groups had no real power: 'I think all the organised ones hinder progress, they combat against each other and unless it's achievable, monitor-able and quantifiable ... they are not really interested in the unquantifiable. The informal networks have no power, no control, so these people just sit in meetings and raise concerns'. Another interviewee felt uneasy that partnership working was more about supporting proposals, rather than tackling a cause or matching assets to needs. JCP wanted people to be more mobile looking for work between North and South Yarmouth, yet they only worked close to their headquarters in North Yarmouth and were unaware of the support needs across the river. Two community facilities delivered through the SRB 5 scheme (the health centre and children's centre) were constantly seeking inward investment. SRB had expected the health centre to be viable and self-sustaining, but JCP could not sign a contract to rent space. The children's centre had offered affordable childcare for parents wishing to return to work, but charges increased substantially and priced locals out, as one interviewee explained: '... set up 12 months ago it did very, very well and it is successful as an SRB project, but it seems to have left the community behind again. I think I am probably right in saying that it now provides more childcare out of the area rather than in the area but in terms of success it is still a viable business'. The organisations were making efforts to fill gaps, but the network chose passivity over collective risk-taking, lobbying and organising more action.

Actual Contacts in SC
A lead agency (N6) had a central network position (see Fig. 6.3), but its relations were not reciprocated, thereby reflecting its contractual and monitoring role. A facility (N4) with the highest reputation amongst the entire network population for work in SC provided an additional snowball contact (N2). This asset attracted organisations to it and built an inward-looking local identity, but there were weak local links surrounding unemployment support.

Case E—Independents

The reputational approach identified a network sample of ten borough-wide organisations associated with unemployment. Half of these had offices or facilities located in the CN wards, while the remainder were situated in SC. They offered specialised services in arts, careers support, community support, education, job support, training support, voluntary sector grants,

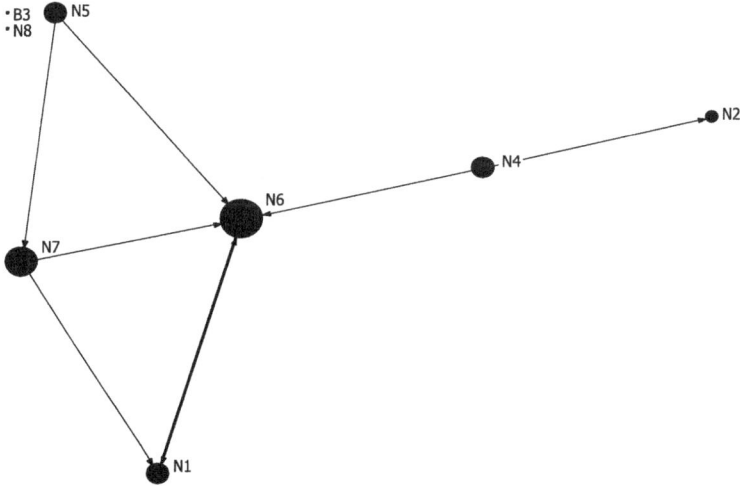

Fig. 6.3 Actual contacts in Case D (SC) relating to the unemployment issues in SC

community support, youth offending and youth careers advice. Half were mainstream provision; the others received grants and NRF.

Actual Contacts

Organisations were connected largely by pendant ties, meaning one contact only (see Fig. 6.4). This suggests an incoherent framework for unemployment action, as organisational collaboration and dependency between organisations was lacking. Contacts operated in networks close to their organisational interests. One interviewee participated in various learning networks. The agency with the highest network centrality ($N = 6$) had only two reciprocated ties; one of these wanted money but they could only allocate spending in certain ways. An interviewee supporting unemployed clients on probation had no formal links with JCP and used informal methods for job searches, including walking round building sites. Another mainstream service was invited to numerous networks, but as one interviewee mentioned, its participation had to 'make it worth it rather than just

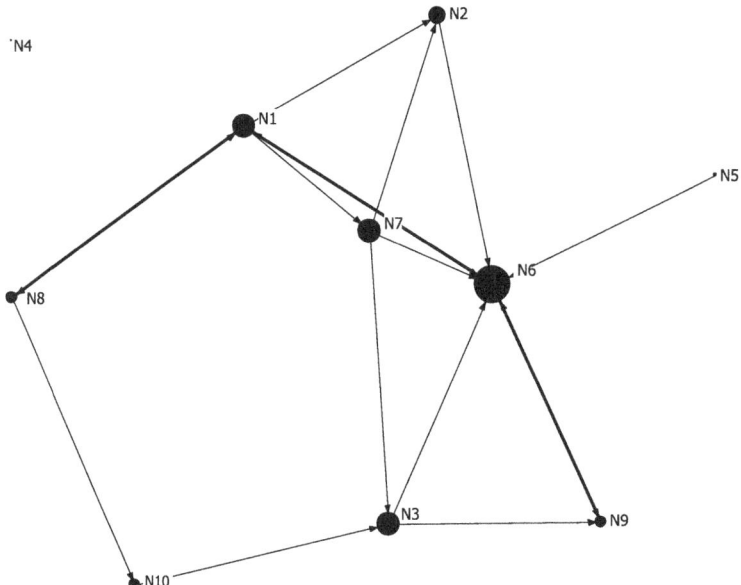

Fig. 6.4 Actual contacts in Case E (BW) relating to the unemployment issues in SC

attend a meeting. We could spend all our days in meetings really because to really be involved takes a lot of staff time'. Multiple groups were the problem, commented another. One contact cited the SRB as an example of a big network linked to funding that did not involve their organisation until it was too late. Another interviewee believed the LSP and the NRF were using a top-down approach and were not good at engaging the voluntary sector: 'I think we have gone slightly network mad in Yarmouth and how effective they are is very questionable. The LSP is supposed to be y'know a very key body but I know that a lot of people question the effectiveness of that in terms of being an accountable body'. One contact said that the income-maximisation working group discussed unemployment but that no positive outcomes or measures emerged. Another said that community empowerment representation did not deal with unemployment and focused instead on health, a community building, safety and welfare. A network of advice and guidance providers monopolised the client referral system according to one contact, and there was rivalry amongst those seeking contracts at different times in the year. One interviewee believed the LSP's Learning Forum competed with an existing learning network that had tried to exclude the statutory sector. Unemployment sustained the tourism industry and low-paid workers, suggested another, but the community partnerships did not know how to address these issues: 'Government talk about worklessness and liveability and things like that so they don't want to talk about it. I mean the economic forum doesn't want to talk about unemployment; they want to talk about job creation. Other people talk about income maximisation. Nobody wants to talk about unemployment'.

Factor Comparison

This section considers how five factors (see Fig. 1.1) impacted on Great Yarmouth's network culture.

Central Environment

The Conservative-led borough council adopted a multitude of governance frameworks mandated by central funding streams. These included: Adult and Community Learning/Personal and Community Development Learning; Investing in Communities Programme; LEGI; NRF; Neighbourhood Management; Safer and Stronger Communities; Sure Start; and WNF. The GOE had oversight of the Neighbourhood Renewal Strategy and the

GYLSP annual performance review.[19] The GYLSP and six other LSPs in Norfolk reported to the Norfolk County Strategic Partnership (NCSP), which had a board, a management group, lead officers, outcome champions and other countywide policy sector partnerships. The NCSP managed the Local Area Agreement for the GYLSP.[20] NCC produced a county regeneration strategy and business plan for an 'Investing in Communities' programme managed by EEDA. The government also required local authorities to combine their efforts to improve economic prosperity and tackle worklessness, skills, housing and transport through Multi-Area Agreements (DCLG and DWP 2007: 17). The Economic Forum (Case A) reported upwards to the GYLSP executive, NCSP and GOE, yet reporting accountabilities did not focus downwards. GYLSP meetings increasingly focused on coordinating NCC Local Area Agreements with round-table discussions to develop action plans and contribute to regional outcomes (GYLSP 2006a). Six officials pointed out that GYBC coordinated a Great Yarmouth 2020 Vision, with forums and a rationale for making bids, long before the LSP. The revised vision states that GYBC produced a Neighbourhood Renewal Strategy 'at the request of government' and that this 'resulted from a change in government' (GYLSP 2002: 3). The GYLSP website[21] subtly disparaged government policy and stated that the LSP contributed to government targets to provide 'us' with better services, stating that 'areas of deprivation' was the 'jargon of politicians', and implying that performance management exercises were not pragmatic: '… in April 2004, the LSP was subjected to a Performance Management Framework exercise by GO-East rather than the annual accreditation process'. The WNP (Case B) delivered by JCP was expected to work closely with the LSP, yet momentum was lacking. The WNP expressed concern about policy inertia for helping Yarmouth people engage in the construction of the outer harbour project, yet it did not represent unemployment on the Economic

---

[19] GOE was the Eastern Regional Government Office that supported local authorities and partners in the East Anglia region to develop LSPs and community strategies, rate partnership performance against six indicators: strategic, inclusive, action-focused, performance managed, efficient, and learning and development, and deliver an improvement plan.

[20] The Local Area Agreement was a three-year agreement (2008–11) between NCSP and central government. It aimed to deliver Norfolk's Sustainable Community Plan, and 75 negotiated local and national indicators set against outcomes, measures baselines and targets (see details on Norfolk Ambition website, http://www.norfolkambition.gov.uk/norfolkambition/home.asp, accessed on the 7 February 2009).

[21] http://www.gylsp.org.uk/lsp/; this link is no longer available.

Forum (Case A) or work strategically with the council. Nevertheless, Case A was aware of the Learning Forum's concern that not enough local people had the right skills for the incoming infrastructure projects. Moreover, unemployed people lacked access to finance to start businesses that would take on the work (GYLSP 2006c: sec. 4).

Central Policy Rhetoric: 'Joined-up Government'
Interviewees were asked about the relevance of 'joined-up government' as it pertained to their work; 60 % confirmed its relevance. Regional strategists emphasised its importance for delivery mechanisms and departmental strategies, more than local actors. Two respondents were unfamiliar with the term, including a private sector training provider and a mainstream operational manager linked to the prison service, while 51 % of interviewees were dissatisfied with the lack of joined-up government. One Case A member based outside the borough was meant to influence joint working, and enable business to be productive and competitive, yet they did not mention Case A or attend meetings, and believed that partnerships obscured delivery: '… sometimes I think because of the plethora of partnerships and partnership approach it's complicated delivery arrangements. It's not clear to see who is actually doing the delivering which benefits, and what has an impact on unemployment'. Another senior member of Case A reflects candidly on learning and skills policy: 'I think too often they become monitors and checkers of things (em) rather than really helping to shape things, and just add to the list of people who are checkers as opposed to those who get on and do things. I would put ourselves on the list of monitors and strategists in that sense and I think there are just too many of those'. One interviewee from the private sector believed regional and local governments had the appearance of being coordinated but lacked vision. A manager said the DWP should join up departments, but they did not because the Department of Trade and Industry (DTI), Office for the Deputy Prime Minister (ODPM) or the Treasury influenced funding priorities. Another suggested the problems had escalated: 'They really don't operate in a joined up way. We've just had a situation where an East region group for international trade was meant to be part of another group and working on this for six months. Suddenly someone nationally said "oh not sure about that" and put a kibosh on it … each government department have their own agenda and they do not seem to be able to work together especially at a local level. It's getting worse for instance in the business support area,

you've got the LSC, obviously you've got Business Link, and everything is being put through the Regional Development Agency. They are pulled and pushed by different agendas at national government level which is stopping them being able to do what they want to do, so by the time it comes to the local level it is completely fragmented and not very satisfactory'.

Another interviewee said the DTI granted freedoms to decide on regional priorities, but they clashed with DEFRA, the ODPM and the DfES, who imposed additional targets on their work. As stated: 'In one sense we are supposed to have this free rein to decide what the priorities for our region is, but we've got one hand behind our back from central government who say "oh by the way we want you to achieve these targets as well" ... Looking upwards from the region it is difficult and a frustration'. Conflicting messages conveyed by the government about which different management approaches the borough should adopt dismayed one official, who commented: '... we've got parts of government telling us that the future lies with Gershon[22] and efficiency, and joining things up, and making them bigger so we use our full spending power and muscle. On the other hand, we've got the Neighbourhood Renewal Unit and so on saying break things down, Neighbourhood Management is the future and y'know you need to devolve and centralise and delegate. But as soon as you start breaking services up into smaller parcels and handing them down to smaller organisations then the costs goes up and y'know we are left to reconcile those apparent conflicting (em) and there's lots of other examples of that'.

One interviewee from the voluntary sector despaired because the local authority did not understand their needs yet expected the sector to deliver welfare through Local Area Agreements. Furthermore, frontline service providers felt exasperated because departments' grant-making reporting systems did not relate, and one who supported people with few opportunities stated: 'I keep hearing words like joined up government and joined up thinking, but I see nothing but a shambles around me ... what's coming through is the total lack of policy coming from this government'. A few interviewees were disappointed that a joint event to discuss family tax credits or childcare tax credits with the Inland Revenue and JCP had not taken place. Senior officials believed uncoordinated initiatives impaired health and education. One manager considered that joined-up government recognised the barriers to education and skills training, but

---

[22] Refers to the Gershon Efficiency Review of public spending carried out by Peter Gershon in 2003-04 who was tasked to make recommendations for the Comprehensive Spending Review 2004, and achieve £21 billion spending cuts by delivering joined-up government.

wanted parents to foot the bill. An executive said no government would join up without civil service reform. The messages of frustration were consistent. A private sector representative stated that businesses were loyal for reasons other than central policies, such as a major business not relocating because the managing director kept a boat on the Norfolk Broads. Moreover, Yarmouth had a tradition of rivalry and resistance between economic elites, overambitious councillors and local governments.

Network Structure

To access central government funding Case A was obliged to apply network structure, Case A institutionalise participation, support economic development narratives, and coercive instrumental alliance-building to support central goals. Case B had central government directives but the manager shaped the governance process through a loose network structure of preferred ego-contacts. In this way, the structure supported the programme itself rather than local needs. Cases C, D and E had no network identity or governmental coordination, and reputational actors worked apart or shaped governance as voluntary independent structures. Organisations associated with Cases D and E had low capacity to construct ties around neighbourhood unemployment issues, and reflected the fragmentation of funding streams. Some organisations could only survive through self-interests and lacked the time, capacity and resources to connect, but the disinclination to work across boundaries was related largely to funding structure, institutional practices and limited job roles.

Area-Based Factors

A study of seaside economies between 1971 and 2001 noted the growth in tourism employment, particularly for women and part-time workers (Beatty and Fothergill 2003). Numbers of benefit claimants also increased, and there was a small amount of in-migration for job prospecting and cheap rental accommodation (2003: 35–40, 93–7). In-migration can also drive growth, in that retirees with spending power contribute to the economy, but also increase spending on public services. The study suggested that the real rate of unemployment is much higher than recorded, as older men on sickness benefits (frequently former manual workers) and female carers looking after family or the home would like to work. Thereby, jobless characteristics are no different from other areas (2003: 80). Location matters, as some seaside towns in wealthier regions benefit from a spillover effect, whereas weaker economies such as Great Yarmouth need 'policies to foster job creation' (2003: 8). The present study asked network actors about their understanding of the unemployment problem in CN and SC. A category and coding technique was used to analyse interview transcripts. Table 6.3 presents the percentage of local knowledge in the networks directed towards the wards or spread borough-wide. One local worker stated that historic legacies were reflected in house prices, which separated manual workers in Cobholm from professionals in Southtown. A voluntary sector manager referred to the 'good solid working-class community' of SC, whereas CN had a transient population. A regional contact tried to conceal their lack of local knowledge by emphasising the borough: 'I think people tend to talk about a Yarmouth problem rather than an actual ward level'. Indeed, strategic managers with resource and policy influence in Cases A and E possessed less local knowledge. However, a strategic view is critical since Yarmouth needs better transport infrastructure to attract industries for job creation, as a regional strategist explains: '… the major barrier is location. In our organisation, a big emphasis is on attracting companies from all around the world, now if you have a brand like Cambridge, that's quite easy to attract because there's familiarity. But, if you are on the periphery of the region that doesn't have say the best logistics and transport networks it's difficult to attract those types of industries to those areas'. Job creation depended on marketing programmes, but two important representatives did not attend an important AGM of the Great Yarmouth Marketing Initiative, and one interviewee considered their lack of commitment an example of joined-up working not working. Notably, Case A had limited ward knowledge but more borough-wide strategists and resource influence (correlate Table 6.2 with Table 6.3).

The second stage of the transcript analysis identified 24 unemployment factors, which are organised here into four categories: locality, structure, capacity to work and behavioural issues. Frequency of statements were totalled as percentages per case. All cases linked unemployment to in-migration, but more Case A actors associated ethnic minorities and economic migrants with housing problems in flats and bedsits in CN than in SC. One interviewee from the public sector commented that business leaders wrongly perceived that some people had travelled to Yarmouth from Liverpool with the sole purpose of claiming benefits. Some dismissed the benefits system for being too lenient and blamed unemployed people for not finding work. Yet, low job

Table 6.3 Contacts in case networks aware of the unemployment issues in CN, SC or borough-wide only (%)

Cases	A	B	C	D	E
	EF %	WNP %	CN %	SC %	BW %
CN	8	40	33	36	21
SC	12	20	33	27	14
BW	80	40	33	36	64
Total[a]	100	100	100	100	100
Case total[b]	N = 24	N = 4	N = 4	N = 8	N = 10
Interview total N = 47					
Citation total N = 50					

CN Central and Northgate wards, SC Southtown and Cobholm wards, BW Borough-wide only.
[a]Total citations are more or less than 100 due to rounding
[b]Includes members with a dual role or dual-network memberships

density was scarcely mentioned, and jobless people cannot overcome structural problems, recognised by some interviewees as linked to job type, industry decline and low-pay work. Case D was concerned with high unemployment in SC. Case A interviewees raised concerns about capacity-to-work issues, particularly skills attainment, and to a lesser extent education, young people and childcare. One interviewee stated that workforce learning had increased but needed to be balanced with job opportunities; if not, people would have to commute. More business representatives framed behavioural issues negatively. For example, one interviewee believed unemployed people were 'unemployable' as they had an attitude problem, and claimed, '… it's rather sad that the majority of people that are unemployed in those areas don't actually want to work'. Another suggested that local unemployed people were lazy. However, Case A did not discuss constructive job interventions or the provision of services to alleviate depression, such as counselling, social activities, sports or arts. Ironically, one of the complainants said the Economic Forum needed kicking. This implies that networks can be unworkable or attitudinally lazy. Therefore, the next section explores the organisations associated with assisting the unemployment issues in CN and SC, and whether high- or low-reputational organisations participated in case networks (see methods, Appendix 2).

Network Agency: Organisational/Participants

In CN, 115 organisations were associated with supporting unemployment. Citation frequency scores were low to high (1–38). In SC, 111 organisations were associated with supporting unemployment. Citation scores were similarly low to high (1–36). Forty-seven per cent of organisations had one citation only. Further analysis linked the high- or low-reputational organisations to case networks.

Contacts with a High Reputation

The top tier reputational scores were similar to those of LBTH. JCP scored the highest ($N = 38$ in CN, and $N = 36$ in SC). JCP had resource power and authority to reach out through a small number of training provider contractors and manage the WNP (Case B), but they did not work across policy boundaries or support local economic development. Great Yarmouth Further Education College was the next highest scorer ($N = 18$ in CN and $N = 21$ in SC) and the highest scorer in Case A; they had ideas and organising ability and seized funding opportunities. Thereafter, the WNP had $N = 13$ nominees for its work in CN. A senior GYBC officer assumed the WNP budget was twice its actual size and that it was linked with prominent actors. This information deficit reveals the council's disconnectedness with the DWP and local JCP. Similar scores were attained for voluntary sector organisations with a borough-wide reach, ranging from community learning and information, enterprise support and a furniture project, to a youth association. A health and resource centre located close to a main road in SC was a beacon and a base for flagship initiatives, scoring $N = 10$ in SC. Some organisations supporting poorer clients were less visible as they were located in backstreets. Nominations for statutory organisations with youth training policy agendas, from Connexions, which coordinated youth issues, to the LSC and schools, were much lower. Organisations obtained only slightly higher scores in ward clusters where their headquarters were based, and a youth association and a learning and information scheme had premises in SC but a higher reputation in CN, which suggests outreach activities were effective.

Contacts with a Low Reputation

A high number of low-reputation organisations were associated with cases. In Case A, a third or more of the contacts lacked a reputation for tackling neighbourhood unemployment. Some had limited interest, responsibilities or capacity to deliver effect change for unemployed people. The low-reputation regional actors could justify their roles by putting in place the mechanisms of governance to support central government initiatives. Some borough-wide organisations steered self-interests outside networks in their preferred localities. One had a high overall reputational score ($N = 12$), yet none when applied to its respective network Case D. One executive had the highest out-degree score (ties from someone to someone else) $N = 12$, but no in-degree score for assisting unemployment. An authority representing port employees could block regeneration, although its employment support role was unclear; it had no reputation score, but it did have in-degree and out-degree scores for connectivity. BL and the LSC had higher reputation in Case A, whereas low reputation business leaders blamed their reduced roles on government.

Network Processes

The government expected networks to operationalise its 'joined-up working' policy. Hence, actors were asked about their understanding of this term and its relevance to their work. The following results were yielded from the interviews.

Joined-up Working: Associations and Meanings

Responses were organised into three categories: (i) general associations; (ii) value associations; and (iii) structural associations (see methods, Appendix 2). Most respondents were familiar with 'joined-up working', but a private sector training provider had not heard the term. Twenty per cent had reservations, and more Case A members had negative perceptions of partnerships than those in the other groups. A selection of views follows. A private sector interviewee said that Yarmouth had worked logically before government came to influence the local government map and 'talk a good talk' in terms of joined-up working. A senior official said the rhetoric of joined-up working lacked evidence to back it up, but that they complied with central government and 'talked the talk', including in Local Area Agreements. One interviewee claimed public/private partnerships like 'Yarmouth *Inte*great project [had] not worked because … the public sector chaired … if you talk to anyone in the tourism industry they go [boom] there wasn't an understanding to allow the entrepreneurship of the private sector to come through … it was like tick boxes …'. The interviewee went on to note: 'what the private sector can add is innovation, entrepreneurship and looking outside the box ... the public sector find that difficult so that's why in partnerships joined up working doesn't work because the private sector just get fed up and walk away'. Whilst sectoral differences hampered joined-up working, the benefits to clients and neighbourhoods attracted little consideration and comment.

Certain values and principles were associated with joined-up working, including 'working together' agenda-sharing, having a common focus and working for the greater good, rather than pursuing selfish goals. A majority of local workers and service providers believed joined-up working referred to working at ground level, avoiding duplication or working in isolation, while others identified specific problems. One executive noted that when it came to helping the client, information sharing or resources did not follow: '... in reality there is very little joined up working, most people (em) pay lip service to it and it is in essence a whole variety of individual organisations pontificating schemes whatever (em) that really do little towards joined up working'. A frontline worker believed networking was: 'talking, but that ain't working together'. A voluntary sector manager was surprised at the number of isolated organisations: 'At the moment it is not joined-up because everyone is more or less doing their own thing from whatever their charitable objectives are'. Networks did not always help the poorest organisations, as a worker serving to rehabilitate the neediest citizens observes: 'There's all sorts of organisations trying to bring people together ... I don't go to a tenth of the meeting invites because I'm constantly battling for funding [...] I know other people in similar positions have the same problems'. A mainstream provider complained that partnerships overlapped and targeted similar clients, but that joined-up working did not reach wards or was not cost-effective. A private sector provider said JCP promoted joined-up working to give a non-biased impression but in reality they did not encourage people to get together. Definitions were less operational-focused and more emotive, ranging from ideology to self-delusion, and cynicism to frustration.

Nevertheless, the concept of joined-up working was highly relevant to respondents' work. They repeatedly used phrases like 'extremely relevant'. Several associated it with partnerships. A strategist associated it with getting their messages out through networks. A manager saw it in terms of a hierarchy. A private sector representative cynically believed joined-up working had only one-way relevance – for government to confer with business and tick a box; it did not mean that the government listened.

Perceptions of who promoted joined-up working were organised into three categories: (i) organisations; (ii) networks/partnerships and initiatives; and (iii) strategic frameworks. JCP had the highest citation score for promoting the concept, but only weakly, at 19 % ($N = 9$). Coordinating bodies, such as the EEDA and the LSC achieved low scores of 6 % ($N = 3$) and 2 % ($N = 1$) respectively. These results hint towards no single organisation leading effectively joined-up working policies. Only 30 % of informants and more borough-wide actors associated two case networks with the concept; Cases A (EF) and C (WNP). Low ratings for promoting joined-up working correlated with their low connectivity. The LSP and the community plan were cited by 28 % of informants as a strategic tool for joined-up working. Only 6 % ($N = 3$) mentioned EEDA's 'Investing in Communities' initiative. A key informant explained how EEDA sought a coordination role and planned to join up solutions and stakeholders from top to bottom and work out how the voluntary sector does things. They would then get JCP, LSP and the LSC to progress collectively to achieve regional significance. Yet EEDA's modest reputation results suggest they had overinflated their joined-up regional governance role. One official said EEDA directed them not to work on a local basis but to focus on better coordination through the IIC (Investing in Communities) and the county, and that this had been a 'painful experience'. In this sense, the regional perspective of joined-up working meant joining upwards, and the neighbourhood perspective began to lose its relevance.

Conclusion

The east of England seaside town of Great Yarmouth has struggled for decades to provide sufficient tourism jobs and year-round work to overcome high levels of unemployment. Despite a public funding surge to grow green industries, the economy and private sector investment remains sluggish. Policymakers want people to be more motivated, educated and skilled, yet job competition excludes much of the potential workforce and entry-level jobs with progression routes are scarce. Fruitless job-searching means individuals eventually turn inward; depression, anxiety and self-harm are high amongst the borough's long-term unemployed.

Policy narratives and network culture in the case sites of CN and SC differed according to local culture and history. CN's physical seafront regeneration programme for the tourist industry lacked focus in relation to housing needs or job creation. During the summer months, part-time low-paid work in CN attracts a transient population and contributes to weak social networks and housing problems. Some migrants accept exploitation but this undermines the outlook for people needing job stability to support family life. In SC, the voluntary sector projected notions of community, local labour history underpinned expectations for local jobs, and some families experienced generational unemployment. Hence, a 'troubled families' scheme was established based on coalition government policy.

None of the five network cases profiled could deal effectively with unemployment. Network structure intensified network management, from regional to county to local level, but neighbourhood unemployment kept on rising. Networks developed around institutional/organisational needs rather than those of the neighbourhood. Poor organisations had survival goals and monitoring outcomes for funders, not local network governance goals. JCP had the highest centrality score across all organisations, but few ties were reciprocated and their uneven support role, based on their needs rather than community needs, did not match expectations. In short, governing networks reinforced a hierarchical approach to consensus-building to avoid conflict over policy content, had little control over the economic impact of national decisions, and roles and responsibilities for dealing with unemployment problems were ad hoc. In subsequent years, funding has intensified to concentrate local organisations on targeting employability deficiency and unemployment's consequences, not its causes; thereafter, strategy and delivery have separated or sidelined local providers. Recently, stakeholders have attempted to legitimise local responses and pool budgets for small-scale neighbourhood provision, but job targeting has had much less attention. In 2016, unemployment in Great Yarmouth remains higher than the national average and numbers of ESA/ IB claimants have reached the highest level since records began. The next chapter explores in greater detail the perceptions of network performance and outcomes.

Bibliography

Beatty, C., & Fothergill, S. (2003). *The seaside economy: the final report of the seaside towns research project, CRESR*. Sheffield: Sheffield Hallam University.

Beatty, C., Fothergill, S., & Wilson, I. (2008). *England's seaside towns: A 'benchmarking' study. A research report*. London: DCLG.

ComeUnity (CU). (2010, May). *Outcomes report 2006–2010*. GYBC and GYLSP. Retrieved from http://www.comeunity.info/

Connexions. (2006). *Not in Employment, Education or Training (NEET): Statistical data*. Retrieved from http://www.dcsf.gov.uk/14-19/index.

DCLG, & DWP (2007). *The working neighbourhood funds.* London: DCLG.
Department for Communities and Local Government (DCLG). (2006). *Strong and prosperous communities: The local government white paper.* Cm 6939-I. London: DCLG.
Department for Communities and Local Government (DCLG). (2008). *Communities in control; Real people, real power: The local government white paper.* Cm7427. Norwich: The Stationery Office.
Department for Communities and Local Government (DCLG). (2010a). *Index of deprivation 2010.* Retrieved from https://www.gov.uk/government/statistics/english-indices-of-deprivation-2010
Department for Communities and Local Government (DCLG). (2010b, December). *National evaluation of the local enterprise growth initiative programme.* Amion Consulting.
Dewson, S. (2005). *Evaluation of the working neighbourhoods pilot: Year one. DWP research report No. 297.* Leeds: Corporate Document Services.
Dewson, S., Casebourne, J., Darlow, A., Bickerstaffe, T., Fletcher, D., Gore, T. & Krishnan, S. (2007). *Evaluation of the working neighbourhoods pilot: Final report. DWP research report No. 411.* Leeds: Corporate Document Services.
Fox, P. (2008). *Stocktake of the Great Yarmouth Local Strategic Partnership: A consultation paper.* From the interim local government re-organization adviser to the GYBC. Retrieved from http://www.communityconnections.org.uk/
Gauci, P. (1996). *Politics and society in Great Yarmouth 1660–1722.* Oxford: Clarendon Press.
Great Yarmouth Borough Council (GYBC). (2004, October 12th). *Minutes of the policy development committee.*
Great Yarmouth Borough Council (GYBC). (2011). *Great Yarmouth economic development strategy 2011–2016.*
Great Yarmouth Borough Council (GYBC). (2012a, May). *Borough profile (facts and figures).*
Great Yarmouth Borough Council (GYBC). (2012b, December 7th). *Employment and skills group minutes.*
Great Yarmouth Borough Council (GYBC). (2013a, December 4th). *Employment and skills group minutes.*
Great Yarmouth Borough Council (GYBC). (2013b, 17th October). *Economic Reference group minutes.*
Great Yarmouth Borough Council (GYBC). (2014a) *Borough profile (facts and figures) 2014.*
Great Yarmouth Borough Council (GYBC). (2014b). *Scrutiny committee minutes,* 6 March 2014, Appendix 1. Measuring the Impact of Welfare Reform November 2013.
Great Yarmouth Borough Council (GYBC). (2014c, June 10th). *Yarmouth area committee minutes.*

Great Yarmouth Borough Council (GYBC). (2014d, June 11th). *Employment and skills group minutes.*
Great Yarmouth Borough Council (GYBC). (2014e, July 22nd). *Full council minutes.*
Great Yarmouth Borough Council (GYBC). (2014f, February 19th). *Employment and skills group minutes.*
Great Yarmouth Local Strategic Partnership (GYLSP). (2002). *GY Local Strategic Partnership Great Yarmouth 2020 vision: The revised path to the future.* Great Yarmouth.
Great Yarmouth Local Strategic Partnership (GYLSP). (2005, September 15). *Minutes of the Great Yarmouth, Economic Forum.*
Great Yarmouth Local Strategic Partnership (GYLSP). (2006a, January 20). *Minutes of the LSP Executive.*
Great Yarmouth Local Strategic Partnership (GYLSP). (2006b, March 9). *Minutes of the LSP, Economic Forum.*
Great Yarmouth Local Strategic Partnership (GYLSP). (2006c, February 09). *Minutes of the LSP Economic Forum.*
GYURC (Great Yarmouth Urban Regeneration Company 1st East). (2007a). *Area action plan preferred option January 2007.*
GYURC (Great Yarmouth Urban Regeneration Company 1st East). (2007b). *Area action plan preferred options consultation: Consultee responses September 10, 2007.* Retrieved from http://www.great-yarmouth.gov.uk/1steast
Great Yarmouth Local Strategic Partnership (GYLSP). (2008, September 11). Minutes of the Economic Forum.
Great Yarmouth Local Strategic Partnership (GYLSP). (2009, September 25). Meeting of the Great Yarmouth LSP delivery executive. Great Yarmouth.
LEPN (Local Enterprise Partnerships Network). (2014). *Building local advantage: Review of Local Enterprise Partnership area economies in 2014.* Athey Consulting Ltd, www.atheyconsulting .co.uk
Lynch, K. (2015, December 03). New staff arrive in Great Yarmouth for wind farm project. *Great Yarmouth Mercury.* Retrieved from http://www.greatyarmouthmercury.co.uk
Meeres, F. (2007). *A history of Great Yarmouth.* Chichester: Phillimore.
NALEP. (2013). *New Anglia Local Enterprise Partnership economic profile 2013.* Wymondham, Norfolk.
NALEP. (2014a). *New Anglia Local Enterprise Partnership strategic economic plan (April 2014).* Wymondham, Norfolk: NALEP.
NALEP. (2014b, November 20). *New Anglia Local Enterprise Partnership board.* Agenda Item 5, Managing Director's report. BTAdastral Park, Ipswich.
NALEP. (2014c, January). *European investment strategy.*
NALEP. (2015, March 20). *New Anglia Local Enterprise Partnership Board.* Agenda Item 10, Business Performance Reports-March 2015. Beacon Park, Great Yarmouth.

Norfolk County Council (NCC). (2013, April 17). *Report to Norfolk Health and Wellbeing Board.* Norfolk Joint Health and Wellbeing Strategy – Responding to the priorities. Item 6.

Norfolk County Council (NCC). (2014a). *NCC public health outcomes framework summary.*

Norfolk County Council (NCC). (2014b, May 6). *Report to Norfolk Health and Wellbeing Board, Norfolk Joint Health and Wellbeing Strategy 2014–2017.* Item 5.

Norfolk County Council Strategic Partnership Board (NCCSP) (2009, January 14). Minutes of the joint scrutiny panel.

O'Gorman, S. (2011). *Engage and enable: Project evaluation.* NIACE November 2011.

Office for National Statistics (ONS). *Labour market profile - Great Yarmouth.* See https://www.nomisweb.co.uk/reports/lmp/la/1946157234/report.aspx?town=great yarmouth

Office for National Statistics (ONS). *Neighbourhood statistics.* See https://www.neighbourhood.statistics.gov.uk/

Office for National Statistics (ONS). (2014, March). *Local labour market statistics.* Retrieved from http://www.nomisweb.co.uk

Office for the Deputy Prime Minister (ODPM). (2000). *The English indices of multiple deprivation 2000.*

Selby, P. (2008). *Net impact evaluation of the Department For Work And Pensions Working Neighbourhoods Pilot.* Working paper No. 51. A report of research for the DWP.

Taylor, M. (2003). *Public policy in the community.* Basingstoke: Palgrave Macmillan.

Work and Pensions Committee (WPC). (2010, December). *Youth unemployment and the future jobs fund.* First Report of Session 2010–11, Volume II, additional written evidence. London: Stationery Office.

Work and Pensions Committee (WPC). (2014, January). *The role of Jobcentre Plus in the reformed welfare system.* Second Report of Session 2013–14, HC479. Norwich: Stationery Office.

WP (Work Programme). (2013). Work programme statistics website. Retrieved from www.gov.uk/government/collections/work-programme-statistics-2

CHAPTER 7

Network Impact: Performance and Outcomes

Network impact and urban regeneration are not easily measurable, countable or cross-checkable and outcomes are subject to chance, antecedent conditions or leakage from other areas; there is little control over who benefits ultimately from such programmes and cost benefits are not precise (see issues raised by Tyler 2011: 46). However, if network rhetoric is not to mislead, more understanding is required about network performance. As Chaps. 5 and 6 demonstrate, during Labour's network governance campaigns interviewees desired to work together, yet connectivity inside networks was low despite the policy rhetoric of 'joined-up working'. Hence, this penultimate chapter surveys interviewees' perceptions of network performance in case areas and correlates results with case networks. It then cautiously assesses case network outcomes at three levels, using indicators based on network performance variables. Finally, it identifies interviewees' preferred outcomes. Findings suggest networks with high reputations for place-based policy coordination or rebalancing the economy had low effectiveness scores, as actors esteemed networks that furthered their own niche organisational self-interests. The final chapter of the book draws together the investigative strands, and comments on the factors influencing network impact.

PERCEPTIONS OF NETWORK PERFORMANCE

Three network performance variables were examined: reputation, effectiveness and ineffectiveness. First, interviewees were asked to hand-draw a mind map of the formal and informal networks they associated with unemployment problems in the local authority ward clusters (which had been operating for at least three months). Survey results are somewhat impressionistic, and rely on interviewees recalling accurately relations between organisations and networks to which they may not belong. Indeed, many networks were not well known. Second, to acquire more understanding about interviewees' belief systems and disposition, they were asked if the networks were effective. Third, they were asked to explain if and, if so, how networks hindered progress. Findings suggest actors use different criteria to evaluate network reputation and network effectiveness. For example, Local Strategic Partnerships (LSPs) had high reputational scores but low effectiveness scores. The results yielded are discussed below.

Network Reputations

Eighty-five interviewees drew mind maps of networks associated with unemployment, and one preferred to list networks. Some maps reveal complex environments, while others show one network. Table 7.1 organises network nomination results for the London Borough of Tower Hamlets (LBTH) at four geographic levels: two ward clusters, borough and London/regional. Total nominations are $N = 94$ networks and citation frequency range is 1–15. Table 7.2 organises network nominations for

Table 7.1 Networks associated with unemployment in the LBTH—by contacts associated with supporting unemployment issues

Citation frequency	SB wards	EIL wards	Borough-wide	London/region
1	(N = 17)	(N = 13)	(N = 22)	(N = 6)
2	(N = 4)	(N = 3)	(N = 6)	
3	(N = 1)	(N = 1)	(N = 2)	(N = 1)
4			(N = 5)	
5		(N = 2)		
6			(N = 2)	
7		(N = 1)		
10		(N = 1)	(N = 3)	
12	(N = 1)	(N = 1)		
13			(N = 1)	
15		(N = 1)		
Network total (N = 94)	23	23	41	7
Citation total (N = 240)	40	76	115	9

SB Spitalfields and Bethnal Green South Wards.
EIL East India and Lansbury Wards.

Great Yarmouth Borough Council (GYBC) at the four geographic levels mentioned above. Total nominations are $N = 116$ and frequency range is 1–23. Further analysis relates reputation scores to network cases, including participants nominated as network organisations. High-reputation scores correlated with political steering, funding power, institutional resources, formal structure, and image-building. In LBTH, high-scoring networks were Case A (Creating and Sharing Prosperity Group, or CSPG), Local Area Partnership (LAP) 7, associated with Case D, and Case C (Poplar Area Neighbourhood Partnership, or PAN), coordinated by a housing body which safeguarded its local leadership and independence from LAP 7, its competitor. In GYBC, the LSP achieved the highest score, followed by the Economic Forum (Case A), and three other LSP forums (Environmental, Social and Learning), thereby indicating overlapping or confused responsibilities. Small networks attained low scores reflecting diversity, resources and reach. Then again, the Working Neighbourhood Pilot (WNP) (Case B) had mid-to-low reputation scores in LBTH and GYBC respectively, and high-level funding. However, scores require

Table 7.2 Networks associated with unemployment in the GYBC—by contacts associated with supporting unemployment issues

Citation frequency	SC wards	CN wards	Borough-wide	County/region
1	(N = 8)	(N = 6)	(N = 35)	(N = 12)
2	(N = 1)	(N = 1)	(N = 9)	(N = 5)
3	(N = 1)	(N = 1)	(N = 6)	(N = 4)
4	(N = 1)		(N = 4)	(N = 1)
5	(N = 1)	(N = 3)	(N = 2)	
6			(N = 1)	
7		(N = 1)		
8			(N = 3)	(N = 1)
9			(N = 1)	
10			(N = 1)	
11	(N = 1)		(N = 1)	
13			(N = 1)	
14			(N = 1)	
18			(N = 1)	
22			(N = 1)	
23			(N = 1)	
Network total (N = 116)	13	12	68	23
Citation total (N = 359)	33	33	247	46

CN Central and Northgate wards.
SC Southtown and Cobholm wards.

interpretation; for example, 11 Great Yarmouth interviewees believed that a particular community partnership (represented in Case D) tackled unemployment, yet an informant said the partnership meetings did not discuss this issue, thereby suggesting that a quarter of interviewees were misinformed. Regional networks were three times more prominent in GYBC than in LBTH, and more borough-wide networks represented GYBC wards; this reflects the funding focus of participants.

Network Effectiveness

Tables 7.3 and 7.4 organise nomination results per case for effective networks into two categories: 'networks, partnerships, initiatives or lead agent', and 'cluster type', which includes structural type; that is, formal, informal or combination, time-operating or ceased, and scale of impact. In Table 7.3, Tower Hamlets ($N = 36$) interviewees nominated 31 different networks as tackling unemployment effectively, while three interviewees regarded all networks as ineffective. Responses suggest 25 networks were ongoing, five were time-limited, and one had ceased operating. Fifteen networks were perceived to be formal, seven informal, and nine formal/informal. The latter descriptor had different meanings, for example, some interviewees said they were neither especially formal or informal, while others said they were sometimes formal or informal as needed. Fifteen had a borough-wide focus, 11 were local, three covered East London and two were London-wide. The most frequently mentioned networks were the TH Employment Consortium ($N = 4$), Employment Solutions ($N = 3$), Case A ($N = 3$), Case B ($N = 3$) and the Ethnic Minority Enterprise Project ($N = 3$).

Similarly, in Table 7.4, Great Yarmouth interviewees ($N = 44$) nominated 39 different networks as tackling unemployment effectively, while two interviewees regarded all networks as ineffective, and one member of the Economic Forum (Case A) had no local knowledge. Responses suggest 31 of the nominated networks were ongoing, seven were time-limited and five ceased operating. Twenty-three networks were perceived as formal, 11 were informal, and nine were formal/informal. Twenty-seven had a borough-wide focus, seven were local, and five were county-level. The most frequently mentioned networks were a group of electronics businesses ($N = 3$), Training cluster ($N = 2$), Education Action Zone ($N = 2$), WNP ($N = 2$), Jobcentre Plus (JCP) training provider network ($N = 2$),

Table 7.3 Actual networks perceived to be effective in tackling unemployment—identified by network contacts in the LBTH

Network focus	Type	Case A (CSPG)	Case B (WNP)	Case C (PAN)	Case D (EIL)	Case E (SB)	Citation total
Network/partnership/ initiative/lead agent							
Employment support	FI/O/BW	2	–	1	1	–	4
–	F/O/BW	3	–	–	–	–	3
–	F/TL/L	1	1	–	1	–	3
–	FI/O/BW	1	1	1	–	–	3
Ethnic minority	FI/O/L	1	–	–	–	2	3
Local regeneration	FI/O/L	1	–	1	–	–	2
Arts	I/O/LW	–	–	–	–	1	1
Business	F/O/EL	1	–	–	–	–	1
–	FI/O/LW	1	–	–	–	–	1
Community centre	FI/O/L	1	–	–	–	–	1
Child support	F/O/BW	1	–	–	–	–	1
Community empowerment	F/O/BW	–	–	–	–	1	1
Community legal support	FI/TL/BW	1	–	–	–	–	1
Community representation	F/O/BW	1	–	–	–	–	1
Education and employment	F/O/L	–	–	1	–	–	1
Employers	F/O/EL	–	–	–	–	1	1
Jobs support	F/TL/BW	1	–	–	–	–	1
Learning	F/O/BW	–	–	–	–	1	1
Local decision-making	F/O/L	–	–	–	1	–	1
Neighbourhood enterprise	F/O/BW	–	–	1	–	–	1
Professional development	FI/O/BW	1	–	–	–	–	1
Voluntary sector initiatives	F/CO/BW	–	–	–	–	1	1
Youth support	F/O/BW	1	–	–	–	–	1

(*continued*)

Table 7.3 (continued)

Network focus	Type	Case A (CSPG)	Case B (WNP)	Case C (PAN)	Case D (EIL)	Case E (SB)	Citation total
Cluster type							
Education:	I/O/EL	–	1	1	1	–	3
Employment:	F/TL/BW	1	1	–	–	–	2
Ethnic youth employment:	I/O/L	–	–	1	1		2
Community self-help:	I/O/L	–	–	1	–	–	1
–	I/O/L	–	–	–	–	1	1
Job support:	I/O/BW	–	–	1	–	–	1
–	I/TL/L	–	–	1	–	–	1
Youth employment:	FI/O/L	–	–	1	–	–	1
None considered effective		–	1	2	–	1	4
[a]Case total		19	5	13	5	9	51

CSPG Creating and Sharing Prosperity Group (LSP), WNP Working Neighbourhood Pilot (private sector managed), PAN Poplar Area Neighbourhood Partnership, EIL East India and Lansbury Wards, SB Spitalfields and Bethnal Green South Wards CSP Creating, sharing prosperity [a]Includes members with dual role and membership. *F* formal network, *I* informal network, *FI* formal and informal network, *BW* borough-wide, *EL* East London, *L* local, *LW* London-wide.

Note: Data point perception in 2005: *O* ongoing, *TL* time-limited, *CO* ceased operating

and Single Regeneration Budget (SRB) ($N = 2$). Case A (Economic Forum, or EF) was mentioned once.

In both local authorities, high-reputation LSP and WNP networks and Case C (PAN) in LBTH had low effectiveness scores (raw data results were compared in Tables 7.1 and 7.2 with 7.3 and 7.4). Networks that were perceived to be effective were not overseeing local coordination or joined-up working goals. Rather, interviewees self-nominated networks supporting self and organisational interests that facilitated personal control over structure, networks that were enterprising, self-determined, collaborative, and which utilised participants' niche worldview and expertise. Business leaders in Case A (GYBC), for example, nominated a network serviced by the GYBC Economic Development Unit supporting 20–30 local electronic companies of all sizes. However, the network did not directly influence unemployment, other than to sustain or create niche jobs. Cluster-type networks loosely joined up actors with similar values, interests and skills, matched to funding opportunities. Four of the informal network clusters

Table 7.4 Actual networks perceived to be effective in tackling unemployment—identified by network contacts in the GYBC

Network focus	Type	Case A (EF)	Case B (WNP)	Case C (CN)	Case D (SC)	Case E (BW)	Citation total
Networks/partnerships/ initiatives/lead agent							
Electronics businesses	F/O/BW	3	1	–	–	–	4
Education	F/CO/BW	2	–	–	–	–	2
Employment support	F/TL/L	2	–	–	–	–	2
–	F/O/BW	1	–	–	–	–	1
–	F/O/BW	–	–	–	–	1	1
JCP Training providers	FI/O/BW	–	–	–	1	1	2
SRB	F/CO/L	1	–	1	–	–	2
Arts	FI/TL/BW	1	–	–	–	–	1
Business	F/O/C	1	–	–	–	–	1
Children's centre	F/O/L	–	–	–	–	1	1
Childcare support	I/O/L	–	–	–	1	–	1
Community safety	F/O/BW	1	–	–	–	–	1
Economic Forum	F/O/BW	1	–	–	–	–	1
Enterprise support	F/O/BW	1	–	–	–	–	1
Environment	F/O/BW	–	1	–	–	–	1
Energy sector	F/O/C	1	–	–	–	–	1
Health	FI/O/L	–	–	–	1	–	1
–	F/O/BW	1	–	–	–	–	1
Information and advice	FI/O/BW	–	–	–	–	1	1
–	FI/O/C	–	–	–	1	–	1
–	FI/O/BW	–	–	–	–	1	1
Learning	F/O/BW	1	–	–	–	–	1
–	I/TL/BW	–	–	1	–	–	1
Learning providers	F/O/BW	–	–	–	–	1	1
Mobility support	F/TL/BW	–	1	–	–	–	1
Outreach	FI/O/BW	–	–	1	–	–	1
Regeneration	I/O/BW	1	–	–	–	–	1
Women's support	I/O/BW	–	–	1	–	–	1

(*continued*)

Table 7.4 (continued)

Network focus	Type	Case A (EF)	Case B (WNP)	Case C (CN)	Case D (SC)	Case E (BW)	Citation total
Young people	FI/O/BW	–	–	–	–	1	1
Youth support	I/O/BW	–	–	–	–	1	1
YMCA	I/O/BW	–	1	–	–	–	1
Cluster type							
Training:	FI/TL/L	1	–	–	1	–	2
–	I/O/BW	–	–	–	1	–	1
–	I/O/BW	–	–	–	–	1	1
Business support:	I/O/C	1	–	–	–	–	1
Economic investment:	I/CO/C	1	–	–	–	–	1
Job brokerage:	F/TL/BW	1	–	–	–	–	1
Training brokerage:	I/O/L	–	–	–	1	–	1
–	I/O/BW	–	–	–	1	–	1
None considered effective		1	–	–	–	1	2
Lacked local knowledge		1	–	–	–	–	1
[a]Case total		24	4	4	8	10	50

EF Economic Forum (LSP), WNP Working Neighbourhood Pilot (Job Centre Plus), CN Central and Northgate wards, SC Southtown and Cobholm wards, BW Borough-wide only. JCP Jobcentre Plus, SRB Single regeneration budget [a]Includes members with dual role and membership. *F* formal network, *I* informal network, *FI* formal and informal network, *BW* borough-wide, *C* county, *L* local.

Note: Data point perception in 2005: *O* ongoing, *TL* time limited, *CO* ceased operating

in Table 7.3 mention the Tower Hamlets College, a funding opportunist. Some nominations appeared to use guesswork. For example, two interviewees nominated the WNP without knowing how often it met; one thought it had £6 million in funding, while the other assumed it involved several prominent contacts, which was not the case.

Network Ineffectiveness

Results for perceived network ineffectiveness were similar in both local authorities; however, half the interviewees declined to comment. Inevitably, public sector professionals are cautious when passing judgement and criticising central or local initiatives. Hence, results interpretation includes a health warning, as competitive environments restrict interviewees from open debate, but concealed opinion undermines network evaluation. Overall, 54 % of interviewees cited ineffective network performance ($N = 20$ in LBTH and $N = 26$ in GYBC). A third of Case A members were critical of the LSP and WNP. In LBTH 60 % of Case C members had the highest complaints against state-led networks; not surprisingly they saw the LSP as wanting to usurp them, yet stated that the WNP (Case B) needed local expertise to achieve targets. Case E had the highest overall dissatisfaction with networks, perhaps reflecting local politics and their perceived marginalisation. In GYBC, 58 % of Case A members were dissatisfied with networks and 33 % criticised state-led networks, including Case A. 50 % of Case E members considered network structure, membership and content problematic; a small number specified LSP networks and the community empowerment network in connection with this.

A theoretical deficit is evident; network governance and network management arrangements did not service organisations' interests in the networked policy processes. Specific problems raised by interviewees were crudely counted and were related to: alliance-building ($N = 11$), problem-solving ($N = 11$), network management ($N = 10$), coordination ($N = 7$), decision-making ($N = 7$), connectivity ($N = 5$) and self-determination ($N = 3$). A few complained that networks lacked political sensibility about local economic and social injustice. In sum, ineffective networks appear to maintain the status quo and power division and are less able to utilise actors' preferences.

Case Network Outcomes

This section assesses case network performance at three outcome levels: (i) neighbourhood; (ii) network; and (iii) organisational/participant (see the assessment criteria in Chap. 1). It draws on qualitative assessment: observations, fieldwork, document analysis, impressions accumulated from interviews, secondary sources and data quantified through social network analysis (SNA); however, the method requires caution (see data analysis, appendix 2).

Table 7.5 summarises the results for LBTH cases A1–E1, and GYBC cases A2–E2. Six outcome indicators (see italic subheadings in the table) and associated optimal/suboptimal empirical referents guide the analysis. A legend (at the foot of the table) explains the coding of results and variable rating scale (rows and cells), from above average (++) to below average (–) where evidence is lacking. Judging network outcomes in different settings is more

Table 7.5 Six outcome indicators and thirty empirical referents for assessing case network outcomes at three levels of analysis: neighbourhood, network and organisational/participant in two local authorities 2004–5

	LBTH					GYBC				
Cases:	A1	B1	C1	D1	E1	A2	B2	C2	D2	E2
Neighbourhood level										
Alliance-building										
[a,b]Local group representation	(−)	(−)	(++)	(−)	(++)	(−)	(−)	(−)	(++)	(++)
[a,b]Problem representation	(−)	(−)	(+)	(−)	(−)	(−)	(−)	(−)	(+)	(−)
[c]Community assets	(−)	(−)	(++)	(−)	(+)	(−)	(−)	(++)	(++)	(++)
[a,b,c]Resource-sharing	(−)	(−)	(+)	(−)	(+)	(−)	(+)	(−)	(+)	(+)
Decision-making										
[a,b,d,e]Service cohesion	(−)	(−)	(−)	(−)	(−)	(−)	(−)	(−)	(−)	(−)
[a,b]Skills support	(−)	(+)	(+)	(−)	(+)	(−)	(−)	(+)	(+)	(+)
[a,b,d]Economic growth plan	(+)	(−)	(+)	(+)	(−)	(++)	(−)	(−)	(−)	(−)
[a,b,d]Jobs/business starts	(−)	(−)	(−)	(−)	(−)	(−)	(−)	(−)	(−)	(−)
[a,b]Projects/joint schemes	(+)	(+)	(+)	(+)	(−)	(+)	(+)	(−)	(+)	(−)
Problem-solving issues										
[a,b]Information/advice services	(−)	(+)	(++)	(−)	(+)	(−)	(+)	(+)	(+)	(+)
[a]Benefit take-up	(+)	(−)	(+)	(−)	(−)	(−)	(+)	(+)	(−)	(+)
[a]Personal support/counselling	(−)	(−)	(+)	(−)	(−)	(−)	(−)	(+)	(−)	(+)
Network level										
Network governance										
[a,b]Democracy/discussion	(+)	(−)	(++)	(+)	(−)	(+)	(−)	(−)	(−)	(−)
[b]Joint working	(−)	(−)	(++)	(−)	(−)	(−)	(−)	(−)	(−)	(−)
[c]Weak ties	(++)	(+)	(+)	(++)	(++)	(++)	(++)	(++)	(+)	(+)
[c]Strong ties	(−)	(−)	(+)	(−)	(−)	(−)	(−)	(−)	(−)	(−)
[c]Clique-forming	(+)	(+)	(+)	(+)	(−)	(+)	(+)	(−)	(+)	(−)
[a,b]Conflict	(+)	(++)	(−)	(++)	(+)	(−)	(+)	(−)	(−)	(−)
[a,b]Leadership	(+)	(−)	(++)	(−)	(−)	(+)	(−)	(−)	(−)	(−)
Vertical coordination	(++)	(++)	(+)	(++)	(−)	(++)	(++)	(−)	(−)	(−)
Network management										
[e]Problem analysis	(−)	(−)	(−)	(−)	(−)	(−)	(−)	(−)	(−)	(−)
[e]Strategic planning	(++)	(−)	(−)	(+)	(−)	(++)	(−)	(−)	(−)	(−)
Policy implementation	(+)	(−)	(−)	(−)	(−)	(+)	(−)	(−)	(−)	(−)
[a,b]Information-sharing	(++)	(−)	(++)	(+)	(−)	(++)	(−)	(−)	(−)	(−)

(*continued*)

Table 7.5 (continued)

Cases:	LBTH					GYBC				
	A1	B1	C1	D1	E1	A2	B2	C2	D2	E2
[c]Hierarchy/bureaucracy	(++)	(++)	(−)	(++)	(−)	(++)	(++)	(−)	(−)	(−)
[c]Image-building	(++)	(++)	(+)	(++)	(−)	(+)	(+)	(−)	(−)	(−)
[c]Network funding	(++)	(++)	(+)	(++)	(−)	(+)	(++)	(−)	(−)	(−)
Organisational/ participant level										
Self-interest										
[a,b]Self promotion/ reputational	(++)	(++)	(++)	(++)	(−)	(++)	(++)	(+)	(+)	(+)
[a]Power assertion	(++)	(++)	(++)	(++)	(−)	(+)	(++)	(−)	(−)	(−)
[a]Skills utilisation	(−)	(−)	(+)	(−)	(−)	(−)	(−)	(−)	(+)	(−)

Key: LBTH cases: A1 (CSPG), B1 (WNP), C1 (PAN), D1 (EIL), E1 (SB)
GYBC cases: A2 (EF), B2 (WNP), C2 (CN), D2 (SC), E2 (BW)
Legend: (++) = significantly above average; (+) = average; (−) below average; (−) = significantly below average
Evidence: [a]interview transcripts, [b]minutes, [c]Social Network Analysis, [d]documentation, [e]observation

an impressionistic processes than a statistical one; however, an audit trail of evidence links to qualitative and quantitative data (see key at the bottom of the table). The following discussion selects evidence at three levels of analysis.

Neighbourhood-Level Outcomes

Three outcome indicators guide the analysis of neighbourhood-level network outcomes: alliance-building, decision-making and problem-solving along with 12 empirical referents.

Alliance-Building

Alliance-building in localities involves four relational activities that steer collaboration rather than competition:

1. *Local group representation* was low in formal cases, and all participants had weak connectivity around unemployment dimensions.

2. *Local issue representation* was weakly coordinated between interest groups, local advocates and front-line workers. Elites in formal cases had limited local knowledge and therefore, were not representative of neighbourhood unemployment.

3. *Community assets* provided focal points for residents to access community services, information, advice and training. Community buildings had benefited from network liaison but stable funding based on local needs

assessment was lacking. Sustaining assets depended on windfalls and personalities in networks (see Cases C1 and D2).

4. *Resource-sharing* in the social economy was constrained by the market economy. Networks had a role to encourage resource-sharing, but organisational flexibility was lacking. For example, Jobcentre Plus (JCP) was unable to contract provision at a South Yarmouth resource centre with an SRB legacy to self-sustain (see Case D2). In North Yarmouth, the WNP (Case B2) shared a community facility that competed for client referrals and strained joined-up working. Some Tower Hamlets contacts in SB (Case E1) shared in-house resources for survival, but funding territorialism was still evident.

Decision-Making
Five decision-making goal-sets were critical for unemployment assistance.

1. *Service cohesion* was impossible, as a proliferation of short-term provisions confused clients and providers. Networks promoted cohesion of new programmes rather than pulling mainstream services together and reviewing needs against past performance. GYLSP issued four sets of minutes per forum (Economic, Social, Learning and Environmental Forums) to coordinate better decision-making, but the process was cumbersome.

2. *Skills support* for unemployed people was often poorly represented. Networks did not systematically review skills gaps, training provision, local labour needs and local cultural perspectives. Case D2 showed flexibility, by instigating a training course directly supporting online applications to a local factory.

3. *Economic inward investment* plans were not integral to cases. In Great Yarmouth, the Urban Regeneration Company (URC) 1st East was distant from Case A2. In Tower Hamlets, community economic planning (2002–6) prioritised local networks' capacity-building so as to engage with LSP priority-setting (LBTH 2002: 20, 33). Case C1 pursued economic and social investment to strengthen housing and regeneration participants' neighbourhood assets.

4. *Job creation/business starts* were represented weakly across all cases. Tower Hamlets LAPs lacked job creation objectives and Neighbourhood Renewal employment floor targets were not met (LBTH 2008). Business start-up support for unemployed people had minimal targeting. A Local Economic Growth Initiative (LEGI) bid supporting business start-ups and loans did not align delivery to the Local Area Agreement or LSP, despite Case A2 having oversight of enterprise policy (GYLSP 2006: 64). GYLSP

Learning Forum raised concern about the lack of targeted job creation at the outer harbour and the need for training to help jobseekers compete for work at the Norwich regeneration area. Case A2 agreed to take matters forward, but a meeting with JCP, the college and the Learning and Skills Council (LSC) was cancelled and progress delayed.

5. *New projects and joint schemes* were pursued ad hoc by single organisations seeking short-term funding. Case A1 over-reported members' achievements as Case A1 outcomes; for example, the acquisition of the development site for the London Mosque, a spatial development consultation to shape a 'Unitary Development Plan to 2016', the building of a Somali Bank and 359 organisations supported by training (see the LBTH Community Plan 2004–2005). One Case A1 member petitioned to establish a borough-wide 'community hub network' for third sector organisations and public agencies and subsequently received WNF.

Problem-Solving
Three variables are decisive in addressing the immediate problems of unemployed people.

1. *Information/advice services* are essential for moving unemployed people to employment. Cases lacked authority over information pathways. In Great Yarmouth, JCP and the Inland Revenue were criticised for not delivering community information events. Careers advice and guidance providers used separate networks to develop progression routes for young unemployed people. JCP provided a government-funded Sure Start programme with a free phone service and jobs information board, but a nearby mental health centre and a centre supporting young women received no support. Case A2 did not coordinate information services, and when the GYLSP Social Forum raised issues pertaining to obstacles to advice services, different language needs, and front-line staffing issues, the outcome was a non-decision (GYLSP 2005a: 30). Nationally and locally, the level of information provision was constantly changing, and the range of employment and training services were unknown. Hence, a working group in 2010 recommended research with the purpose of mapping services in LAP areas (LBTH 2010).

2. *Benefit take-up* requires specialist advice since the welfare system can lead people into debt and health problems. In 2003/2004, Case A1 aimed to reduce debt and welfare benefits dependency; instead, it reported on a 'successful' benefit take-up campaign. This outcome is misleading given that it relates to the Community Legal Services Partnership. Conversely,

Case A2 passed benefits advice responsibilities to GYLSP Social Forum, although actual CAB provision was continually under-resourced.

3. *Personal development and counselling* provides support to maintain the self-confidence of jobseekers. It is essential, as unemployment induces poor health and downstream costs on the NHS. The cases ignored these issues or transferred to ambitious funding bids. Following the demise of Great Yarmouth WNP, JCP's 'Building a better future' Neighbourhood Renewal Funding (NRF) bid in 2006/2007 wanted a £500,000 health management programme to offer a new way of working between the NHS, social services and voluntary and community sector to engage the 'hard to reach', and address unemployment, worklessness, health, crime and environmental issues. A working group in Tower Hamlets focused on employment equality to increase the number of Bangladeshi workers in NHS roles (LBTH 2010: 25–6), and the Tower Hamlets Primary Care Trust (PCT) linked jobs, training and health care information on a website, but in 2013 the Tower Hamlets Clinical Commissioning Group replaced the PCT and the unemployment theme was dropped. Meanwhile, Bart's Health NHS Trust 'Community Works for Health Pathway' assisted 100 East Londoners to gain administrative and clinical support roles in 2014. Innovative programmes clearly have life-changing potential and require a jobs horizon.

Network-Level Outcomes

Network processes relate to network-level outcome indicators: network governance and network management, and 15 empirical referents.

Network Governance
Eight variable referents support network governance.

1. *Democracy/transparency* assists open communication and information flow to influence ideas, issues and reduce ambiguity. Discussion of unemployment related to incoming programmes in the formal Cases A1 and A2 and the informal Cases E1 and C2. Case C1 meeting minutes had an open communication style for exploring problems and options. Case D1 organised a subgroup to discuss unemployment and spending priorities in LAP 7, but action plans lacked detail and follow-up work was delayed.

2. *Joint working activities* involve issue-exploration and appropriate linkages. Local governance in Case C1 resembled a housing body staff meeting that others could influence for joint working. Area-based time-

limited interventions were introduced with little local consultation, and were intended to steer joint working; however the expectations associated with imposing partnership working were unrealistic in short time-scale.

3/4/5. *Weak ties, strong ties, cliques*—the outcomes were similar; cases had more weak ties than strong ties, or cliques influenced policy and resource allocation. WNPs (Cases B1 and B2) resorted to an ad hoc expenditure amongst weak ties. Strong ties and cliques amongst Great Yarmouth business elites progressed economic interests separate from the social agenda. Community activities depended on influencing elites; in this way, social enterprise support reached the agenda, although with marginal outcomes in terms of job creation.

6. *Conflict* often occurs when expectations differ or another party interferes; it is integral to development work and may result in positive outcomes. LSPs used network governance and funding pathways to avert conflict and policy contestation. Nevertheless, LSP officials wanted LAP 7 to usurp the PAN network (Case C1), and deflected the resultant conflict by claiming actors needed to understand LAP goals. The GYLSP learning forum imposed upon an existing learning network led to tensions, according to some interviewees.

7. *Leadership* refers to network capacity to coordinate unemployment policy. In some of the cases, local leaders steered major funds for neighbourhood projects which also promoted their organisations; such as, a regeneration agency, the college and a housing body. In all cases, unemployment leadership continued to defer to JCP. Case E1 had limited leadership for local growth. Leadership was also lacking in Cases A1–E2.

8. *Vertical versus horizontal coordination* preoccupied LSP networks implementing performance management goals. Case A2, for example, struggled to horizontally coordinate an LSP self-assessment, including 29 Local Area Agreements, and vertically coordinate local public service agreements and Norfolk County Local Area Agreements in separate policy areas while simultaneously reporting to NRF, the LEGI and the Government Office for the East of England (GOE).

Network Management
Network management consists of seven variables.

1. *Problem analysis* requires follow-up work and active network management. Yet problems churned year-on-year in minutes, reports and scrutiny reviews (see LBTH 2010: 16–20). Case C1 had not conducted a local needs survey. Cases could not redress the deficit of entry-level jobs or work placements for unemployed graduates. Community volunteering placements could not guarantee transition to employment.

2. *Strategic planning* requires strategic network behaviour and detailed objectives for tackling unemployment. Case A2 members increasingly reported upwards to legitimise county and regional strategies and initiatives. In 2005–2006, the neighbourhood action plans for EIL LAP 7 delegated the 'outcome' of 'reduce unemployment' to Case B1, while SB LAP 2 ignored unemployment and the outcome of 'better coordination of training' was deferred to Case A1.

3. *Policy implementation* inside networks often more resembled a publicity stunt than actual planned workload. Between 2002 and 2005, Case A1 pledged to reduce unemployment, increase employment rates and tackle poverty by decreasing debt and welfare benefits dependency. But pledges fell behind schedule or were dropped, such as 'training opportunities leading to jobs'. The network even vowed to 'increase tourism', despite there being no tourism representative. In reality, outcomes were more modest. Flagship Neighbourhood Renewal policies demanded network governance but implementing joined-up working required major changes.

4. *Information-sharing* had overloaded LSP agendas. Networks handled so many information presentations and work updates, they lost focus. A stock take of the GYBC LSP reported that some presentations 'have appeared to be "overt funding pitches"' (Fox 2008: 9). Local Area Agreements gave impressions of information sharing but target setting did not connect to a network-driven work plan.

5. *Hierarchical bureaucracy* was transferred through network management, to appease upper-tier funding institutions and policy coordinators, with little downstream accountability. Network structure expanded to service programme bureaucracy. For example, NCC proposed the establishment of a Resources Advisory Group to handle Great Yarmouth's LEGI and WNF (GYLSP 2008a).

6. *Image-building*, sometimes referred to as 'communications' is a priority management activity of state-funded networks. The LBTH LSP used pervasive marketing, a website and publications to legitimise the network infrastructure and investments. The council-funded weekly newspaper, *East End Life*, delivered to 90,000 homes, promoted broadsheets on the THP that told residents they lived in 'LAPs 1–8' (East End Life 24 January–6 February 2005: 11). In 2014, LBTH was found to have defied the Local Audit and Accountability Bill and publicity code, which had been put in place to curb newspaper propaganda and unfair competition, and the DCLG minister investigated East End Life for allegedly promoting the

mayor, Moreover, the Secretary of State instigated an investigation into 'Favouritism on the part of the Mayor/Authority [such as grant making and property transactions] towards certain sections of the community [in LBTH]...for the purposes of gaining political favour (PwC 2014: 3, 76).

7. *Network funding* as a process for managing redistribution was not satisfactory. Judgements about the LSP performance depended on NRF distribution; groups had to adapt to unrealistic funding timescales and bureaucratic processes. Hence, the LSP funded 'capacity-building' training to help small groups participate in funding rounds and ensure compliance with programmes. Nonetheless, THLSP LAPs were preoccupied with constitutional matters and a funding underspend meant local needs were not adequately identified. Although NRF emphasised new ways of delivering and sustaining community activity, only 35 per cent of NRF projects in Great Yarmouth were mainstreamed according to the final NRF report (GYLSP 2008b).

Organisational/Participant-Level Outcomes

Networks assist those at the organisational/participant-level to pursue self-interested outcomes.

Self-Interests

The disposition to self-interest in network outcomes usually takes one of the following three forms:

1. *Self-promotion/reputation-building* enables organisations to expand neighbourhood services. Cohorts lose out if representative organisations lack credible reputations in programmes loyal to government policy objectives. In the studied cases, competent players in funding games gained temporary funding assistance (e.g., from NRF and WNP), but funding incentives to develop an overview were lacking.

2. *Power assertion* is present in a network when the members' strength of funding, information or resources dominates the policy agenda or overshadows that of another party. The WNP (Cases B1 and B2), for example, used its spending power to commission whatever innovation was available, from a community bus refurbishment to salary costs for a Portuguese newspaper (GYLSP 2005b). Power assertion does not necessarily aim for shared governance or service cohesion in the long term; rather, it supports the shortest route to the earliest self-interest gain.

3. *Skills utilisation* in networks assists to maintain participant's commitment to specific goals. Funding opportunists utilised their bid-writing skills to draw down NRF expenditure. Economic representatives in Case A2 complained that state-centric interests limited their input. Informal contacts in Case D2 utilised their skills to operate services from a community health centre, including job advice for minorities, occasional training, business start-up advice, community development and childcare. However, the precariously funded participants were not primarily representing unemployed people; rather, participants used networks to demonstrate various skills in order to embed security for themselves.

Preferred Network Outcomes

This section analyses a preference survey based on concrete examples of network outcomes that assisted interviewees to achieve an unemployment policy goal in the ward clusters. From the codified transcripts, there are 79 statements of outcomes, analysed from 86 interviews. Results were organised at three levels (neighbourhood, network and organisational/ participant) per local authority and network and outcome type, and citation frequency. Of those interviewed, 8 % were unable to provide an example (four from LBTH and three from GYBC). The results are largely associated with networks perceived to be effective (see Tables 7.3 and 7.4).

Similarities and Differences

Interestingly, only one interviewee mentioned the outcome of 'connecting people to employment'. More examples were focused at borough-wide than at ward-based level. Some outcomes were ambiguous or in transition, while some actors struggled to provide an example, thereby implying operational slippage and inappropriate network membership. Several outcomes had weak association with local unemployment, which suggests actors lacked experience or capacity for understanding unemployment needs, or that their roles were misaligned, invalid or under-utilised. Some examples described indirect support, such as establishing a learning centre or developing a sub-network for ideas.

Citation results for neighbourhood-level outcomes were $N = 12$ (LBTH) and $N = 13$ (GYBC). The former had more mention for training courses and outreach, and work placements and information in the latter. Results for the network-level outcomes were $N = 11$ (LBTH) and $N = 11$

(GYBC). Network discussion and bringing agencies together were mentioned more often in the former, while the latter mentioned coordinating the electronics sector, referral procedure and network development. Results for organisational/participant outcomes were $N = 12$ (LBTH) and $N = 20$ (GYBC). Voluntary sector representation, funding acquisition and staff training were cited in the former, and the latter mentioned contracts, new assets, funding support for private sector jobs and business representation.

Overall, funding acquisition ($N = 5$) and staff training ($N = 3$) were mentioned more often than IT training for job applications ($N = 1$) or supporting links between school, employers and vocational training ($N = 1$). More outcomes supported organizational- and network-level outcomes in GYBC, particularly those associated with funding, possibly because funding is perceived as difficult to attain outside of cities and participants depended on new projects for job security. Jobs scarcity in GYBC also led to more outcomes for information and guidance, work placements and business start-ups. Representatives of electronics companies chose grants for staff training on the basis they helped to sustain jobs in Yarmouth. One interviewee avoided giving a straight answer, but finally offered the example of a network that they had been developing for two years, but still had not launched: '... the (em) learning systems group for the East. We would establish this, and it will happen whether or not people came on board, so that one is much more of a formal process in the sense that we would drive it to existence'. Further, a GYBC officer could offer only an example of a network in development, which was in the process of establishing a URC. Another network in development involved a Norwich-based organisation supporting people with transport barriers for which the nominee intended to commission services and declare their board membership in the tendering process.

One official said that LAP 7 had brought actors together but, to date, had not delivered anything concrete. A voluntary sector representative chose as an example a European Social Fund (ESF) network, now defunct, that had delivered business enterprise support. However, the unstable network policy field had disconcerted the actor, and the following long excerpt is justified on the grounds of its descriptive power:

> ...my impression is ... if I go out there now and say ... let's come together and form a partnership and develop some project ideas they will probably respond positively and come together to one meeting, or two meeting [*sic*]... but ... if in five months they realise there is nothing concrete on the horizon

in terms of funding opportunities it will probably disperse. Because for all the regeneration effort, for all the unemployment and employment activities that happens in the area for the funding that comes through, there is still a distinct lack of strategic coming together, and strategic thinking needs to be directed, needs to be guided. The local authority is not doing that and the voluntary sector is not well organised enough to do that. The funding that is coming to the area for these activities from LSC, from LDA [London Development Agency], from [the] local authority, from Europe, are isolated, they're not coming joined-up, they are coming in their own way and they have all got their own criteria and systems and procedures, so people are responding to them in their own ways. It's not exactly conducive to people working together for a long time. JCP may come in and force a couple of organisations to work together, and another partnership will come in and force the partnership to go separate ways, that's what's happening so it is difficult.

Evidently, preferred outcomes were nested within the limits of funding strategies, over which participants had little control. These results are not a scientific summary; however, they demonstrate a concentration of self-interested priorities and pressure points arising from local culture. The combined results show that preferred outcomes were focused at network and organisational level more than at neighbourhood level.

Conclusion

This chapter surveys the perceived network performance, case network outcomes and actors' preferred network outcomes in case areas. Reputation scores were highest for state-mandated formal networks with large controlling structures for place-based policy coordination, high visibility, power resources and institutional backing. However, network effectiveness scores did not correlate with high-reputation networks. Rather, actors chose self-selected networks that utilised their skills and expertise, increasing their prospects for self-promotion and personal power. Half the interviewees mentioned ineffective networks, while the remainder were unable to comment or make a judgement. In both case areas, state-led networks had low effectiveness scores and fell short of expectations.

The analysis of case network outcomes draws on indicators and empirical referents. Neighbourhood-level outcomes supported positive activities, from projects that expanded information provision to targeted employment support and provision of community assets, although often without a legacy from which to run them. Outcomes depended on organisations' ability to promote themselves in formal networks and compete for funding, an unsatisfactory process given the lack of neighbourhood advocates in governing networks and many dead-weight unemployment representatives with low local knowledge. Network-level outcomes ranged from the more successful, such as image-building self-referencing networks, separating professionals by job status in hierarchical governance tiers for policy coordination, and reputational positioning to assist national and regional programmes, to the less successful, which displayed a lack of formation and the transfer of responsibilities. Organisational-level outcomes focused less on joint working as policy influence was slight, given that quangos functioned above these organisations, and the funding streams induced competitive behaviour necessitating the promotion of self-interest, power assertion and reputation-building. In Great Yarmouth, the private sector withdrew from governing networks as the public sector dominated. In Tower Hamlets, business leaders were distant, and the voluntary sector felt used by the public sector and the private sector providers.

Case networks could not transform local economies. Central initiatives funded networks to implement dogmatic neo-liberal policy that increased competition between local actors and organisational compliance with national priorities; suitable jobs were unavailable and job creation remained insufficient. Socially suboptimal network outcomes include problem avoidance, disruption of local relations and increased planning to track the multiple short-term projects promoting departments and institutional reform that confused workers and citizens alike. There was no uniform direction for tackling social and economic barriers to work.

Finally, participants' preferred outcomes reflected self-nominated interests in wide-ranging niche networks as opposed to networks coordinating place-based policy. None of the concrete examples could significantly reduce unemployment rates. Future networks could select performance criteria applicable to policy and outcomes desired at three impact levels: neighbourhood, network and organisational/participant. The final chapter draws conclusions from the results in order to directly answer the research questions.

Bibliography

East End Life. (2005, 24 January–6 February). *Local area partnerships.* Tower Hamlets: Municipal Newspaper, p. 11.

Fox, P. (2008). *Stocktake of the Great Yarmouth local strategic partnership: A consultation paper.* From the interim local government re-organization adviser to the GYBC. Retrieved from http://www.communityconnections.org.uk/

Great Yarmouth Local Strategic Partnership (GYLSP). (2005a, December 8). *Minutes of the Social Forum.*

Great Yarmouth Local Strategic Partnership (GYLSP). (2005b, September 15). *Minutes of the LSP Economic Forum.*

Great Yarmouth Local Strategic Partnership (GYLSP). (2006, March 9). *Minutes of the Economic Forum.*

Great Yarmouth Local Strategic Partnership (GYLSP). (2008a, September 11). *Minutes of the LSP Economic Forum.*

Great Yarmouth Local Strategic Partnership (GYLSP). (2008b). *Great Yarmouth Neighbourhood Renewal Fund final report on 2006–2008 achievement.* Great Yarmouth.

London Borough of Tower Hamlets (LBTH). (2002). *Community economic development plan, 2002–06.* Retrieved from http://www.towerhamlets.gov.uk/

London Borough of Tower Hamlets (LBTH). (2008, May 6th). *Report of the scrutiny working group evaluating Neighbourhood Renewal Funding.* Overview and Scrutiny Committee Agenda Item 9.3., pp. 105–144.

London Borough of Tower Hamlets (LBTH). (2010). *Overview & scrutiny committee, 9 March 2010, supplemental agenda 9.1 report of the scrutiny review working group on reducing worklessness amongst young adults 18–24.*

PWC. (2014). Best value inspection of London Borough of Tower Hamlets report 16 October 2014. PricewaterhouseCoopers LLP. Retrieved from http://pwc.co.uk

Tyler, P. (2011). What should be the long-term strategy for places with patterns of decline and underperformance. In P. Lawless, H. G. Overman, & P. Tyler (Eds.), *Strategies for underperforming places* (pp. 36–51). SERC Policy Paper 6. London: SERC.

CHAPTER 8

Conclusions: Modelling Suboptimal Network Outcomes

This final chapter draws together evidence, and delivers a pessimistic line of argument. Difficulties for networks are inextricably linked to weak economic thinking and to institutions that are detached from the local labour market realities. Our findings could galvanise the creation of counter-resolutions for socially optimal network outcomes.

This study defines the 'network impact' concept and orientates networks in the British policy process. It claims that the literature pays more attention to concept understanding at the organisational/participant- and network-level than at the level of neighbourhoods and citizens. Since the early 1990s, landmark governance literature has normalised network ideation of governance interaction for democratic participation. To accommodate governance, network theories have expanded in disciplines from political science and social policy to public policy. Yet scholars have shown weak problem-solving crossover, have lost sight of labour realities, have avoided direct questions of economic democracy and have ignored classic accounts of imperfect policy implementation (see Pressman and Wildavsky 1973; Hanf and Scharpf 1978; Hogwood and Gunn 1986: 198–209; also Lane 1993, Chap. 4; Papadopoulos 2003). In effect, governance roles and networks support 'job clubs' in which 'outsource specialists', 'experts' and 'representatives' deliver welfare-to-work, while sustainable jobs are eroded in the deregulated labour markets and labour matters lack network representation.

Ostrom (1990: 7) claimed that scholars of collective action 'would rather address the question of how to enhance the capabilities of those involved to change the constraining rules of the game to lead to outcomes other than remorseless tragedies'. Stoker (1991: 261–8) identified future models of local governance as hierarchical, market and network mechanisms enacting policy through participatory democracy and organisational cooperation. The Rhodes model views policymaking analysis and policy change at three explanatory levels: intergovernmental relations (macro-level), policy network relations (meso-level), and actors' network behaviour (micro-level) (see Rhodes and Marsh 1992a: 12). Kooiman (1993: 2) distinguishes governing, governance, governability and 'meta governance' for strengthening primary relations for public purposes across sector domains. The Dutch school emphasises the prospects for policy networks steered by public management processes (Kickert et al. 1997). Danish scholars deliberate interactive network governance, as applied to employment policy (Sørensen and Torfing 2008; Damgaard and Torfing 2010). Davies (2011: 151) regards network governance as 'spearheading neoliberalism'. Yet even critical texts lack substantive network impact research on unemployment (with exceptions; see Moore et al. 1989). As Moore et al. (1989: 134) comment, 'effects should be seen not only in quality of jobs created, but in broader political, social and possibly ideological terms'.

Likewise, networks have only been able to handle internal labour market problems in a piecemeal way. Four theories remain relevant (Rosenbaum et al. 1990). Segmented labour market theory suggests that primary and secondary labour markets diminish the chances of certain groups, such as women, minorities and youth. Human capital theory refers to the work-entry efforts that people make to improve their capability and work chances, but this depends on external market forces, information flow and policy reform. Signalling theory concerns the limited information and empirical evidence employers use to 'signal' skill shortages and select job applicants, as despite norms, such factors as age, qualifications and schools attended may be misunderstood or used to discriminate. Finally, network theory concerns the use of institutional linkages to improve job entry conditions, such as cooperation between the state, employers and trade unions. However, while governance operations and growth policies add context and impetus, theories applied have not led to plentiful jobs across all parts of Britain.

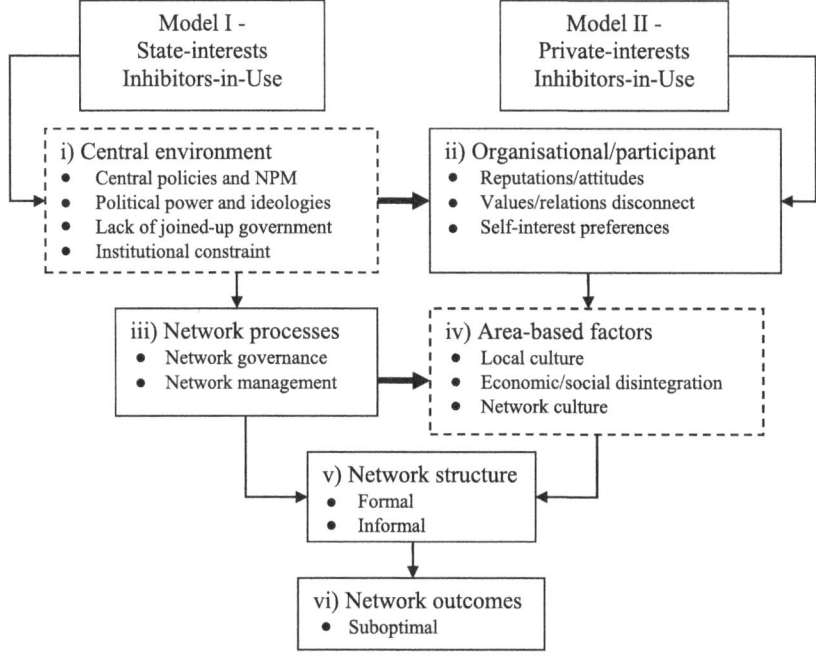

Fig. 8.1 Two explanatory models of factors and key variables that impede network outcomes

Gateways to Suboptimal Network Outcomes

Two 'gateways' model suboptimal network outcomes, see Model I (exogenous type), State Interests, and Model II (endogenous type), Private Interests in Fig. 8.1. The models interrelate in a sequence and depict causal processes and causal direction of inhibiting factors and variables.[1] The 'central environment' and 'area-based factors' offer plausible accounts and predict network structure and network ineffectiveness. Models are

---

[1] The models in Fig. 8.1 resembles exogenous and endogenous group characterisation (see Dunleavy 1988: 42–7). The former refers to a fixed type of membership shaped by external factors. The latter refers to like-minded people sharing similar preferences and is self-selected membership.

policy- and context-dependent; only two local authorities and one policy area were tested, thus they cannot be completely accurate as a theory of outcomes.

Model I

This model resembles the rational scientific management tradition. The state coerces networks to deliver central goals. Hierarchical power flows from the 'central environment', represented by the thick one-directional arrow, to the 'organisational/participant level'. Participants, coerced to play the rules of the game, gain reputation and resources, but joined-up working may be falsely represented and obscure the real needs in localities. The 'network processes' and 'area-based factors' are no less influential in the causal mechanism and manipulate 'network structure'; however, to a greater extent the former is affected by the 'central environment' and the latter depends on the capacity and behaviour of 'organisational participants'. The thick one-directional arrow flowing from the 'network processes' to the 'area-based factors' represents the intentional power of the state to manipulate localities instrumentally through decision-making structures, network governance and network management. Network structure (formal or informal) is not a stand-alone predictor of failure but an outcome of inhibiting factors, as findings indicate:

Model I Factors
   (i)   *The overplayed central environment*: central policies instigate network change down a vertical chain of influence and departmental competition tightly controls political preferences. State-led interventions and funding streams ensure network structures and processes maintain capitalist hegemony, privatise welfare and endorse self-help and growth models. Institutional bureaucracy, however, stymies employers' engagement in central skills interventions, apprenticeships and learning systems.
   (ii) *The organisational/participant level (a narrow outlook)*: actors appear together in high-reputation networks but mainly work apart, adapting organisational interests to funding policy. Strategists are powerful on one level but weak on others. Values are not always shared between organisations or complementarities recognised. Socio-economic responses are rare. Attitudinal orientations, cognitive filters and 'belief traps' skew representations – some

business figureheads assume unemployed people are 'work-shy'. Participants focus on jobseekers' inadequacies, not jobs supply or coordinating firms.

(iii) *Network processes depoliticise governance*: network governance evades politically contentious positive change. Over-hyped network capacity masks joined-up government problems. Separating elites from community roles prevents cross-talking, and deters political games and decision-making veto. Funders depoliticise policy implementation and shape resource allocation details aside from organisations so as to avoid conflicts of interest should any of them bid in future funding rounds. Funding themes change network direction, but none have proven adequate to address unemployment and persistent economic uncertainty.

(iv) *Area-based factors, difficulties and neglect*: the networks barely considered how local culture, history and the quality of local services reflect the limits of the market. Funding rounds with short lead-ins substitute for a lack of local planning and distract from environmental limitations or a weak local economy. Stakeholders coerced into centralised networks have a minimal interest in local unemployment dimensions. Outcomes are uneven or favour borough-wide responses. Community assets are the exception and are a legacy of former networks or the influence of different policy fields.

(v) *The extent network structures support localities*: a continuum of formal state-mandated network structures governed and institutionalised relations since the 1980s. These governing networks aimed to decrease sector reliance on local authorities, professionalise and bureaucratise neighbourhood networks and increase reliance on central agendas, from coercing policy innovation to giving power as a reward. The spotlight on self-referencing grand structures could not help the elites or underprivileged groups to overcome weak connectivity.

(vi) *Network outcomes*: network-level image building masks weak unemployment representation, problems coordinating elites and local problem avoidance. Information overload, continuous reporting to different structures and pre-determined network funding reduces participants' policy influence. Short-term initiatives confuse workers, organisational roles and citizens alike. Imposing network structure to steer and expand national and institutional priorities through power assertion creates conflict. Network funding to privatise community services, reduced smaller organisations political voice. Organisations adapt to the competition culture by increasing opportunistic and isomorphic behaviour, and delivering programmes, some with values they do not always support.

Model II

Organisational participants drive the causal pathway in this model. The one-directional arrow flows downwards from the 'organisational/participant level' to the 'external factors'. Actors progress interests in self-determined networks that are less dependent on central network processes.

Model II Factors

(i) *Organisational/participant level priorities*: contrary to Model I, actors shape self-organising networks reflecting their area of concern, niche interests, expertise and contextual circumstances, supporting self-affirming roles, agency, autonomy and values. They do this to anticipate opportunism, control competition and focus on outcomes they prefer.

(ii) *Difficulties arising from the perceptions of area-based factors*: professionals tend to avoid neighbourhoods with environmental degradation, conflict or 'politics' and dissuade others from concerted network action, as place association can prejudice or damage career reputations. Indeed, career promotion is usually rewarded to those who transfer from micro to macro level. Hence, neighbourhoods are vulnerable to neglect or uneven development and under-privileged organisations defending neighbourhoods often have limited network participation and reduced visibility.

(iii) *Difficulties arising from the multiplicity of network structures*: actors identify structures beyond case network typology, including 'network organisations', meaning self-reliant community entities matching services to local need. Actors prefer network pluralism, but this relies on ad hoc leadership and makes coordination difficult. Nevertheless, the state dismisses community leadership in a central mandated structure.

(iv) *Network outcomes*: self-maximisation, reputation maintenance, opportunities for job security, rewards and recognition, competition and specialisation produce low tolerance, a smaller worldview and inflexibility. Then again, the creative individualist impulse in networks is the lifeblood some localities depend on.

Implications

Actors desired to work together in self-affirming networks to sustain private interests and organisational reputations. Localities were of less interest, and, in a pronounced sense, they were victims of self-referencing networks.

These ranged from the state-interested corporate networks to the self-interested pluralistic networks. Neither type sufficiently addressed unemployment. Networks align themselves to the cultural, sectoral differentiation and neighbourhood problems, but policymakers and network participants need to consider policy limits. Many actors in neighbourhoods have limited opportunities to participate in networks and represent their clients' support needs in relation to the different dimensions of unemployment. Resource-rich networks are not accountable to the unemployed but to their central funding tenure, and thus actors have a tendency to protect their private interests, institutions, contracts and reputations. As such, governance networks require geocentric orientation (reconciling multiple worldviews and values). This is no easy task, but the problems identified in this book provide a template for understanding how central funding, joined-up governance policies and institutions' 'tunnel vision' can hamper outcomes, and why actors often work in a silo mentality or prefer self-determined networks.

Epilogue to the Fieldwork

The grand aim of improving and sustaining economic performance in English regions by 2008 and reducing the persistent gap in growth rates was initially a joint target of HM Treasury and the Department of Trade and Industry (DTI), alongside English regional assemblies. Responsibilities later transferred to new departments (see Chap. 4), but the continuous reform that has taken place suggests departments were assigned problems they could not solve, and that their lack of joined-up government led to a coordination problem on the ground. Institutions and organisations also failed to operate through case networks for two reasons: (i) a crisis of competence (adequate authority but lacking competence), and (ii) a crisis of authority (adequate competence but lacking authority to implement or participate in a policy domain) (Alexander 1995: 176 citing Lehman 1975: 95–6, 303). Subsequently, government requested that local authorities work harder to reduce unemployment through employment subgroups and employment strategies, but with no additional funding. Actors participating in these subgroups were interviewed for this study. Certainly, growth was stronger in both regions compared to others; however, ward statistics beyond the fieldwork period show that *despite* the presence of these networks, unemployment increased in the case local authorities, then decreased from 2012 but ESA claimants increased (see Tables 5.1 and 6.1).

From 2005, 'unemployment' became a remit of neighbourhood management approaches, and case areas organised local provider coalitions to deliver the Working Neighbourhood Fund (WNF). In 2010, the Conservative and Liberal Democrat coalition government changed the network map again, replacing Labour's national targets, Regional Development Agencies (RDAs) and Local Strategic Partnerships (LSPs) with the 'business-led' subregional Local Enterprise Partnerships (LEPs). Local authorities remain subservient to regional governance. The 2007, 2010 and 2015 Indices of Multiple Deprivation (IMD) rank the London Borough of Tower Hamlets (LBTH) as the third most deprived local authority district in the country for extent of deprivation, whilst Great Yarmouth Borough Council (GYBC) moved from thirtieth position to the fourth most deprived area for local concentration of deprivation, and shares eighth position for proportion of working-age adults in employment deprivation (DCLG 2010; 2015).

Using the example of Great Yarmouth, local governance eroded further when WNF providers were replaced by Prime Providers who lack accountability to the Council's Economic Reference Group (formerly Case A, the 'Economic Forum') or the voluntary sector-led Employment and Skills Group (currently inactive). The New Anglia Local Enterprise Partnership (NALEP) secured Enterprise Zone status for Great Yarmouth and Lowestoft and Regional Growth Fund (RGF) has been used to regenerate unused land for renewable energy development. The EZ reportedly created 1,895 jobs to February 2016 (includes construction), and the most recent NALEP report suggests 16 EZ businesses are supporting 350 jobs (NALEP 2016: item 5, sec 2.: 4). Despite repeated enquiries in 2015, however, the NALEP has not disclosed how many jobs were taken up by Great Yarmouth residents. Reliance on growth to redress unemployment was perhaps naive, as only three Great Yarmouth businesses accessed match funding investment grants from the NALEP Growing Business Fund, despite a £3 million underspend defrayed until the end of 2015 for the LEP area. The NALEP Growth Hub has only delivered 4 percent of intensive support in Great Yarmouth, compared to 24 percent in South Norfolk. The National Audit Office (NAO) has already criticised RGF expenditure on 'projects that offered relatively few jobs for the public money invested' (NAO 2012: 9). The Committee of Public Accounts (CoPA) considers that government initiatives overseen by LEPs have fallen short of expectations (see Fig. 2, CoPA 2014: 12). Nevertheless, the NALEP, whose annual running costs are circa £800,000, expects a £2 billion wind turbine project off the southern Norfolk coast from 2017 to create 3000 jobs. Furthermore, it aims to deliver 'world-class' industry, institutions, facilities, locations, and skills clusters, from agri-tech to green energy. Yet much less is said about services for citizens on the ground, such as social investment in a well-trained caring profession.

In Tower Hamlets, new transport links have boosted growth, the employment rate has improved and 43.6 % of the Tower Hamlets working-age population (aged 16-64) are qualified to degree level and above (LBTH 2014: 4). Nevertheless, only 0.8 % of these had achieved apprenticeship and 15.6 % of Tower Hamlets working age residents had no qualification according to the census in 2011 (2014: 4). In May 2013 25,650 people in Tower Hamlets claimed out-of-work benefits, including 12,260 Employment and Support Allowance (ESA) or Incapacity Benefit (IB) claimants, of whom 46 % have mental health problems. From the 21,000 residents who would like to work, 10,346 are Jobseeker's Allowance (JSA) claimants, 4250 are unemployed but not claiming benefits, and 6300 are economically inactive. Tower Hamlets residents would need 13,300 jobs to meet the London employment rate (LBTH 2011: 34). Although Tower Hamlet IMD employment indices improved from thirty-eighth rank in 2010 to seventy-sixth in 2015, income deprivation increased from tenth rank in 2010 to sixth in 2015, and first for proportion of children (39.3 %) and older people (49.7 %) living in income-deprived households (DCLG 2010; 2015). The Tower Hamlets Community Plan 2015 seeks a 'fair and prosperous community',' yet the community economic impetus is not a strong campaign to redress the benefits trap and housing costs, or 'heed residents' concerns about the impact on existing communities of the large number of high value high-rise homes being built' (THP 2015: 06).

The Tower Hamlets Partnership (THP) continues to drive community plan delivery groups (a children and families partnership board; a community safety partnership board; an economic task force; a health and well-being board; a housing forum) and a network of subgroups for local partnership work (2015: 23). Community ward forums and community champion coordinators seek ways to support citizens, from digital inclusion to involving them in decision-making in health services. A cross-partnership of providers intends to pilot localised integrated employment services responding to the benefit cap and barriers to employment, from housing and health, basic skills, money management to affordable childcare (2015: 11, 28). Arguably, partnerships between institutions and providers for trust and reputation-building are not robust economic solutions for citizens on the ground, and scope for job creation remains limited.

Government has spent billions on speculative business activity; subsiding firms, equipping industry, supporting thousands of SMEs, investing in high value skills, automated technologies, transport improvements, new college facilities and housing plans. ALMPs, behaviour change programmes for people with 'complex needs' and 'radical' schemes targeting NEETs are constantly reinvented. As this book is going to press, the number of people unemployed in 2016 is higher than when this study began, over a decade before. The unemployment rate fell to 4.8 % (1.60 million ILO measures), but the claimant count increased 14,000 from the previous quarter (Apr-June 2016) to 803,300, and, 9,900 more than for a year earlier (ONS 2016). Between May-Oct 2016, the unemployed claimant count increased in three of four case study ward clusters (ONS Neighbourhood Statistics). NEETs have also increased, compared to the previous quarter and previous year (Jul-Sept 2015). Job vacancies decreased between Nov–Jan 2016 and Jul-Sept 2016, from 763,000 to 752,000 or 2.1 unemployed heads per job vacancy. 4.79 million people are self-employed and numbers are rising, the trend responds to structural change in the labour market and a form of disguised unemployment, as many self-employed people want more hours. Since the Work Programme launch in 2011, 1.56 million people completed time on the scheme but 65% returned to JCP, and only a quarter were still in work after two years (DWP 2016). 434,000 people have been unemployed for more than one year. Hence, we need to look beyond faceless statistics and count the real cost of unemployment impacting people at neighbourhood level. Networks could support system change for more jobs and equitable job structure; otherwise, unemployed people will be forever encouraged to take jobs with poor conditions and which prop up capitalist systems of injustice (Crisp 2008: 183).

Investigative Limitations and Future Research Directions

Network accounts are inevitably selective and incomplete, and findings in other policy areas and locales may differ. A comparative study of the shift towards governance in education policy and local economic development policy in England and France, for example, has confirmed that place history and culture differentiate governance outcomes (Cole and John 2001). This book argues that governing style and network culture differ according to civic culture, but rather government policy narratives dominate the governance landscape. The role of LEPs, for example, appear uncertain given the drive for English devolution, such that local authorities in Cambridgeshire and Peterborough, and Norfolk and Suffolk have been asked to form two new East Anglia combined authorities tied to an elected mayor and a finance package for overseeing transport, strategic planning and skills training. Notably, Norwich City, Great Yarmouth, Breckland and North Norfolk have rejected proposals. Yet governance problems appear secondary to the role of the deficient unemployment policy that networks handle. Weaknesses in the orthodox economics should not be ignored. More case studies in the policy field in different locales would increase the possibility of generalisation.

The book did not privilege structure, agency or process, one above the other, and theoretical synthesis is relevant. Future studies, therefore, could build on the factor approach. The central environment factor questioned the state's influence in network inter-relations and support for social and economic regeneration. Policy worked to minimise the effects of unemployment rather than its economic causes, and local networks, interest groups and unemployed people lack influence over macro-economic policies. Analysing network structure provides an important reality check for monitoring who participates in networks, their reputations, inter-relations and preferred structures. Network structure changes over time; hence, longitudinal studies can help to track power shifts in strong and weak ties combined with the use of qualitative and quantitative methods. Organisational/participants' competitive behaviour conflicts with actors' inherent desire for collaboration and support for joined-up working. Elites in this study, for example, did not establish strong ties. A nuanced study might explore these facets amongst classes of employees without losing sight of outcomes and attitudes towards beneficiaries. Schneider et al. (2003: 152, 155) considered how environmental policy networks bridged the ideological cleavage between pro-development business groups and slow-growth environmental interest groups; they also call for more systematic research on the impact of local networks on agency outcomes. Network processes highlight coordination deficiencies caused by network-centric governance and network management of private governance. The research might return to earlier concerns in the literature over governance accountabilities, such as networks being a form of private government, not responsible government (Rhodes and Marsh 1992b: 203). Future studies should determine how policy and network processes might harness better outcomes for citizens. Finally, area-based factors conjoin with other factors to consider how ward-based culture shapes local networks. Local workers facing real-world problems will inevitably question the authenticity of network studies that overlook this dimension.

Final Comments

Two factors appear to undermine overall network performance and indicate mismatches between Whitehall and actors on the ground. A health warning applies since these are mostly qualitative impressions. First, the central environment maintains unemployment to keep markets functioning and prevent inflationary pressure on wages. Networks embody the ideology of market economies, and their relative failure to achieve outcomes for combating unemployment might be considered to reflect this. The conundrum is indicative of the economic challenges of integrating a capitalist system of production with a socialist system of distribution and speaks directly to the empirical puzzle. Institutions, unable to transcend their structural problems, resort to self-referencing policy responses to legitimise their roles, or they combine 'strategies of self-interest with interpretations of the public interest' (Burns et al. 1994: 74–5). In the process, the citizen and neighbourhood focus decreases.

Second, the local culture, stemming from history, socio-economic conditions and politics, shapes the network culture and predicts structure, agency and outcomes. Future networks, whether place-orientated (resource-poor) or state-orientated (resource-rich), would need to audit the network conditions that are making neighbourhood outcomes difficult to achieve and consider how participants might apply their self-interest, resource capacity and positional influence more effectively at the level of neighbourhood problems. Yet cases suggest that professionals coerced into sustaining network hierarchies have limited connectivity, and that voluntary networks closer to the ground sustain core services rather than joined-up working. Network governance fills gaps around the table, but weak alliance-building impacts on unemployment representation and network processes can distract from neighbourhood-level outcomes. Short-term network initiatives were stimuli for a hotchpotch of trial-and-error welfare support or supply-side provision and central publicity. Yet their networking efforts did not remedy the long-term problems, and a lack of legitimate governance and a quick-fix approach to expenditure created conflict. Actors' preferred networks were policy-sector specific, exploratory or involved peer support, although governing and contractual networks were of less interest.

Labour aimed to bring 'Britain together' and tackle 'silo working' through network governance and joined-up working policy. Its network campaign brought forward organisations with allegiances to central policies. The Conservative and Liberal Democrat coalition government (elected in May 2010) usurped Labour's network infrastructure and put business-centred networks at the forefront of local growth initiatives. These structures remain a 'work in progress' under the Conservative government (elected in May 2015), with a greater emphasis on funding apprenticeships and reducing youth unemployment. However, networks cannot remedy persistent unemployment without unifying collective interests with self-interest and constructing responses other than blaming, as even within the framework of Conservative ideology, the community responsibility agenda does not communicate with Britain's labour market. Future studies should continue to ask why the impact of networks has been negligible for citizens in places with weak economies, but also in neighbourhoods adjacent to some of the richest districts in the country. Industrial policy for economic growth and technological advancement cannot provide enough jobs for everyone who wants to work. Self-interests tarnish the capitalist system. Market turbulence depletes contributions to the public purse. A work on-demand economy, increasingly coordinated through digital technology lacks employment rights in sight of a social contract. The working poor need the right to earn a decent living, moreover, unemployed people need jobs not mandatory work schemes tied to a welfare state industry. To this end, employment policy and governing networks need a seismic shift to address unemployment.

Bibliography

Alexander, E. R. (1995). *How organizations act together*. Luxembourg: Gordon and Breach.
Burns, D., Hambleton, R., & Hoggett, P. (1994). *The politics of decentralisation*. Basingstoke: The Macmillan Press.
Cole, A., & John, P. (2001). *Local governance in England and France*. London: Routledge.
Committee of Public Accounts (CoPA). (2014). *Promoting economic growth locally, sixtieth report of session 2013–14*. HC 1110. London: The Stationery Office.
Crisp, R. (2008). Motivation, morals and justice: Discourses of worklessness in the welfare reform green paper. *People Place & Policy Online, 2*(3), 172–185.

Damgaard, B., & Torfing, J. (2010). Network governance of active employment policy: The Danish experience. *Journal of European Social Policy, 20*(3), 248–262.

Davies, J. S. (2011). *Challenging governance theory: From networks to hegemony.* Bristol: Policy Press.

Department for Communities and Local Government (DCLG). (2010). *Index of deprivation 2010.* Retrieved from https://www.gov.uk/government/statistics/english-indices-of-deprivation-2010

Department for Communities and Local Government (DCLG). (2015). *Index of deprivation 2015.* Retrieved from https://www.gov.uk/government/statistics/english-indices-of-deprivation-2015

Department for Work and Pensions (DWP). (2016). Quarterly Work Programme national statistics to Mar 2016. London: DWP.

Dunleavy, P. (1988). Group identities and individual influence: Reconstructing the theory of interest groups. *British Journal of Political Science, 18*(1), 21–49.

Hanf, K. I., & Scharpf, F. W. (Eds.) (1978). *Interorganizational policy making: Limits to coordination and central control.* Beverly Hills: Sage.

Hogwood, B. W., & Gunn, L. A. (1986). *Policy analysis for the real world.* Oxford: Oxford University Press.

Kickert, W. J. M., Klijn, E.-H., & Koppenjan, J. F. M. (Eds.) (1997). *Managing complex networks: Strategies for the public sector.* London: Sage.

Kooiman, J. (Ed.) (1993). *Modern governance: Government-society interactions.* London: Sage.

Lane, J.-E. (1993). *The public sector: Concepts, models and approaches.* London: Sage.

London Borough of Tower Hamlets (LBTH). (2011, April). *Tower Hamlets employment strategy.*

London Borough of Tower Hamlets (LBTH). (2014) *Understanding skills and qualification levels in Tower Hamlets.* Research briefing. June 2014. Tower Hamlets.

Moore, C., Richardson, J. J., & Moon, J. (1989). *Local partnership and the unemployment crisis in Britain.* London: Unwin Hyman.

NALEP. (2016, April 20). *New Anglia Local Enterprise Partnership Board. Agenda Item 5, Managing Director's Report April 2016.* Centrum, Norwich.

National Audit Commission (NAO). (2012). *Regional growth fund.* HC 17. London: The Stationery Office.

Office for National Statistics (ONS) (2016, November). *UK Labour Market: Nov 2016.* Statistical Bulletin.

Ostrom, E. (1990). *Governing the commons: The evolution of institutions for collective action.* Cambridge: Cambridge University Press.

Papadopoulos, Y. (2003). Cooperative forms of governance: Problems of democratic accountability in complex environments. *European Journal of Political Research, 42*(4), 473–501.

Pressman, J. L., & Wildavsky, A.B. (1973). *Implementation: How great expectations in Washington are dashed in Oakland or, why it's amazing that Federal programs work at all, this being a saga of the Economic Development Administration as told by two sympathetic observers who seek to build morals on the foundation of ruined hopes.* Berkeley: University of California Press.

Rhodes, R. A. W., & Marsh, D. (1992a). Policy networks in Britain: A critique of existing approaches. In D. Marsh & R. A. W. Rhodes (Eds.), *Policy networks in British government* (pp. 1–26). Oxford: Clarendon Press.

Rhodes, R. A. W., & Marsh, D. (1992b). New directions in the study of policy networks. *European Journal of Political Research, 21,* 181–205.

Rosenbaum, J. E., Kariya, T., Settersten, R., & Maier, T. (1990). Market and network theories of the transition from high school to work: Their application to industrialized societies. *Annual Review of Sociology, 16,* 263–299.

Schneider, M., Scholz, J., Lubell, M., Mindruta, D., & Edwardsen, M. (2003). Building consensual institutions: Networks and the national estuary program. *American Journal of Political Science, 47*(1), 143–158.

Sørensen, E., & Torfing, J. (2008). Introduction: Governance networks research: Towards a second generation. In E. Sørensen & J. Torfing (Eds.), *Theories of democratic network governance* (pp. 1–21). Basingstoke: Palgrave Macmillan.

Stoker, G. (1991). *The politics of local government* (2nd ed.). Basingstoke: Macmillan.

THP (Tower Hamlets Partnership). (2015). *Community plan 2015.*

Erratum to: The Impact of Networks on Unemployment

J.M. Hurst

Erratum to:
J.M. Hurst, *The Impact of Networks on Unemployment*,
DOI 10.1057/978-1-349-66890-8

The Author information and some content was printed in error in October 2016 and has been removed from the book and the content updated in December 2016.

....

The online version of the updated original book can be found under DOI 10.1057/978-1-349-66890-8

....

J.M. Hurst, *The Impact of Networks on Unemployment*,
DOI 10.1057/978-1-349-66890-8_9

Appendix 1: UK Unemployment Rates and Trends

Claimant count historical estimates 1881–2015 (experimental statistics)

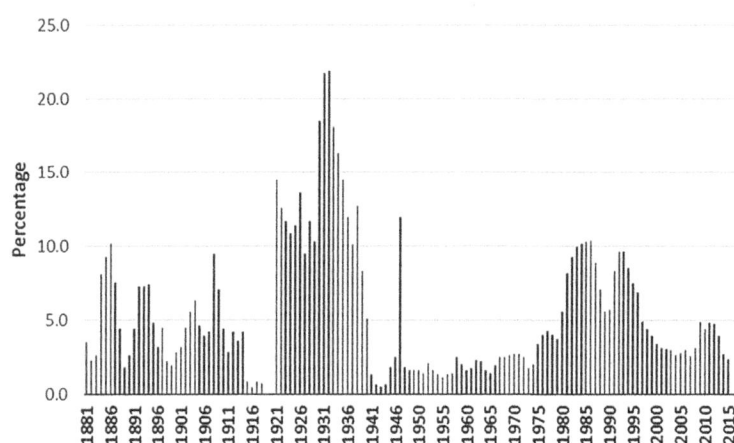

Source: ONS (http://www.ons.gov.uk).

Appendix 1: UK Unemployment Rates and Trends

Unemployment rates in the UK 1973–2015 (percentage of total labour force)*

Year	Rate	Year	Rate
1973	3.7	1997	6.9
1975	4.5	1998	6.2
1980	6.8	1999	6.0
1981	9.6	2000	5.4
1982	10.7	2001	5.1
1983	11.5	2002	5.2
1984	11.8	2003	5.0
1985	11.4	2004	4.8
1986	11.3	2005	4.8
1987	10.4	2006	5.4
1988	8.6	2007	5.3
1989	7.2	2008	5.7
1990	7.1	2009	7.6
1991	8.9	2010	7.9
1992	9.9	2011	8.1
1993	10.4	2012	8.0
1994	9.5	2013	7.6
1995	8.6	2014	6.2
1996	8.1	2015	5.4

*Standardized to accord with the ILO definition of unemployment.
Source: ONS (http://www.ons.gov.uk).

Labour Force Survey: ILO unemployed over 12 months in the UK 1992–2015 aged 16–64*

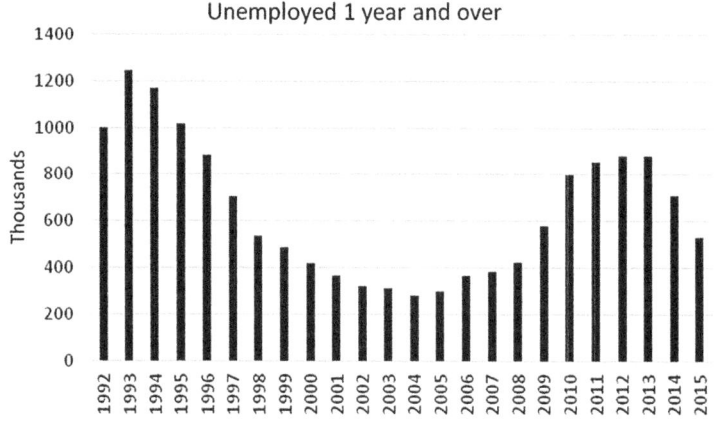

*Seasonally adjusted
Source: ONS (http://www.ons.gov.uk).

APPENDIX 1: UK UNEMPLOYMENT RATES AND TRENDS 299

Labour Force Survey: ONS: All unemployed in the UK 1992–2015 aged 16–24*

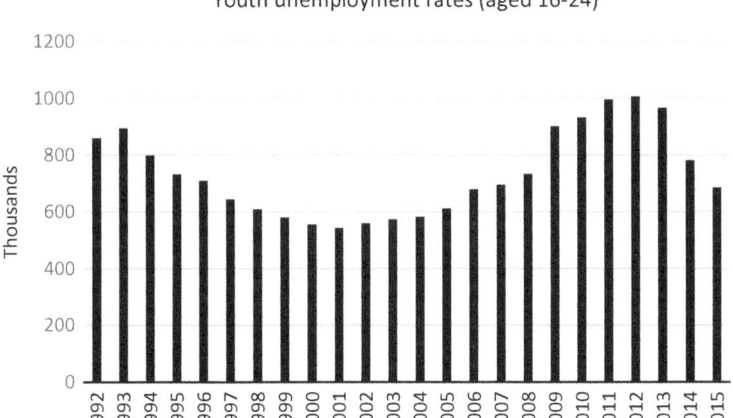

*Seasonally adjusted
Source: ONS (http://www.ons.gov.uk).

Labour Force Survey: ONS part-time workers: Could not find full-time job*

*Seasonally adjusted
Source: ONS (http://www.ons.gov.uk).

Appendix 2: Research Design and Methodology

The research design primarily focuses on comparative case studies and public management. The research strategy focuses on the local conditions in order to delimit (i) why networks have not resolved the problems of neighbourhood unemployment and (ii) what type of outcomes networks achieved (and why they vary). Each case is a single unit of bounded phenomena in different contexts with several properties and 'factors reflecting historical processes and human agency' (Ragin 1987: 71; Gerring 2004: 341; Baxter and Jack 2008: 550). The units of analysis are different types of formal and informal networks associated with unemployment, and interviews with network participants. The data collection supports the fieldwork to uncover networks, as reported in Chaps. 5 and 6, and the data analysis assists with overcoming bias, and increase accuracy of results. Case studies involving the network population defy measurement by one procedure or method. Mixed methods aim to generate enough observations to support causal inference about the conditions a factor (independent variables) is present the direction of influence and the outcomes that occur (dependent variables) (see Fig. 2.1). Three approaches offer guidance (see Blatter and Blume 2008). Co-variance assesses the relationship between causal factors and casual effects, supported by data triangulation. Causal process-tracing explores causal mechanisms, 'smoking gun' factors or unanticipated variables to check reporting bias, or findings based on chance. Congruence analysis uses empirical observations to explore the strength and breadth of theoretical propositions, avoid theoretical exclusivity, reduce normative judgements, support different explanations and compare accounts. Deductive theories formed prior to research were refined or developed inductively as the research unfolds (May 2001: 32–3). However, theory replication across small-N cases should not be generalised. Methods for uncovering network structures attract criticism, as network attributes are unstable, difficult to replicate, switch focus and terminate on programme completion or when the political scene changes (Provan et al. 2007: 511). The cause-and-effect relationships (attribution) between factors are 'difficult to measure in any precise manner' (Yin 1994: 110; Klijn et al. 2010: 1065). Actors' multiple objectives or single goals in decision-making processes change over time (2010: 1065–6). Factors producing outcomes in one type of network may not be generalisable to another. Some outcomes are not directly measurable, or may contain too much or insufficient data. Randomised experimental trials are expensive to conduct, narrow the field of exploration and are not suitable for pluralistic situations. Demonstrable impact also takes time to develop or emerge, or results may be contestable without a control condition for testing neighbourhoods without networks addressing unemployment, or, where unemployment levels are low. People may find jobs in spite of networks, and 'there is rarely any way to determine what would have happened to clients in the absence of intervention' (Lipsky 1980: 49). In lieu of methodological limitations, the book includes as many cases as possible with which to test and reassess theories against patterns of empirically based qualitative and quantitative evidence. This involves revising framework propositions as findings emerge, explaining different cases and comparing causal factors and outcomes in the same way for each case (Geddes 2003: 172).

CASE SELECTION CRITERIA

A three-stage sampling strategy was designed to identify case sites and ward clusters, and network cases. Stage 1 selected two local authorities from the IMD (ODPM 2004) with (i) Neighbourhood Renewal Funding (NRF) to operate a Local Strategic Partnership (LSP), and (ii) Department for Work and Pensions (DWP) funding to operate a 'Working Neighbourhood Pilot' (WNP) in specific ward clusters.[1] From 88 NRF-funded local authorities 12 were selected to run WNPs.[2] Two local authorities matched the criteria: the London Borough of Tower Hamlets (LBTH), and the Great Yarmouth Borough Council (GYBC). Stage 2 identified two ward clusters from within each local authority with persistently high unemployment. One ward cluster must operate the WNP. The ward clusters selected in LBTH were East India and Lansbury (EIL), and Spitalfields/Banglatown and Bethnal Green South (SB). The ward clusters selected in GYBC were Central and Northgate (CN), and Cobholm and Southtown (CS). Stage 3 applied positional, decisional and reputational methods to uncover case networks representing ward clusters (Knoke 1994: 280–1). Where a network begins and ends is a sampling problem, and time-intensive to research, thus boundary specification strategies are critical for deciding

---

[1] The English Indices of Deprivation measure areas and wards in England with high levels of poverty on seven domains: education, skills and training; employment; geographical access to services; health deprivation and disability; housing; income; and crime.

[2] NRF allocated £1.875 billion between 2001-06 to 88 local authorities. 86 local authorities were allocated £1.05 billion between 2006-08, mainstreaming support was intended thereafter.

who to include or exclude, and then contacting whole network populations (Alter and Hage 1993: 298; Marsden 2005: 9; Isett et al. 2011: 163, 166). The positional method identifies formal networks listed in official documents or websites with high visibility and governance remits. Whilst easier to identify, formal network membership quickly goes out of date, and subordinates often attend meetings in place of named actors on a list. Latent networks may outperform in terms of connectivity or coordination but their boundaries are less evident or are 'artificially created by the researcher', or the sample may depend on the perception of the issue or dominance of network types over another (Berry et al. 2004: 546; Rhodes and Marsh 1992: 187). The decisional approach asks public actors with leadership roles, local decision-making influence or hierarchical positions in a programme to identify their ego relations (Wasserman and Faust 1994: 42). Similar methods are the position generator and named generator approaches (Lin 1999: 38). Some researchers ask interviewees to state whom they are in touch with from a predetermined list (see Milward and Provan 1998: 390). But interviewees may fear sanctions and over-report on contacts (Isett et al. 2011: 166). Likewise, researchers cannot assume that actors with visibility are 'where the action is' (Wasserman and Faust 1994: 179). Hence, the reputational method is a simple 'snowball' technique that allows participants to determine network boundaries perceived relevant to a situation. Thus, interviewees were asked to nominate organisations associated with assisting the unemployment issues in the ward clusters, name their actual contacts and draw a perceptual map of networks by recall. When organisations reached three nominations, the lead member was contacted and interviewed. The cut-off point with this method is often arbitrary and biased towards strong ties, and also requires flexibility, as overlaps occur between data collection, analysis and coding (Wasserman and Faust 1994: 34). Moreover, interviewees can incur recall, memory lapse and biases during interviews and missing data in network analysis can distort results (Marsden 2005).

Networks Uncovered

Not all of the case networks uncovered were neighbourhood-focused; some represented ward-clusters, or wards from a borough-wide position. The LBTH networks includes one formal borough-wide network, two formal ward-focused networks and two informal ward-focused networks ($N = 5$). The GYBC networks includes one formal borough-wide network, three informal ward-focused networks and one informal borough-wide network ($N = 5$). The *positional method* uncovered two sub-networks of the LBTH LSP and GYBC LSP. The other positional case links to a housing association in the LBTH EIL wards. All three had websites, network membership lists, network administrators or coordinators and meeting minutes. Three cases uncovered by the *decisional method* were associated with three officials; two had expenditure decisions associated with the WNP, the other official had coordination responsibilities and a decision-making role in NRF. Four cases used the *reputational method* to examine connectivity between organisations associated with ward unemployment. In one ward cluster, organisations associated with unemployment had no connectivity, and this case represents the counterfactual.

Research Subjects

Qualitative semi-structured interviews in case areas involved professionals ($N = 86$) who had associations with networks targeting unemployment or who were selected through the fieldwork. Subjects included sub-government officials, executives, and senior managers from the regions, local authority officials and front-line statutory, voluntary or private-sector providers. Organisational interests included regeneration, education and skills, local business community, employment-training provision, community advice, childcare centres, community colleges, youth services, probation services, housing services, community health centres and arts programmes. Funding sources included central government departments, institutions and quangos, EU programmes, the local authority, housing associations or charities. However, many job roles had limited responsibility for unemployment or no direct contact with unemployed people.

Data Collection

Primary and secondary data-gathering assisted the author to determine (i) central environment infrastructure and policy discourse shaping network environment, (ii) local demographics and provision, (iii) network attributes; size, type of member, locality focus, policy interests, and outcome type; (iv) governance, management and decision-making styles, (v) the network participant's relational patterns, reputation, attitudes and perceptions of effort. Qualitative interviewing techniques gave voice to stakeholders' perspectives, and a quantitative survey conducted during interviews determined relational interactions and additional contacts to interview. Network coordinators verified membership lists. A literature review collated documentation from the following: academic and institutional sources; government and funding regimes websites; internet searches for policy documents, plans, and reports; journals and books; area statistics and demographics; and unemployment policy and data. Local authorities provided reports, strategic planning documents, brochures and e-newsletters. Public interfaces, from town halls, local libraries to Jobcentres, provided local newspapers, public reports and information leaflets. Formal network websites archived minutes and reports, or these were obtained through the Freedom of Information Act.

Occasionally, fieldwork led to unplanned, informal, ad hoc spontaneous conversations with professionals in community venues, which were dated in a notebook. This was not a systematic review of services, but revealed poor communication, duplication of effort or lack of joined-up information flows between institutions, networks and neighbourhoods.

Despite the emphasis on joined-up working and engaging the hard to reach, community notice boards (including local colleges) tend not to promote competing organisations' activities. Interviewees were unaware of some providers who advertised in local newspapers. Informal visits to research sites in each local authority provided insight into differences between ward clusters' culture and history, community provision, housing and transport needs. A walkabout at street-level, whilst not scientific, identified service duplication and neglect, and return visits found new provision had terminated; however, observations should be triangulated to avoid bias. Nevertheless, all data collection techniques carry health warnings (see Foddy 1993: 126–52).

Data-gathering activities occurred over an 18-month period including a four-month set-up phase to ensure case study areas were not research-fatigued or competing with another study, and to map baseline data in the case areas and pilot the interview procedure. Interviews were carried out over one year, from November 2004 to May 2005 in Tower Hamlets, and from May 2005 to November 2005 in Great Yarmouth. Initial fieldwork yielded a wide range of networks and stakeholders and an interview participation rate of 98 % (86 of 87 organisations interviewed). Thirty-nine actors were interviewed in the LBTH; 12 were associated with other case networks, hence the citation total $N = 51$. Forty-seven actors were interviewed in the GYBC; three were associated with other case networks and counted in the citation total $N = 50$. Subsequent network developments have been informally tracked. In 2011, the Information Commissioners Office assisted the author to conclude a one-year investigation into an unpublished report commissioned by one case network.

DATA ANALYSIS

The data analysis involves discourse and content analysis, social network analysis (SNA) and a framework of qualitative indicators to assess network outcomes. The evidence examined includes documents, statistical records, interview transcripts, network structure and behaviour and common themes, theories and phrases. The policy goal 'joined-up working', for example, is largely undefined in the literature, yet its conditions and issue-range are vast, from complete coordination to dysfunctional relations, planning processes to implementation gaps. Interview transcripts shed light on discourse meanings and the management of meanings in practice. Phrase frequencies, patterns and themes were analysed using systematic interpretative and coding techniques rather than a precise measure (Bell 1993). Transcripts were numbered and stored in MS Word computer files, copied into a question-by-question table format, read, reread, and reduced into colour-coded categories with 'higher-order headings'. Data refinements were saved in new files at each stage, providing an audit trail. Final tables totalled citation frequencies in percentages across cases and results compared. Caution also applies to the examination of public documents and programme evaluations, since they must satisfy a political process for managing public relations, and results may be subject to reporting conditions.

Attribute and relational data collated in responses to questions during interviews supports SNA. Attribute data draws on interviewees' perceptions about organisations and networks associated with unemployment support in two ward clusters. Relational data draws on information about the organisations they actually contacted about unemployment issues in the two ward clusters, the frequency of contact, if it occurred through a network, the quality of the relationship and if it had changed over time. They were also asked to draw a map of the local network environment. Results (per local authority) were transcribed into binary data sets in an adjacency matrix and saved in MS Excel. Matrices contain the presence or absence of actors' ties, in a grid format of rows and columns populated in each cell. A number one in a cell represents one tie. A zero represents absent relations. Rows quantify the number of directed ties actors have with others (out-degree). Columns quantify the number of ties actors receive (in-degree). A sum at the end of each column and each row counts alter and ego scores. A similar matrix plotted organisations perceived to assist unemployment. The matrix-assisted methods supported the analysis of centrality scores in ten networks using visualisation software: UCINET 6 (Borgatti et al. 2002) calculates Freeman's centrality measures, and NetDraw 1.0 (Borgatti 2002) computes a visual display of network ties in sociograms. The unit of analysis in sociograms is the organisational participant represented by a node (N). The node size increases in tandem with the number of organisational ties. However, ties may not be reciprocal; hence, the arrow-headed nodes show the direction of the tie and whether the connection is reciprocal or asymmetrical. Participants with few or no ties are the isolated nodes at the edge of the sociogram. Centrality is 'a basic axis for describing network patterns' (Alter and Hage 1993: 162; Kilduff and Tsai 2003: 29–33). The star sociogram, for example, is an optimum egocentric position for the central actor who receives all network participants' attention, but could repress the other actors (see Scott 1991: 9–11). Thus, degree-centrality results show where members' ties could be strengthened (Holman 2008: 539).

Finally, a coding scheme with performance indices analysed data on case network outcomes and preferred outcomes (see a similar approach in a study of political effectiveness, Roller 2005: 210–12; Table 4.24). A coding score judges network outcome as significantly above average (++), average (+), below average (−) or significantly below average (− −). A legend at the bottom of Table 7.5 indicates the type of evidence, from interview transcripts, minutes, SNA and documentation to observation. Scores should be treated with caution, as meanings vary from case to case.

Appendix 3: Interviewing Strategy

Interviewees were contacted about the research purpose and interview procedures in advance, including the estimated interview time, and with a guarantee of confidentiality and citation anonymity. The interview protocol included reading aloud a statement of confidentiality and informal and formal network definitions so that interviewees could provide their own definition if different. Interview length ranged from 15 min to 2 h, but the average was 1 h. Audio-digitally recorded interviews were transcribed verbatim into MS Word files.

A semi-structured open-ended questionnaire was piloted and refined, mainly to capture experiences or behaviour (what people have done or do), opinion or values, (concerning beliefs) and perception of effort (concerning motivation). The Part 1 questions (1–5) focused on the unemployment problem, local knowledge, concept understanding and the relevance of joined-up working and joined-up government, and actors who promoted the policies. The Part 2 questions (6–9) asked about relations and reputations in the policy field, the frequency and quality of communication and network awareness. The Part 3 questions (10–15) enquired into how networks assisted unemployment policy in the ward clusters, the type of network, frequency of contact and outcome. Self-selection or success maximisation provided insight about actors' motivation, their preferred outcomes and those networks perceived as effective. The interview questions follow.

1. What do you think the unemployment problem is in [...] wards and [...] wards?
2. What do you understand by the term 'joined-up working'?
3. How is 'joined-up working' relevant to your work?
4. Is 'joined-up working' promoted in the unemployment policy sector in this area, and by whom?
5. Is 'joined-up government' relevant to your work?

    [*Prompt*: I am doing a mapping exercise of this area and interested in finding out which organisations are linked to unemployment issues, what kinds of formal/informal inter-organisational networks exist to address unemployment issues and which ones you are in touch with.]
6. Can you please list the organisations that are associated with assisting the unemployment issues in the [...] wards and [...] wards?
7. Which organisations are you in touch with, in relation to unemployment issues?

    – How often do you meet, i.e. daily, every other day, monthly, or quarterly?
    – Do these relations occur through formal and/or informal inter-organisational networks?
    – How would you describe these relations, i.e. positive, negative, or neutral?
    – Have they improved or worsened over time?
    – [Answers were logged in table format during the interview.]

8. Please draw a map illustrating the formal and/or informal inter-organisational networks in the [...] wards, and [...] wards linked to unemployment issues (operating for at least 3 months).
9. Please say if these are networks more effective than other networks, and/or they hinder progress.
9a. Can you say what was here before these networks?
10. Can you give me a concrete example of how a network has played a part in assisting you to achieve an unemployment policy goal? For example, did it assist you to get over difficulties working in this

field, enable additional resources or service improvements, increase training places, or create jobs?
11. Is it a formal network + named/informal network?
12. Do you meet regularly? What is the frequency of these meetings?
13. How many organisations are involved, and which ones?
14. Who coordinates it, and why do they?
15. What do you think it has contributed overall, i.e. to the inter-relations in [...] wards or [...] wards, and unemployment policy goals?

Appendix 4: Employment-Related Funding Streams and Provision in Tower Hamlets 2004–2008

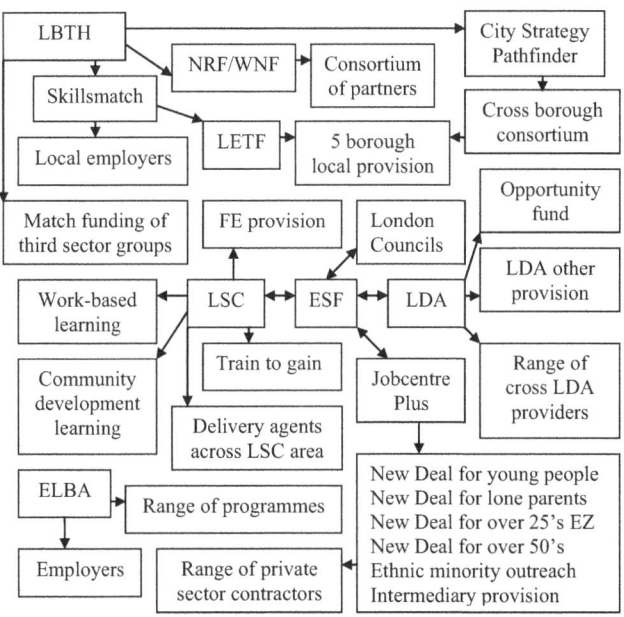

Note: Programmes evolving or existed in 2004–8 (Adapted from source LBTH 2008 Fig. 1: 11).

Appendix 5: Primary Networks in Great Yarmouth (2004-2008)

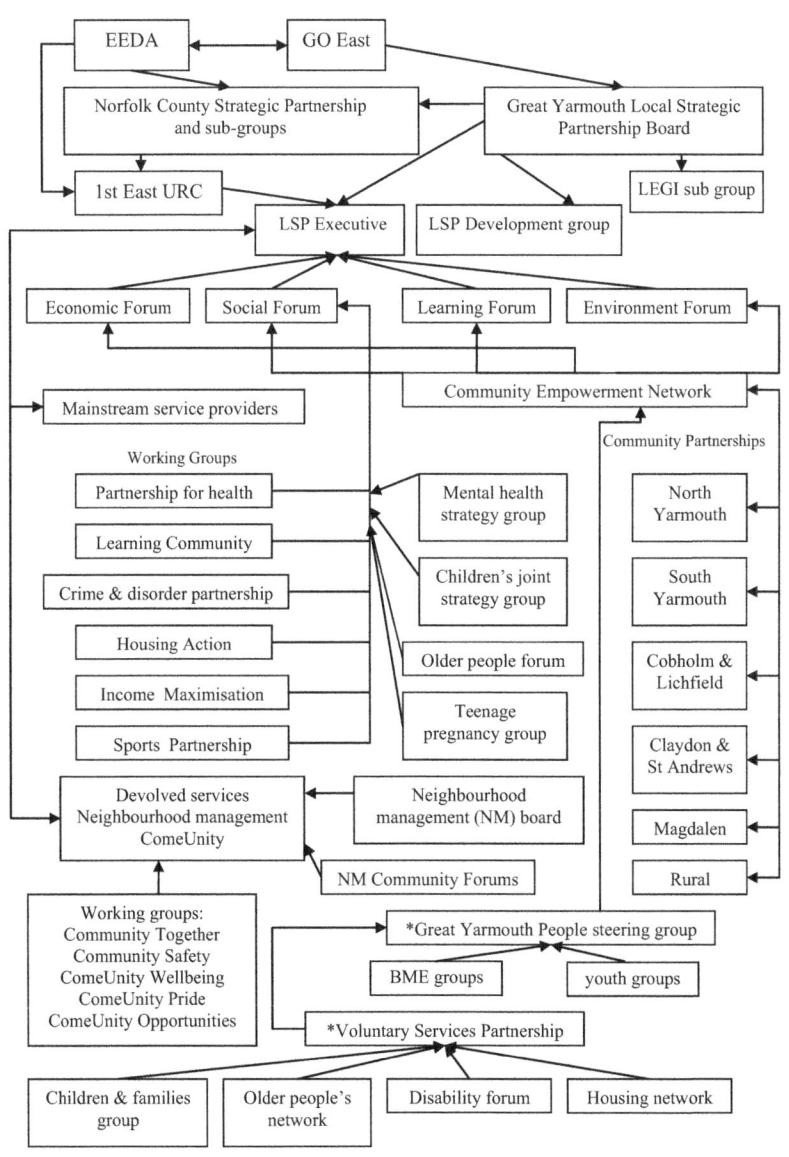

Bibliography

Alter, C., & Hage, J. (1993). *Organizations working together*. Newbury Park, CA: Sage.

Baxter, P. and Jack, S. (2008). Qualitative case study methodology: study design and implementation for novice researchers. The Qualitative Report, 13,(4,), 544-559. Retrieved from http://www.nova.edu/ssss/ QR/QR13-4/baxter.pdf.

Bell, J. (1993). *Doing your research project*. Milton Keynes: The Open University.

Berry, F. S., Brower, R. S., Choi, S. O., Xinfang Goa, W., Jang, H., Kwan, M., et al. (2004). Three traditions of network research: What the public management research agenda can learn from other research communities. *Public Administration Review*, 64(5), 539-552.

Blatter, J., & Blume, T. (2008). In search of co-variance, causal mechanisms or congruence? Towards a plural understanding of case studies. *Swiss Political Science Review*, 14(2), 315-356.

Borgatti, S. P. (2002). *Netdraw visualization software*. Cambridge, MA: Analytic Technologies.

Borgatti, S. P., Everett, M. G., & Freeman, L. C. (2002). *UCINET 6 for windows: Software for social network analysis*. Harvard, MA: Analytic Technologies.

Foddy, W. (1993). Constructing questions for Interviews and questionnaires. Cambridge: Cambridge University Press.

Geddes, B. (2003). *Paradigms and sand castles: Theory building and research design in comparative politics*. Ann Arbor: University of Michigan Press.

Gerring, J. (2004). What Is a case study and what is it good for?' The American Political Science Review. 98(2): 341-354.

Holman, N. (2008). Community participation: Using social network analysis to improve developmental benefits. *Environment and Planning C: Government and Policy*, 26(3), 525-543.

Isett, K. R., Mergel, I. A., LeRoux, K., Mischen, P. A., & Rethemeyer, R. K. (2011). Networks in public administration scholarship: Understanding where we are and where we need to go. *Journal of Public Administration Research and Theory*, 21(suppl 1), i157-i173.

Kilduff, M., & Tsai, W. (2003). *Social networks and organizations*. London: Sage.

Klijn, E.-H., Steijn, B., & Edelenbos, J. (2010). The impact of network management on outcomes in governance networks. *Public Administration*, 88(4), 1063-1082.

Knoke, D. (1994). *Political networks: The structural perspective*. Cambridge: Cambridge University Press.

Lin, N. (1999). Building a network theory of social capital. *Connections*, 22(1), 28-51.

Lipsky, M. (1980). *Street-level bureaucracy, dilemmas of the individual in public services.* New York: Russell Sage Foundation.

London Borough of Tower Hamlets (LBTH). (2008, November). *Tower Hamlets employment strategy 'getting neighbourhoods working'.*

Marsden, P. V. (2005). Recent developments in network measurement. In P. J. Carrington, J. Scott, & S. Wasserman (Eds.), *Models and methods in social network analysis* (pp. 8–30). New York: Cambridge University Press.

May, T. (2001). *Social research: Issues, methods and process.* Buckingham: Open University Press.

Milward, H. B., & Provan, K. G. (1998). Measuring network structure. *Public Administration, 76*(2), 387–407.

Office for the Deputy Prime Minister (ODPM). (2004). *The English indices of multiple deprivation 2004.*

Provan, K. G., Fish, A. C., & Sydow, J. (2007). Interorganizational networks at the network level: A review of the empirical literature on whole networks. *Journal of Management, 33*(3), 479–516.

Ragin, C.C. (1987). The comparative method. Moving beyond qualitative and quantitative methods. Berkeley: University of California Press.

Rhodes, R. A. W., & Marsh, D. (1992). New directions in the study of policy networks. *European Journal of Political Research, 21*, 181–205.

Roller, E. (2005). *The performance of democracies: Political institutions and public policies.* Oxford: Oxford University Press.

Scott, J. (1991). *Social network analysis: A handbook.* London: Sage.

Wasserman, S., & Faust, K. (1994). *Social network analysis: Methods and applications.* Cambridge: Cambridge University Press.

Yin, R. K. (1994). *Case study research design and methods* (2nd ed.). Thousand Oaks, CA: Sage.

Index[1]

A
activation policy. *See also* Jobcentre Plus (JCP); unemployment; welfare benefits
  active labour market policies (ALMPs), 7, 12, 82, 98–101
  defined, 98
  European, 91
  German and Swedish case studies, 98
  OECD's active society, 101
Aftab, I., 172n8
agency. *See also* network; network cases; unemployment
  dimensions of, 52–3
  and European policy, 89
  local limits to, 277–8
Agranoff, R., 3, 9, 15, 26, 27, 31
Ainsworth, P., 172
Alexander, E. R., 32, 287
Ali, R., 163
Alter, C., 21, 25, 31, 32, 53, 54, 56, 304, 312
Amin, S., 87
anarchists, 48
Angelica, Y., 96
Appelby, Y., 101
apprenticeships
  deficit of work placements, 108, 115, 181
  without ESOL (English for speakers of other languages), 171
  funding focus, 104
  in Germany and Scandinavia, 89
  governance, 103
  historic legacies, 248
  levy, 105
  low employers' engagement, 81, 103, 104, 205, 284
  and partnerships in Austria, Germany and Switzerland, 103
  schemes, 112, 113, 150
  take-up, 103
  targets driven by the LSC, 110
  weak progression routes in UK and USA, 89, 103
  for youth, 103

---

[1] Note: Page numbers followed by "n" denote notes.

Apprenticeships, Skills, Children and Learning Act 2009, 113
Asquith, Herbert, 73
Atkinson, R., 83, 129, 131, 134, 137–9, 141, 178
austerity measures, 67, 80

B
Bache, I., 55, 135, 143, 144
Ball, R. M., 142
banking crisis
  credit flow, 67
  poor performance, 67
  post-recessions, 98
  remuneration, 67
Bank of England
  monetary policy, vii, 80, 88
Barker, A., 128, 137
Barnett, A., vii
Barr, N., vii, 88
Bassett, K., vi, 14, 129, 138
Bathmaker, A. M., 101
Batley, R., 135–8
Baumberg, B., 7
Beatty, C., 64, 82, 83, 213, 217, 247
Beazley, M., 141
Bednarzik, R. W., 101, 102
Begley, T. M., 53
Begum, H., 171
Bell, D. N. F., 6, 68
Bell, J., 311
Bell, K., 7
Bell, S., 26, 27
Bellis, A., 111
Bennett, R., 78, 81
Benschop, Y., 52
Benson, J. K., 14, 49
Berclaz, M., 89, 94
Berg, E., 94
Berry, F. S., 49, 50, 304
Beveridge, William, 74, 87, 95

Bevir, M., 15, 50
Big Society, 128
Biggs, J., 163
Bish, R., 9
Blair, Tony, 127, 128, 144, 146
Blanchflower, D. G., 6, 68
Blatter, J., 301
Block, F., 7, 72, 79
Blokland, H., 51
Blume, T., 301
Bogason, P., 6
Boh, W. F., 52, 53
Booth, Charles, 73
Borgatti, S. P., 312
Börzel, T., 4
Boyne, G. A., 26
Brennan, A., 140, 143, 144
Brewer, G., 26
Brewer, M., 84
British Broadcasting Association (BBC)
  on fraudulent provision, 101
  on poor job support, 99, 101
  on race discrimination, 85
Brown, Gordon, 111, 149
Brown, K., 17, 51
Bulpitt, J., 88
Burns, D., 18, 138–40, 174, 175, 292
Burt, R. S., 182
business enterprise support. *See also* Chamber of Commerce; enterprise allowance scheme
  and Business Link, 78, 112, 176, 223, 245
  Business Support Service, 144
  and industrial strategy, 114
  limitation, 106, 164–5
  local, 72, 106, 138, 151, 217, 225, 250, 277, 307
  low participation, 31, 106
  social enterprise, 227, 273
  stakeholders, 109
Business Innovation and Skills Committee (BISC), 152

C

Cabinet (1929), 74, 132
  Unemployment Committee, 132
Cabinet Office, 113, 145, 146, 150
Caines, K., 140
Callaghan, James, 137
Cameron, David, 114, 128
capitalism
  Adam Smith's view, 70–1
  anti-capitalists, 76
  and credit and mortgage abuse, 67
  and general crises, 67, 96
  and gentrification, 173
  government dependency, 87, 89, 96
  hegemony, 284
  impact on working classes, 6
  injustice, 290
  limits of, 7
  Marx view, 48, 70, 73, 96
  representatives, 11
  reproduction of the capitalist order, 95
  structural weakness, 70, 90, 95–6, 136
  tensions of, 73
  and unemployment, 72, 73
  varieties of, vi, 89
  uneven development, 149
Cardoso, G., 11
Casebourne, J., 81n11, 85, 187, 237
Castells, M., 11
Castles, F., 82
Catney, P., 135, 144, 147n8
Centre on Dynamics of Ethnicity (CoDE), 85
Chamberlain, T., 27
Chamber of Commerce, 34, 106, 141, 234, 235
  in Norfolk, 218
Chen, B., 53
Churchill, Winston, 74
Cinalli, M., 90, 109
Citizen Advice Bureau (CAB), 66, 132, 272
City Challenge. *See also* urban regeneration
  excessive monitoring, 141n7
  funding, 141n7
  partnership approach, 141–3
civil service
  benthamites, 134
  call for reform, 134, 140, 246–7
  chartists, 134
  class bias, 134
  entrepreneurs, 134
  historic reform, 134
  IMF cut workforce, 137
  Next Steps reform, 140
  white male enclaves, 134
classical economics, 70–2
  laissez-faire markets, 72, 87
Clinton, Bill., 99
Cole, A., 4, 290
collaboration
  for alliance-building, 269–70
  desire for, 291
  expectations undermined, 141–2
  for governance, x, 142
  interagency, 141, 141n6
  lack of accountability, 14, 140, 196
  network assumptions, 9, 11
  for reputation-building, 289
  for trust-building, 47, 53, 57
  weak, x
ComeUnity (CU), 225–7
Cominetti, N., 83
Commission of the European Communities (CEC), 91
Committee of Public Accounts (CoPA), 100–1, 107, 288
communism and communists
  American, 76
  British, 75

Communities and Local Government Committee (CLGC), 131
Community development project (CDP), 136
Community Plan Action Groups (CPAG), 179, 183, 184, 197
community power, 49. *See also* power relations
Community Programme, 78
competition. *See also* City Challenge; Industrial; New Public Management; Single Regeneration Budget
  civil service, 134
  departmental, 150, 284
  discourages solidarity, 48, 72–3
  encourages opportunism, 286
  EU policy, 91
  funding, 11, 20–1, 141, 144
  global, 115, 150, 215n3
  for jobs, v, 6, 178, 181, 205, 223, 253
  public tenders, 139, 179
  right-wing, 140
  steering, 16, 216
  sub-optimal network outcome, 279, 283–7
  wealth inequality and, 72
Connexions, 113, 176, 197, 219, 223, 250
conservatism
  business leadership, 138
  enterprise culture, 138
Conservative-Liberal Democrat coalition government, 13, 100, 131, 149, 288, 293
Conservative Party, 77, 90, 96, 114, 136, 148, 212
  in Great Yarmouth, 212
Considine, M., 4, 16, 20, 50, 53, 82
Cook, K. S., 19
coordination. *See also* joined-up government
  capitalist, 64
  costs, 31, 52, 53, 56
  deficiency, 5, 107, 136, 137, 139, 141, 148, 153, 198, 291
  deficit: in cases, 167, 179, 267
  EU policy, 89, 92
  and governance, 18, 19, 54
  industrial, 88, 89
  initiative overload, 146, 167
  of labour relations, 89
  limits, 26, 98, 153
  under NPM, 138, 139
  open method of, 91
  place-based, 259, 264, 278
  processes, 55, 183, 191
  role inflated, 253
  strain of pluralist networks, 286
  strategy, 51, 152, 183
  vertical and horizontal, 29, 139, 268, 273
  wasteful, 203
Copeland, P., 91
Cope, S., 30, 138
Cotterill, S., 49
CPAG. *See* Community Plan Action Groups (CPAG)
Craig, G., 135, 136
Crisis (Charity for the homeless), 97
Crisp, R., 81, 86, 128, 129, 148, 151, 152, 290
Cristofoli, D., 3, 24, 25
critical realist perspective, 7
Croall, J., 81
Crouch, C., 11, 77, 102
Crowley, L., 83, 108, 181

D

Daguerre, A., 77, 78, 86, 99
Dahl, R. A., 49
Daly, H. E., 94
Damgaard, B., 25, 282
Daugbjerg, C., 4
Davies, J. S., 3, 11, 17, 26, 50–2, 68n5, 129–31, 144, 282

INDEX 327

DCLG. *See* Department of Communities and Local Government (DCLG)
De Bruijn, J. A., 31, 56
deLeon, L., 30
Delors, Jacques, 90
   and EC White Paper (1993), 90, 94
demand-side policies, 82, 218
Dench, G., 163n2, 174, 175
Denzau, A. T., 6
Department for Business, Energy and Industrial Strategy (DBEIS), 114
Department for Business, Enterprise and Regulatory Reform (DBERR), 112
Department for Business Innovation and Skills (DBIS), 106, 112–14, 149–52, 171, 216, 225
Department for Communities and Local Government (DCLG), 17, 83, 111, 150, 152, 161n1, 169, 183n12, 212, 216, 217, 225, 225n12, 274
Department for Education (DfE), 112–14
Department for Education and Skills (DfES), 111, 196, 197, 246
Department for Innovation, Universities and Skills (DIUS), 111, 112
Department for Work and Pensions (DWP). *See also* Working Neighbourhood Fund (WNF); Working Neighbourhood Pilot (WNP); Work Programme
   background to, 110
   co-commissioning, 111, 177n11
   contracts, 187
   and European Social Fund (ESF), 216
   and full employment, 96
   isolated role, 86, 197, 245, 250
   Prime Providers, 100, 168, 180, 186–7, 196
   schemes, 97, 108, 111–13, 151, 162
   staff and sanctions, 66, 221
   and warning from UK Statistics Authority, 79
   youth contract, 107–8
Department of Employment (DoE), 137, 139, 141, 143
Department of the Environment, Transport and the Regions (DETR), 111n19, 146, 148
Department of Trade and Industry (DTI), 78, 112, 143, 185n14, 197, 224n11, 245, 246, 287
De Rynck, F., 24, 25
DETR. *See* Department of the Environment, Transport and the Regions (DETR)
devolution of power, vi, 104, 113, 114, 148, 150, 291
Dewson, S., 111, 186, 187, 237
DfES. *See* Department for Education and Skills (DfES)
Dickson, T., 77, 137n4
Dietz, R. D., 50
discrimination. *See also* British Broadcasting Association (BBC)
   of African-Caribbean graduates, 85
   cohorts, 69, 84, 85, 115
   ethnic minorities, 69, 144, 164, 165
   housing, 135n1
   lack of support, 17, 144
   lack of voice, 17, 30
   physical and mental health impact, 164
   Race Relations Act 1976, 135n1, 137
   recruitment, 85, 165
   stereotyping, viii, 79
DIUS. *See* Department for Innovation, Universities and Skills (DIUS)
DoE. *See* Department of Employment (DoE)
Douglas, T., 79

Dove, C., 111, 180
Dowding, K., 8, 18, 49
Dowling, B., 3
Dryzek, J. S., 5
DTI. *See* Department of Trade and Industry (DTI)
Dunleavy, P., 33, 283n1
Durkheim, E., 48
Durose, C., vi, 18

E
Easley, D., 19
East End Life, 274
East of England Development Agency (EEDA), 213n1, 214, 215n4, 223, 244, 253
   coordination role perceived weak, 253
Ebers, M., 26
economic. *See also* employment; Keynesian theories; neo-liberal policy; network; network cases; socioeconomic; unemployment; urban regeneration
   alignment to NAIRU, 86–7, 95
   assets, 19, 285
   behaviour, 18–19, 49
   and community, 20, 29, 136
   crisis, 74, 80
   culture, 161
   decline, 20, 77, 88, 108, 136, 138
   decline in Oakland, 26
   demand, and shortfall, 76, 86
   development, 6, 19, 69, 80, 81, 93–4, 112, 140, 142–4, 151, 176, 201, 215, 227–9, 247, 250, 264, 290
   ecology, 90, 94
   equality, 82, 224
   EU targets and coordination, 77, 92, 136
   exclusion of citizens, 51, 85, 93
   exploitation, 72, 224
   feminist literature, 96
   fiscal, 23
   German competitiveness, 90
   global downturn, 77, 175
   governance, 147, 148, 215, 290
   growth, v, vi, viii, 5, 7, 11, 13, 72, 93–4, 111, 114, 128, 134, 136, 143, 149, 150, 224, 227, 268, 270, 293
   history, 70–8, 132
   inadequacy, viii, 135
   incentives to attract firms, 105–6, 215
   inequality, 7, 8, 20, 26, 31n4, 145
   limitation in Europe, vi
   macro, 50, 64, 81, 85–90, 92, 101, 291
   malfeasance, 53
   micro, 81, 92
   migrants, 115, 212, 247–8, 253
   monetary policy, 76, 80, 87, 88, 224
   neo-liberal, 76–8
   OECD influence, 86, 93–4
   performance, 89, 215n3, 287
   place-shaping, v, 4, 20, 23, 239
   planning, 77, 216, 270
   policies, ix, x, 5, 11, 23, 31, 81, 87–90, 92, 101, 132, 291
   recession, 64, 137
   recovery, vii, 12, 80, 88
   requirement, 102
   rules challenged by anarchists, 48
   structure, 8, 12, 48, 136
   sustainability, 94, 214
   transaction cost theory, 53
   uncertainty, v
   views espoused by Marx, 48, 73

Edelenbos, J., 3, 10, 25–8, 52, 302
Edelman, M., 10, 28
Edwards, J., 135–8
Eggers, W. D., 17, 50, 55, 108, 146
Elmore, R. F., 26
emigrant training programme, 74
Emirbayer, M., 49
empirical puzzle, ix, 8, 21, 57, 64, 292
employment. *See also* employment barriers; job creation; labour market
  decline in manufacturing, 78, 78n8, 137n4, 175
  and EU law, 25, 86, 90
  EU policy influence, 77, 89–98
  growth, vi, vii, 74, 164, 217, 247
  historic, 68, 175, 178
  incentives to firms, 97, 105–6, 111, 151
  initiatives, 64, 73–6, 216
  insecure, vii
  Labour's premise of full employment, 77
  local authority desire to lead schemes, 108
  location dependent, 20, 95, 128
  policy, 86, 87, 89, 90, 93, 133, 176
  protection and lack of, 6
  rate, vii, 65n3, 165, 181, 289
  and regeneration, 214
  relations, 11
  separate from social policy, 81
  support of personal advisers, 99, 110n18
  support services in, 106, 151, 185
employment barriers
  benefit trap, 166, 205
  for carers, 96
  for cohorts, 85, 96–106, 165, 178
  domestic, 95
  economic, 95

ESOL (English for speakers of other languages) courses
  oversubscribed and underfunded, 97, 171
  to full employment, 92–8, 107
  human rights perspectives, 93
  institutional, viii, 82
  international growth, 93, 94
  labour flexibility, 86, 87
  location, 84, 164
  policy view, 96–8
  skill deficiency, 100, 101
  structural, 50, 95, 96, 152, 199
  policy view, 96–8
  provision, 138, 139, 153, 168, 177, 201
  for youth, 17
enterprise allowance scheme, 78, 106, 106n16, 133
enterprise zone (EZ)
  anticipated jobs, 217, 227
  definition, 151
  funding incentives, 151, 217
  in Great Yarmouth and Lowestoft, 215, 218, 288
  job outcomes, 152, 288
Entwistle, T., 3
Enum, Y., 169
Ernst, C., 94
Esmark, A., 25, 108n17
European Commission (EC), 68, 90, 91, 93, 94
European Employment Strategy (EES), 25, 91, 108n17
European Exchange Rate Mechanism, 78
European Union (EU). *See also* employment; European Employment Strategy (EES); National Action Plans (NAPs)
  annual growth strategy, 92

European Union (EU) (*Cont.*)
  Britain's entry, 136
  bureaucracy, 91, 142
  community-led economic development, 142
  Europeanisation, 89, 91–2
  European Social Fund (ESF), 92, 100, 112, 148, 151, 168, 179, 216, 225, 277, 317
  Eurozone crisis, 67
  governance, 5
  impact on women, 90
  open method of coordination (OMC), 91
  politics, 23
  programmes, 55, 307
  referendum, vi, 90, 114
  regional development and European Regional Development Funding (ERDF), vi, 55, 91, 92, 148, 151, 213n1, 216
  structural investment funding, 12, 92, 213
  URBAN initiative, 136
  youth unemployment, 17, 68, 99
Evans, G., 130
Evans, J., 127, 129, 130
Everett, M. G., 312

F
Fainstein, S., 172, 173
Fairclough, N., 78
Farmer, R. E. A., 87
Faust, K., 15, 49, 304
Fawcett, P., 4
Field, J., 74, 75
Finegold, D., 102, 105
Finn, D., 99
Fish, A. C., 9, 15, 23, 24, 27, 302
Fitzpatrick, J., 163
Fletcher, D., 83, 86, 99, 110n18

food insecurity and food banks, 84
Fordism, 76
Foreign Office and Commonwealth (FCO), 93
Forstater, M., 80, 93
Foster, S., 171
Fothergill, S., 64, 213, 217, 247
Foucault, M., 12, 18
Fox, P., 228, 229, 274
Freedom of Information (FOI) Act, 14, 167, 309
Freeman, L. C., 312
free-rider problem, 19, 33, 79, 201
Friedman, Milton, 77, 86, 96
Füglister, K., 90, 109
Fuller, C., 78, 81
Furlong, P., 8

G
Gaffney, D., 7
Gage, R. W., 15
Galaskiewicz, J., 48n1, 51, 52
Gamble, A., 88
Gardiner, J., 95, 96
Gardiner, L., 104, 113
Garraty, J. A., vii, 67, 68n6, 70, 74, 76, 80, 87
Gauci, P., 228n13
Gavron, K., 163n2, 174, 175
GDP. *See* Gross Domestic Product (GDP)
Geddes, B., 47, 48, 302
Geddes, M., vi, 3, 15, 17, 26, 27, 30, 142
General Strike, Oakland, USA, 26
George, Henry, 73
Giarchi, G. G., 32
Giguère, S., 4, 10
Gilchrist, A., 15, 27, 55, 134, 135
Ginsburg, N., 83
Girling, J., 72
Giugni, M., 89, 94

Glendinning, C., 3
GOE. *See* Government Office for the East of England (GOE)
Goetschy, J., 77, 91
GOL. *See* Government Office for London (GOL)
Golden, S., 27
Goldsmith, S., 17, 50, 55, 108, 146
Goodship, J., 30, 138
Goodwin, J., 49
Goss, S., 25, 26
governance. *See also* network
  under Blair, 127, 128
  citizen, 17
  community, 13, 17, 147, 223
  corporate, 16
  European, 91
  failure, 4, 17, 26, 50, 52, 287
  flawed, 57
  frameworks, 243
  hierarchical, 13, 279
  informal, 32
  instruments of, 30
  joined-up governance, x, 195, 216, 287
  local, 50, 54, 69, 78, 131, 142, 272, 282, 288
  market, 16
  meta, 4, 5, 282
  models of, 17, 18, 54, 141
  multi-level, 34, 54, 91
  multi-sector, vi
  neighbourhood, 30, 128, 129, 142
  New Public Governance, 16
  orders of, 5
  political, 12–14, 16, 17
  and power, 4, 13, 17, 18, 30, 31, 50, 53, 131, 134, 138, 145, 148, 150, 192, 195, 203, 205, 240, 278, 284, 285
  procedural, 16
  public, 16, 139
  regional, 147–9, 253, 288
  restricted by commercial confidentiality, 100
  roles and responsibilities, 91, 108–14, 181–2, 231–2, 254, 281
  shift to government, 17, 138, 290–1
  of skills shift to employers, 105
  strategic, 17
  structures, ix, 4, 108, 143, 180, 191, 204, 232
  support for apprenticeships, 110
  theory, x, 4, 15, 204
  of unemployment, 108–13
  and the Westminster model, 145
  weak in LEPs, 150, 216, 230
  widens inequality, 204
Government Office for London (GOL). *See* London
Government Office for the East of England (GOE), 215n4, 223, 225, 243, 244, 244n19, 273
Graddy, E. A., 53
Granovetter, M., 18, 20, 49, 53, 165
Gray, A., vii
Greater London Authority, 176n10
great recession, 67. *See also* banking crisis; European Union (EU)
Great Yarmouth Borough Council (GYBC). *See* network cases
Great Yarmouth Local Strategic Partnership (GYLSP). *See* network cases
Great Yarmouth Urban Regeneration Company 1st East (GYURC), 214, 224, 224n10, 270, 277, 319
Green, A., 65, 83, 84
Gregg, P., 63
Greive Smith, J., 88

Gross Domestic Product (GDP), 74, 94, 134, 224
Grover, C., 71, 95
Gunn, L. A., 19, 23n3, 281
GYURC. *See* Great Yarmouth Urban Regeneration Company 1st East (GYURC)

H
Hage, J., 21, 25, 31, 32, 53, 54, 56, 304, 312
Hajer, M., 4, 10
Hall, P., 89
Hall, S., 141, 143, 144
Hambleton, R., 18, 138–40, 174, 175, 292
Hanf, K. I., 3, 9, 17, 18, 21, 24, 26, 52, 56, 98, 281
Hansard, HC Deb, 66, 83
Hardin, R., 19
Harper, G., 171
Harris, J., 78
Harrison, J., vi, 151
Harvey, P., xin1, 75, 93
Hasluck, C., 64, 77, 78, 85, 89, 98, 101, 102, 107, 115, 151
Hastings, A., vi, 142, 144
Haughton, G., 140
Hay, C., vii, 4, 49
Hayek, Friedrich, 77
Heath, Edward, 136
Heinz, J. P., 52
Hennessy, J., 96
Heseltine, Michael, 78, 113, 141
Hill, M., 11, 16, 23n3, 27, 50
Hindmoor, A., 26, 67
Hjern, B., 24, 98
Hobson, John. A., 73
Hofstede, G., 10
Hogarth, T., 80, 97
Hoggett, P., 18, 138–40, 174, 175, 292
Hogwood, B. W., 19, 23n3, 281

Holman, N., 312
Home Office (HO), 110, 136, 179, 196, 225
Hood, C., 147
Hoppe, R., 16, 18
Houghton, S., 111, 112, 180
House of Commons Library, 149
housing
    Action Trusts, 140
    affordability, problems, 10, 30, 104, 107, 166, 178, 218
    the benefit trap, 166, 205, 289
    clearance programmes, 134, 135
    and decent homes standard, 163, 212
    disrepair, 20, 134, 174
    and economic migrants, 248, 253
    and employment support, 179, 188
    ethnic minorities, 143
    and ethnic minorities, 134–5, 135n1, 248
    failure to regulate proprietors, 205
    gentrification, 30, 130, 172, 173
    homelessness, 66, 78, 84, 222
    impact of social housing auction, 167
    institutions, 179
    lack of, 104, 166
    overcrowding impact, 163, 172, 199, 212
    policy, 177
    politics, 140, 174
    right-to-buy schemes, 140
    social landlords, 188
    targets for new build homes, 114, 150–2, 166, 173, 214, 216, 289
Huang, K., 27
Hudson, R., 81
Huggins, R., 106
Hughes, C., 181
Hull, K., 83, 94
human rights. *See also* right to work
    British response, 93
    European, v, xin1, 94

and full employment, 93
UN Human Rights Council, 93
Hunter, D. J., 26
Hunter, F., 49
Hupe, P., 11, 16, 23n3, 27, 50
Huxham, C., 9, 53

I
ILO. *See* International Labour Organisation (ILO)
immigration control, 90, 96
implementation
  case studies, 98
  impairment, 51
  of joined-up working, 205, 311
  and network management, 55
  and policy, 10, 21, 26, 29, 51, 52, 90, 153, 196, 268, 274, 281, 285
  Pressman and Wildavsky's study, 3, 26, 281
  process, 18, 50, 83
  in public administration, 50
  studies, 3, 281
Index of Multiple Deprivation (IMD)
  and case selection, 161n1
  English indices of deprivation, 303n1
industrial. *See also* enterprise zone; skills; trade unions; unemployment
  closure of Docklands, 164, 174
  coordination, 88, 89
  decline, 10, 67, 80, 83, 88, 108, 127, 137, 137n4, 138, 199, 217
  de-industrialisation, 77, 136
  Delors green agenda, 90
  improvement areas, 137, 151, 253
  Industrial Relations Act 1971, 136, 137
  low wage culture, 20, 97
  networks, 34, 64, 134, 137
  oppression, 72, 73, 90
  policy, 29, 90, 114–5, 137n4, 289, 293
  regeneration, 127, 129, 133
  relations, 11, 30, 69, 74, 77, 80, 96
  subsidisation, 6, 11, 87, 97, 103, 115, 290
  training boards, 131, 133
industrial revolution, 72
institutions
  community, 144
  crisis of competence and authority, 287
  of economic development, 81
  European, 65, 91
  financial, 67, 86
  fragmented, 139
  funding streams, 23
  economic impact, 5, 281
  governing, 22
  inflexible, 30, 198, 230, 274
  Matthew Effect, 31n4
  in meta-governance environment, 5
  and network relations, 14, 98, 287, 289, 309
  and network typology, 33
  promoting neo-liberal policy, 82, 86–7, 93
  regeneration, 179
  resource-rich, 13, 193, 204
  social, 73, 87
  social-regulatory, 6
  state, 14, 15, 50
  and structure, 51, 83, 292
  sustain asymmetric power model, 49
  training, 105
  and unemployment, 33, 67, 79, 82–3, 86, 98, 176, 179
  as work barriers, 82
International Labour Organisation (ILO), vii, 93, 94, 102, 168, 181, 290, 298

International Monetary Fund (IMF), 86, 93
   loan to Britain, 137
Iqbal, B., 15
Isaac-Henry, K., 55, 56, 78, 139, 198
Isett, K. R., 3, 15, 24, 33, 304
Islamic identity, and Muslim, 171–2, 205

J

Jackson, A., 140
Janoski, T., 107
Jenkins, K., 140
Jessop, B., 4, 50
Jobcentre Plus (JCP). *See also* job vacancies; New Deal; Work Programme; Working Neighbourhood Fund (WNF); Working Neighbourhood Pilot (WNP)
   business culture, 82, 110n18
   coordination problems, 21, 110, 167, 171, 190, 191, 234, 238, 240, 246, 270
   function of, viii, 110, 112, 176, 223
   joined-up working perceptions of, 189, 204, 228, 233, 250, 252, 253
   leadership, 273
   partnerships, 105, 111, 148, 179, 184, 201, 226, 234, 271, 278
   personal advisers, 99, 99n3, 110n18, 177, 186
   quality of support, viii, 20, 101, 107, 110n18, 151, 168, 176, 177, 226, 240, 270, 271
   reputational scores, 201, 250, 254
   sanctions, viii, 221
   staff reduction, 238, 240

job creation
   in cases, 110, 243, 270, 271, 279
   fragile in low growth areas, 107
   lack of trickle-down, 81, 88, 127, 173, 253
   in LEP remit, 1, 50, 216, 217
   overestimated, 14, 96, 105–6, 152
   place dependent, 30, 129, 138, 248
   policy, 80, 85, 96–8, 107–8, 289
   schemes, 104, 107–8, 115
   and social enterprise, 273
   in weak economies, 88, 248
job dearth, 98, 99, 129, 176, 181, 223, 227, 234, 273
job guarantee, xin1, 99
job quality, 68, 94, 107
job search
   assistance, 73, 173, 238–9, 240
   benefit conditionality, 221, 226
   clubs, 99, 226, 240
   demoralising, 99
   quality of support, 177, 226, 242
   weak ties theory, 20, 49, 165
job vacancies
   and the Beveridge curve, 87
   difficult to fill, 105
   lack of, 21, 104, 181, 223, 234
   number of, vi, 67–8, 290
   quality of, 68, 107, 224
John, P., 4, 10, 13, 17, 18, 144, 290
Johnson, C. L., 55, 56
joined-up government. *See also* Social Exclusion Unit (SEU)
   coordination of departments, 51, 148, 287
   limits to, 30, 148, 197, 285
   policy rhetoric, 196–7, 245–7, 259
Jones, M. R., 144
Jones, P., 127, 129, 130
Judge, D., 77, 137n4

K

Kabeer, N., 172
Karlsson, J. C., xin1
Kassim, H., 16
Kauppinen, T. M., 197n18
Keast, R., 17, 24, 27, 51, 135
Keep, E., 101, 102, 105
Kenis, P., 3, 4, 15–17, 21, 23, 24, 27, 50, 52
Keohane, R., 28, 47
Keynes, John Maynard, 76, 87, 96
Keynesian theories, xin1, 64, 73, 76, 77, 88, 115, 136, 137
Khatri, N., 53
Kickert, W. J. M., 16, 31, 50, 52, 54, 55, 182, 282
Kilduff, M., 8, 21, 48n1, 52, 312
King, G., 28, 47
King, S., 49
Kintrea, K., 83
Kitson, M., 79, 95, 141
Kjaer, A. M., 14
Kleinberg, J., 19
Klijn, E-H., 3, 4, 10, 11, 16–18, 25–7, 28, 31, 49n1, 52, 54, 55, 282, 302
Knoke, D., 15, 53, 303
knowledge economy, 102
Kooiman, J., 13, 282
Koppenjan, J. F. M., 16, 18, 31, 49n1, 52, 54–6, 182
Kortteinen, M., 197n18

L

labour market. *See also* activation policy, active labour market policies (ALMPs); discrimination; neo-liberal policy
  democratic, 72
  deregulation, 224, 281
  emphasis on individual performance, 10, 11, 47, 48, 51, 69, 73, 82, 84, 86, 89–90, 96, 99–101, 113
  flexible jobs, 6, 68, 84, 86, 87
  flexicurity, vii
  ill-prepared for economic crisis, 81
  inequality, 86, 115
  local, 93, 95, 107
  low wages, 6, 81, 82
  national minimum wage, 63, 84, 86
  regulation, 57, 76, 77, 87
  retreat from, 80
  signalling, 103
  statistics, vii, 171
  technological advances, 64, 96
  theories, 282
  wage bargaining, 65, 87, 95
  and weak economic thinking, 281, 293
  weak bargaining power, 6, 57
Labour Party, 75, 77, 134, 163, 174
Lahusen, C., 89, 94
Lane, J-E., 26, 30, 31, 139, 281
Lane, P., 171
Lanning, T., 102, 113
Lansbury, George, 174
Lawless, P., vi, 17, 83, 104, 138, 151
Lawton, K., 102, 113
Leaman, J., 87
Learning and Skills Council for England (LSC)
  association with unemployment support, 176, 223
  background to, 110–12
  corruption, 110
  joined-up working, 203, 253
  lavish spending, 171
  inadequate provision, 148, 171, 189
  and NEETs, 197
  strategic coordination, 177n11, 185n14
  un-joined-up, 196, 203, 228, 278

Le Galès, P., 16, 50
LEGI. *See* Local Economic Growth Initiative (LEGI)
Leitch, S., 102, 107
Lemaire, R. H., 15
LEPN. *See* Local Enterprise Partnerships Network (LEPN)
Levi, M., 19
Lewis, B., 212
Lewis, J. M., 82
lifelong learning, 86, 102
Lindsay, V. A., 31
Ling, T., 145
Link, B. G., 7, 79
Lin, N., 15, 52, 304
Lipsky, M., 17, 302
Liverpool, 20, 129, 224, 248
Livingstone, D. W., 6
Lloyd, P., 140
local economic development
 and governance outcomes, 290
 and Great Yarmouth business leaders, 228
 for inward investment, 81, 151
Local Economic Growth Initiative (LEGI), 224, 224n11, 225, 234, 243, 270, 273, 274
Local Enterprise Partnerships (LEP). *See also* enterprise zone (EZ); New Anglia Local Enterprise Partnership (NALEP)
 background, 13, 92, 288
 and enterprise zone, 151
 Local Growth Fund, 151
 and Regional Growth Fund (RGF), 150–2, 215, 225, 288.
 remit, 149–52
Local Enterprise Partnerships Network (LEPN), 175, 217
Local Exchange Trading Schemes, 81

Local Government Association (LGA), 108, 113, 147
localism, 17, 142, 150, 173, 174, 176n10, 214
Localism Act 2011, 176n10, 214
Local Strategic Partnerships (LSP). *See also* network cases
 background, 13, 146–7, 185n14, 260, 288
 case selection method, 303, 305
 and Community Plan Action Groups (CPAG), 179, 183, 184, 197
 and Creating and Sharing Prosperity Group (CSPG), 167, 176, 182–6, 195, 197, 199, 204, 261, 263–4, 269
 Great Yarmouth LSP, 215, 224, 224n10, 228, 228n14, 232, 234, 237, 244, 245, 270–5
 and Local Area Agreements, 183, 244, 246, 251, 270, 273, 274
 and Local Area Partnerships (LAP), 179, 183, 189, 193, 270, 271, 274, 275
 in map of Great Yarmouth primary networks, 319
 and Multi-Area Agreements, 111, 149, 177n11, 180, 244
 Tower Hamlets Partnership (THP LSP), 183, 232, 270, 274
Locksley, G., 134
London
 Dockland, 20, 173, 177, 178
 Dockland industrial closure, 174
 and Docklands Development Corporation, 139, 164
 employment problems, 181
 Government Office for London (GOL), 165, 196, 197

Greater London Authority (GLA), 147, 176n10
  housing issues, 166
  job competition, 181, 205
  London Assembly, 147, 177
  London Development Agency (LDA), 176, 176n10, 177, 177n11, 185n14, 196, 203, 204, 278
  and London Legacy Development Corporation (LLDC), 173
  manufacturing decline, 175
  Olympics, 177, 178
  Skills and Employment Board, 177
London Assembly. *See* London
London Borough of Tower Hamlets (LBTH). *See* network cases
London Development Agency (LDA). *See* London
London Legacy Development Corporation (LLDC). *See* London
Lowe, R., 134
Lower Super Output Area (LSOA), definition, 161n1, 169, 214, 219, 221, 226
Lowndes, V., 13, 30, 55, 129
Lu, Y., 25
Lynch, K., 218
Lynn, L. E. Jr., 5

M
Maccio, L., 25
MacDonald, K., 128, 137
MacDonald, Ramsey, 132
MacInnes, T., 6, 12n2, 66
MacKay, R. R., 80, 81, 138
Mahoney, W., 139
Major, John, 140
Malthus, Thomas, 70
managerial culture, 47
Manchester
  Chamber of Commerce and Industry, 141
  regeneration, 129
  university anthropologists, 48
Mandell, M. P., 15, 17, 24, 27, 51, 135
Mann, M., 71
Manpower Services Commission (MSC), 78, 137
Marchant, G., 96
Marsden, P. V., 52, 304
Marsh, D., 3, 8, 14, 33, 78, 134, 135, 282, 291, 304
Martin, R., 79, 95
Marx, Karl, 48, 70, 73, 96
Marxism, 73, 115, 129, 136
Matthew Effect, 31n4
Mawson, J., 141, 143, 144
May, T., 302
May, Theresa, 114
Mayhew, B. H., 52
Mayhew, K., 101, 102
Mayhew, L., 171
Mayor of London, 166, 176
Mayor of Tower Hamlets, 167
McAnulla, S., 8, 51, 56
McCarthy, M., 135
McGregor, A., 145, 146
McGuiness, D., 149
McGuire, M., 3, 9, 15, 24, 26, 27
McKinlay, R. D., 82
McLaughlin, E., 84
McNeil, C., 99, 110n18
McVey, Esther, 66, 83
Meadows, P., 177
Meegan, R., 19, 140
Meeres, F., 213n2, 223n9
Mehta, P., 165
Meier, K. J., 15, 23, 25, 31, 55
Merton, R. K., 19, 31n4
Michie, J., 71, 141
Miles, R. E., 31, 198

Miller, P., 142
Mill, John Stuart, 73
Milward, H. B., 9, 21, 23–7, 49, 50, 56, 304
Mitchell, A., 19
Mitchell, J. C., 15
Mitchell, S. M., 26
Mitchell, W. F., xin1, 79, 87, 94, 101
Mizruchi, M. S., 48n1, 51
Moon, G., 129, 131, 134, 137–9, 141, 178
Moon, J., vi, 5, 11, 14, 79, 82, 92, 107, 115, 128, 131, 151, 282
Moore, C., vi, 5, 11, 14, 79, 82, 92, 107, 115, 128, 131, 151, 282
Moore, M., 139
Mourshed, M., 5, 17, 68
multi-factor framework for network investigation
    components, 21–3
    Great Yarmouth, 243–53
    Tower Hamlets, 195–204
Murray, M. J., 80, 92
Muysken, J., xin1, 79, 87, 94, 101

N
Naastepad, C. W. M., 87, 92
NALEP. *See* New Anglia Local Enterprise Partnership (NALEP)
NAO. *See* National Audit Office (NAO)
National Action Plans (NAPs), 25, 91, 108n17
National Audit Office (NAO), 100, 111, 149, 150, 152, 288
National Minimum Wage, 63, 84, 86, 97, 104
National Reform Programmes, 91, 92
National Unemployment Workers Movement, 75
Nazroo, J. Y., 164
NCC. *See* Norfolk County Council (NCC)
NEET (Not in employment, education or training), 66, 108, 197, 219, 290
neighbourhood concept, 29
neighbourhood renewal
    affiliation and influence, 179, 235, 276
    case selection, 303, 305
    job targets not achieved, 234, 270
    and LGA, 147
    and policy action teams, 145–6
    priority areas, 162
    strategy, 145, 147n8, 243, 244
    themes, vi
    vision, 127, 274
Neighbourhood Renewal Fund (NRF)
    allocation, 146, 303n2
    background to, 111, 147, 183n12, 183n13
    Great Yarmouth, 162, 219n6, 272, 273, 275
    priority areas, 146
    Tower Hamlets, 162, 179, 183, 195, 196, 198, 201
Neighbourhood Renewal Unit (NRU). *See also* Social Exclusion Unit (SEU)
    National Strategy, 145, 146, 246
neo-classical
    economists, 70, 88, 89
    ideas, 115
neo-liberal policy, 5, 12, 16, 87, 279
    competition strategies, 48, 279
    economics, 76–8, 89, 93, 95
    flexible market, 11
    fundamentalism, 11
    institutions, 5, 6, 82, 87
    labour market, 12, 95
    preferences in regeneration, 127, 128, 140

INDEX 339

promoted by New Labour, 78
market-induced public services, 18
World Bank policy, 87
network. *See also* collaboration; housing; Local Enterprise Partnerships (LEP); Local Strategic Partnerships (LSP); network cases
central-local urban partnerships, 132, 135–7
closed decision-making networks, 132, 134–5
conceptual framework, 22
conflict, 24
culture two-tier, 13
definitions, 8–9, 15, 313
evaluation and fallibility, 23–4, 259, 267
formal and informal, vii, ix, x, 15, 22–4, 32, 33, 53, 161, 162, 181–2, 191, 193, 198, 228, 231, 232, 240, 242, 259, 262, 264, 266, 269, 272, 276, 278–9, 284–5, 301, 304, 305, 309, 313–15
function, x, 4, 26, 28, 69
governance, v, 4, 5, 16, 17, 25, 50, 134, 281, 285, 287
governing networks, 16, 132–52
impact, v, vi, 4, 5, 8–10, 15, 22, 34, 47, 161, 259, 281, 282
inner-city partnerships, 132, 137–8
instability, 5, 89, 277, 302
inter-organisational contracting networks, 141–2
lack of transparency, 5, 29, 152, 272
management definition, 50
management dimensions, 15, 16, 18, 22, 23, 25, 26, 28, 29, 47, 50, 53, 195
management of LSP infrastructure, 194

management outcomes, x, 4, 147, 201, 254, 267, 273–5, 284, 291
management through instrumentalism, interaction, and institutionalism, 55–6, 198
managers, 31, 55, 56
market-driven networks, 138–40
multi-dimensional impact: symbolic, political, structural, processes, behavioural, and area-based, 9–21
optimal and sub-optimal outcomes, outcome indicators and analytical levels: network, organisational/ participant and neighbourhood, ix, 19, 22–4, 27–9, 30, 279
outcomes impeded, 47, 191, 283–7
parish, v, 12, 71
performance criteria in education, employment, homecare, and mental health services studies, 24–6
performance: reputation, effectiveness, and ineffectiveness, x, 19, 22, 25, 27–9, 31, 49, 52–3, 259, 260, 262–7, 275, 278–9
preferred outcomes, x, 276–8, 279, 312
and public-private partnerships, 142–4
regulation, 32
reputation, dimensions of, 179, 192, 200, 238, 241, 251, 259–62, 284, 286, 287, 291, 303–5, 309, 313
self-governing, 4, 13, 14, 33, 196
society, 10, 11
structure, agency and processes, ix, 51–6, 291

network. (*cont.*)
  theory, viii, ix, 5, 8, 20, 47–56, 267, 281, 282
  trans-European, 90
  trust, 47, 52–3, 56–7
  typology, 8, 33–4, 286
  and weak democracy, 5, 13, 16, 18, 25, 50, 134, 139, 152, 192, 281
network cases, 161–205, 211–54
network cases in Great Yarmouth. *See also* Local Strategic Partnerships (LSP); Neighbourhood Renewal Fund (NRF)
  area-based factors, 247–9
  case comparison, 231–43
  Central and Northgate wards (CN), 218, 218n5, 219–21, 223, 225–7, 228, 231, 232, 235–7, 238, 241, 248–50, 253, 261, 265, 266, 269, 303
  central environment, 243–7
  culture and politics, 222–3
  demographics, 219
  deprivation, 219
  employment context, 217–18
  environment, 221–2
  epilogue to the fieldwork, 287–90
  factor comparison, 243–53
  health problems, 214, 221, 230, 231, 239, 246, 271–2
  industry, 211–13, 221n7, 222, 223, 234, 243, 249, 251, 253
  jobs dearth, 223, 227, 271
  network agency, 250–1
  network culture, 228–31
  network effectiveness, 262, 264–6
  network ineffectiveness, 266–7
  network outcomes, 267–8
  network processes, 251–3
  network structure, 247
  overall policies, 223–8
  preferred outcomes, 276–8
  primary networks in Great Yarmouth, Appendix, 319
  Southtown and Cobholm wards (SC), 214, 218, 218n5, 219–23, 226–8, 231, 232, 235, 236n17, 239, 241, 242, 248–50, 253, 254, 261, 265, 266, 268–9
  unemployment, 219–21
network cases in Tower Hamlets. *See also* Local Strategic Partnerships (LSP); Neighbourhood Renewal Fund (NRF)
  area-based factors, 198–200
  benefits trap, 166, 205, 289
  case comparison, 181–94
  central environment, 195–7
  contentious politics, 174
  culture and politics, 174–5
  demographics, 168–9
  deprivation, 168–9
  East India and Lansbury wards (EIL), 168, 168n7, 169–73, 174, 177–9, 182, 184, 186–90, 196, 198–201, 205, 260, 263, 264, 269, 274, 303, 305
  employment context, 165
  employment-related funding streams and provision, 317
  environment, 172–4
  epilogue to the fieldwork, 287–90
  factor comparison, 195–204
  financial district, 163, 164, 173, 200, 204
  Freedom of Information case, 14, 167, 195, 309
  health problems, 164, 169, 175, 183, 184, 187, 198, 200, 202, 204, 205
  industrial decline, 174, 199
  London employment problems, 181
  network agency, 200–2
  network culture, 178–80
  network effectiveness, 262–4, 266
  network ineffectiveness, 266–7

network outcomes, 267–8
network processes, 202–4
network structure, 198
Olympics legacy, 177–8
overall policies, 176–7
preferred outcomes, 276–8
Spitalfields/Banglatown and Bethnal Green South wards (SB), 168
unemployment, 169, 172
New Anglia Local Enterprise Partnership (NALEP), 214–18, 223–5, 230, 288
New Deal
American, 75, 99
and demoralised clients, 99
Flexible New Deal from 2009, 99, 100, 106n16, 110–12, 186n15, 317
and unemployment reduction, 63
New Public Management (NPM), 16, 69, 138, 139
non-accelerating inflation rate of unemployment (NAIRU), 86–7, 95
Norfolk County Council (NCC), 212, 213n1, 214, 215, 215n4, 219, 221, 223, 225, 226, 228–30, 231, 233, 234, 244, 274
North, D. C., 6
NPM. *See* New Public Management (NPM)

O
O'Connell, R., xin1
ODPM. *See* Office for the Deputy Prime Minister (ODPM)
OECD. *See* Organisation for Economic Cooperation and Development (OECD)
Office for National Statistics (ONS), vi, vii, 6, 65n3, 66, 68, 85, 102, 164, 171, 177, 214, 217, 221, 290, 297–9
Office for the Deputy Prime Minister (ODPM), 111, 111n19, 161n1, 168, 169, 179, 189, 196, 218n5, 219, 224n11, 245, 246, 303
O'Gorman, S., 230
oil shocks, 67, 77, 88, 136
Oliver, A. L., 26
Olson, M., 33
Olympics. *See* London; network cases in Tower Hamlets
ONS. *See* Office for National Statistics (ONS)
Open Method of Coordination (OMC), 91
Organisation for Economic Cooperation and Development (OECD), 5, 7, 65n2, 82–4, 86, 93, 98, 101, 102
Osborne, George, 96–7
Ostrom, E., 19, 52, 282
Oswald, A. J., vii
O'Toole, L. J., Jr., 4, 9, 15, 17, 18, 23, 26, 31, 52, 55, 56
Owen, Robert, 73
Özler, S., 71

P
Pacione, M., 20, 136
Page, E., 50
Page, S., 24, 26, 54, 141n6
Painter, C., 55, 56, 78, 139, 198
Painter, J., 17
Papadopoulos, Y., 4, 50, 128, 281
Parkinson, M., 137
Patel, J., 5, 17, 68
Peck, J., 12, 82, 95, 99, 101
Pedrazzi, L., 25
Percy-Smith, J., 68, 84
Perkins, N., 26

Perri, 6, 11, 30, 31, 139, 140
Peters, B. G., 12, 13, 147
Phelan, J. C., 7, 79
Philpott, J., 95
Pierre, J., 14, 17–19, 54
Pill, M., 130
Polanyi, K., 12
political science, ix, 18, 48–50, 281
Pollin, R., 92
Pollitt, C., 51, 138
Poor Law
   1834 Amendment Act, 72
   replaced by Local Government Act 1929, 74
Poplar HARCA, 173, 174, 179, 188, 189, 196, 201
Porter, D. O., 24, 98
poverty. *See also* network cases
   child, 111, 163, 164, 169
   Child Poverty Action Group, 135
   classical economists view of, 70
   'deserving' and 'undeserving poor', 7, 71
   discourse depoliticised, 17
   division of labour attributed to, 71
   and education, 17, 101
   and ethnic groups, 171
   increase, 7, 65, 68, 71, 79
   intervention, EU, 140
   and landownership, 73
   and LSP intervention, 146
   national reform programme (NRP), 92
   regeneration response to, 128
   population theory of, 70
   superficial responses to, 10, 128, 139
   targeting, 111
   theory, 135
   trap, 135
   uneven capitalist development, 10, 83, 135, 136
   and welfare, 79
   working poor, 73, 97
Powell, M., 3
Powell, W. W., 8
Power, A., 134
power relations. *See also* New Public Management (NPM)
   assertion, 29, 269, 275, 279, 285
   asymmetric power model, 49
   balance of, 138
   bargaining power, 6, 13, 14, 55
   community power, 49
   diffusion, 24
   between funder and contract holders, 139
   govern by network, 145
   hierarchical, 284
   maintained by institutional leaders, 205
   managerial, 141
   negotiated, 147
   over or with, 13, 131
   power sharing, 14, 229, 232
   prioritise central roles, 10, 148, 204
   and privilege positions, 17
   and resource-dependencies, 14, 49
   restructured through governance, 18
   for self-promotion, 29, 275, 278
   tracking, 291
   transaction, 11
   Westminster model, 17, 145
Pratchett, L., 13
Prescott, John., 148
Pressman, J. L., 3, 26, 281
privatisation of public services, vi, 15–16, 69, 77, 86, 100–1, 284, 285
Provan, K. G., 3, 4, 9, 15–17, 21, 23–7, 49, 50, 52, 56, 302, 304
public administration, ix, 8, 48, 50, 92
Pugalis, L., 149
Putnam, R. D., 20

Q
quangos (quasi-autonomous non-governmental organisations), vi, 140, 147, 148, 184, 279, 307

R
Rai, S., 13
Rahman, L., 163
Ramsden, M., 78, 81
Reese, L., 21
Rees, J., vi
Regional Assemblies, 147–9
Regional Development Agencies (RDA), 13, 143, 147–9
Regional Growth Fund (RGF). *See* Local Enterprise Partnerships (LEP)
Reid, B., 15
research design and methods, 301–12
Rhodes, J., 140, 143, 144
Rhodes, R. A. W., 3, 4, 14–18, 30, 33, 50, 137, 139, 198, 282, 291, 304
Ricardo, David, 70
Richards, D., 4, 15, 17, 49, 51, 77, 131, 134, 136, 140, 145, 147, 148
Richards, S., 30
Richardson, J. J., 5, 11, 14, 18, 79, 82, 92, 107, 115, 128, 131, 151, 282
Richardson, L., 19
Rick, J., 81n11, 85
right to work, v, xin1, 75, 84, 90, 93, 94n12, 109
Rigney, D., 31n4
Ritchie, H., 81n11, 85
Robson, B., 130
Roller, E., 81, 312
Roosevelt, Franklin, D., 75, 76
Rose, N., 142
Rosenbaum, J. E., 282
Rosenfeld, R. A., 21
Rouse, J., 53, 55, 56, 78, 139, 198
Rowthorn, R. E., 78, 78n8

Rutt, S., 165
Rydin, Y., 130

S
Saffron, K., 164
Sako, M., 102
Salman. S., 171
Say, Jean-Baptiste, 70
Scharpf, F. W., 3, 6, 26, 51, 52, 56, 91, 281
Schneider, M., 291
Scott, J., 48n1, 312
Selby, P., 237
self-employment
　business start-up support, 78, 81, 106, 106n16, 112, 144, 184, 200, 224, 225, 270, 276
　in case areas, 164, 214, 290
　increase in, vii, 65, 65n3
Serious Fraud Office (SFO), 110
SEU. *See* Social Exclusion Unit (SEU)
Shaikh, A., 67
Shephard, A., 84
Shortell, S. M., 26
Sigala, M., 111
Silver, H., 79n9
Simpson, L., 93
Single Regeneration Budget (SRB). *See also* urban regeneration
　background, 142–4
　in Great Yarmouth, 228, 222n8, 229, 239, 240, 243
　performance, 264, 265
Skelcher, C., 5, 32, 49, 137, 139, 141, 142, 144, 145
skills. *See also* apprenticeships; joined-up government; UKCES
　barriers, 67, 79, 84, 85, 88, 100, 102, 105, 107, 167, 172, 199, 200, 205, 245
　and basic education, 101, 222, 249, 289

skills (*cont.*)
  blaming, 7, 12, 79, 101, 105
  in Cuba and India, 103
  employability training, 101
  ESOL, 97, 100, 101, 171, 240
  funding devolution, 113, 291
  governance, 105, 110–12, 148, 197, 215, 216, 234, 288
  individual competitiveness, 82, 113, 115, 176, 181, 223
  institutions, 81, 82, 98, 102, 105, 110, 176, 179, 223, 230, 284
  investment, 81, 98, 114
  and job scarcity, 7, 101, 102, 107, 177, 181, 217, 218
  Leitch review of, 102, 107
  for life courses underfunded, 97
  mismatch, v, vii, 17, 67, 69, 98, 107, 245
  and OECD policy, 82, 101, 102
  overqualified, 102
  planning, lack of, 103, 107, 111, 112, 150, 196, 234, 270, 271
  policy, 81, 82, 102, 129, 147, 231, 244, 245, 289, 290
  provision, continuous reform, 81, 109–14
  and urban regeneration priorities, 129
  Sector Skills Councils, 105n15
  skills support, 29, 100, 102, 185n14, 225–7, 268, 270
  supply-side fundamentalism, 101, 102
  support, 29, 226, 268, 270
  surplus or shortage, 102–4, 105, 227
  training for employability, 78, 99–105
  utilisation in networks, 276, 278
  and volunteering, 81, 217, 226, 227
  Wolf report, 84, 103
Small, M. L., 166
Smith, Adam, 70–1
Smith, G., 8, 53, 139
Smith, M., 141
Smith, M. J., 17, 77, 134, 136, 140, 147, 148
Snow, C. C., 31, 198
Social Exclusion Unit (SEU), 63, 83, 111, 127, 145, 146, 147n8
socialist theorists, 72–3
social movements, 32, 76, 135
social network analysis (SNA)
  in cases, ix, 182, 239, 267
  and data analysis, 311, 312
  history of, 15
socioeconomic, 161
  policy rift, 11–12, 31n4, 81, 88, 161
sociometry, 48
Somers, M., 7, 72, 79
Sørensen, E., 4, 5, 16, 25, 30, 50, 55, 282
Sorrentino, C., 101, 102
Speenhamland allowance, 12, 70–2
Spicker, P., vii, 79n9
SQW, 164, 167, 169, 178
Stahl, G. O., 55
Steijn, B., 3, 10, 25–8, 302
Stewart, J., 71, 95
Stewart, M., 141, 146
stigmatisation theory, 7, 83
Stoker, G., 16, 50, 134, 138, 282
Stone, C. N., 13, 14, 30
Storm, S., 87, 92
Streeck, W., 6
structural sociology, 48–9
Suder, K., 5, 17, 68
Sullivan, H., 5, 29, 30, 32, 49, 98, 129, 137, 137n3, 139, 142, 144, 145, 147
Swyngedouw, E., 4, 18
Sydow, J., 9, 15, 23, 24, 27, 50, 302

T
Tabb, W. K., 70, 86
taxation. *See also* welfare benefits

Council tax protest, 174
evaders, 7
fear of, 79, 89
incentives, 22, 97, 99, 105, 106, 114, 151
industrial subsidies, 97
Keynesian policy, 88–9
land, 73
local, 71, 72n7
low corporate, 80, 80n10
New Homes Bonus, 150
poll tax protests, 77
revenue, 80, 88
of savings, 73
tax-raising power, 49, 86, 87
Taylor-Gooby, P., 77, 78, 86, 99
Taylor, M., 5, 18, 20, 27, 29, 49, 50, 55, 98, 139–41, 147, 233
Taylorism, 76
Teisman, G. R., 11, 17
Ten Heuvelhof, E. F., 31, 56
ter Haar, B., 91
Thatcher, Margaret, 77, 82, 86, 138, 140
Theodore, N., 12, 95, 99, 101
Therborn, G., 6, 77
Tofarides, M., 144
Tomaney, J., 148
Torfing, J., 4, 5, 16, 25, 30, 50, 55, 282
Tower Hamlets ChangeUp Consortium (THCUP), 180
Tower Hamlets Community Organisations Forum (THCOF), 180
Tower Hamlets Partnership (THP). *See also* network cases in Tower Hamlets
  background to, 183, 183n13, 195
  and Creating and Sharing Prosperity Group (THP CSPG), 167, 176, 195
  and Excellent Public Services (THP EPS), 184
  focus, 176, 179, 180, 183, 184, 189, 191, 198, 204, 274, 289
  and Local Area Partnership 2 (THP LAP 2), 193, 274
  and Local Area Partnership 7 (THP LAP 7), 189, 189n16, 191, 196, 261, 272–4, 277
trade unions, 6, 7, 11, 56, 75, 77, 78, 86, 88, 90, 91, 97, 105, 113, 134, 136, 137n4, 150, 213, 282
Training and Enterprise Council (TECs), 78, 106n16, 110, 144
Treasury, 68, 74, 86, 91, 99, 128, 130, 132, 224n11, 245, 287
  and welfare policy, 86, 99
Triantafillou, P., 25, 87, 98, 108n17
Trust for London, 164, 181
Tsai, W., 8, 21, 48n1, 52, 312
Tsang, E. W. K., 53
Tuite, M. F., 3
Turok, I., 130
Turrini, A., 3, 24, 25
Tyler, P., vi, 128, 138, 140, 141, 143, 144, 148, 151, 259

U

UKCES. *See* United Kingdom Commission for Employment and Skills (UKCES)
UKSA. *See* United Kingdom Statistics Authority (UKSA)
underemployed or part-time workers, overqualified, vii, 6, 65, 68, 93, 96, 102, 173, 181, 215n3, 219, 253
unemployment. *See also* Jobcentre Plus (JCP); industrial; network cases; welfare; welfare benefits
  activation policy, 7, 12, 20, 82, 91, 98–101
  area based, vi, 50, 115, 198, 247, 285
  blaming, 7, 12, 69, 79, 135, 293

unemployment (*Cont.*)
  cabinet responses to, 74, 113, 131, 132, 145, 146
  causes of, 50, 67–8, 70, 99, 167
  definitions, 64–5
  demand-deficient, 94
  depoliticized, 17, 138
  dimensions of, 64, 68, 269, 285, 287
  economic dimension of, 80–1
  and European policy, 89, 90, 92
  false reporting, 66
  gender issues, 194
  governance structures, ix, 108–14
  and group dimensions of, 85
  history of, 70–9
  impairs mental and physical health, viii, 66, 69, 83–5, 97, 164, 169, 175, 187, 200, 214, 221, 230, 231, 246, 271, 272, 289, 303n1, 319
  inadequate responses to, 143
  individual dimensions of, 84–5
  institutional dimensions of, 82–3
  and macro-economic policy, 87–9
  measures of, 65–6, 79
  neighbourhood dimensions of, 83–4
  policy, 85–98
  policy definition, 86
  policy interventions, 98–108
  political dimensions of, 82
  rates, 63, 65, 65n2, 67, 80, 87, 90, 95, 164, 165n6, 169, 171, 177, 230, 279, 290, 297–9
  and recession, 10, 67–9, 77, 80, 88, 97, 98, 137, 141, 169, 219
  representation, x, 5, 14, 29, 69, 93, 152, 153, 168, 184, 200, 227, 229n15, 243, 268, 269, 284, 285, 292
  skills and training to counter, 101–5
  social dimensions of, 81
  structure/agency narratives, 51
  types of (cyclical, frictional, long-term, seasonal, structural, technological), 79–80
  youth unemployment, 5, 14, 17, 68, 84, 99, 100, 103, 113, 165, 165n6, 169, 171, 175, 181, 199, 201, 293
Unemployment Assistance Act 1932, 75
UNISON, 113
United Kingdom Commission for Employment and Skills (UKCES), 83–5, 105, 105n15
United Kingdom Statistics Authority (UKSA), 66, 79
Urban Aid Programme (UP), 135–9, 141, 142
urban regeneration. *See also* enterprise zone; Great Yarmouth Urban Regeneration Company 1st East (GYURC); single regeneration budget (SRB)
  City Challenge, 254
  commercial, 127, 172
  community, 128, 179
  context, 127–8
  cross-cutting policy, 129, 144–6
  culture, design or events-led, 130
  definitions, 130–1
  and EU, 92, 142, 230
  gentrification, 30, 129, 130, 173
  and governing networks, 17, 33, 127, 132–52
  high-income officials, 13–14
  high-profile, 17, 152
  inadequate social regeneration, 142
  incentives to firms, 105–6, 151
  neighbourhood, 129, 130, 132, 142, 145, 152, 153
  outputs, 128
  place-shaping, 19–21
  Policy for the Inner Cities, 137

political process of, 13, 127–33, 152
programmes evaluation, 111, 128, 143, 178, 187, 224–5, 237
property-led, 81, 129, 138, 139, 177
and Race Relations Act 1976, 135n1, 137
Victorian tradition revisited, 71, 138, 141
Urry, J., 20

V
Vaattovaara, M., 197n18
Van den Brink, M., 52
Van Dooren, W., 24, 25
Vangen, S., 53
Verba, S., 28, 47
Versteeg, W., 4
Voets, J., 24, 25

W
Wadsworth, J., 63
Wagenaar, H., 10
Wahhab, I., 111, 112, 180
Walker, F., 27
Walker, R. M., 26
Walsh, K., 89
Ward, H., 144
Ward, K. G., 144
Warren, R. L., 54
Warr, P., 84
Wasserman, S., 15, 48n1, 304
Weaver, M., 175
Webb, Beatrice, 74
Webb, Sidney, 74
Weber, M., 48, 51
welfare. *See also* activation policy; privatisation of public services; Speenhamland allowance; welfare benefits
administration, 72, 102–10
advice services, 166, 199
attack of recipients, 78, 79
and capitalism, vi, 12, 73, 89
carbon costs, 6
corporate, 7, 80, 150
dependencies, 64, 81, 86, 101, 109, 128, 166, 271, 274
food insecurity and food banks, 84
industry, v, 115
in-work, 12
and moral attitudes, 7, 79
Nordic model, 98
partners, 17
paternalist, 135
policy, 5, 82, 86, 99
politics of, 82, 86, 95
reform, vi, x, 72, 82, 89, 96, 99, 166, 216, 218
retrenchment, 75, 77, 99
social, xin1, 11, 12, 75, 75n9, 89, 195
state, 81, 84, 86, 134, 135, 293
structures for full employment, 95
Treasury role in, 86, 99
to-work, 12, 79, 86, 99–101, 164, 167, 197, 230, 281
welfare benefits. *See also* activation policy; housing; Jobcentre Plus; New Deal; Work Programme
benefits trap, 166, 205, 289
Employment and Support Allowance (ESA), viii, 65, 100, 170, 171, 219–21, 223, 287, 289
Enterprise Allowance, 78, 106, 106n16, 133
history of, viii, 12, 71–2, 74–6, 144

welfare (*cont.*)
  Incapacity Benefit (IB), 65, 100, 164, 170, 219, 220, 289
  Jobseekers Act 1995, 144
  Jobseekers allowance (JSA), viii, 6, 12n2, 84, 100, 104, 106, 106n16, 107, 111, 114, 167, 169, 171, 186n15, 219, 221, 226, 289
  off-benefit flow outcomes, viii, 221
  rise of means testing, 75, 78, 166
  in rundown neighbourhoods, 83
  rise of means testing, 75, 78, 166
  sanctions, viii, 7, 12, 12n2, 57, 65, 66, 75, 78, 82, 96, 115, 169–71, 216, 218, 220, 221
  take-up, 29, 184, 208, 268, 271
  undeserving association, 7, 71, 97
  Universal Credit, viii, 113, 170, 220
  workfare schemes, 97, 101
  working tax credits, 12, 63, 71, 84, 97, 99, 164, 166, 173, 240, 246
Whitehead, M., 9, 17
Wildavsky, A. B., 3, 26, 281
Wilkinson, F., 71, 79, 95
Wilks, S., 137
Williams, N., 106
Williamson, O. E., 53
Wilson, D., 54, 142
Wilson, Harold, 135, 137
Wilson, T., 104, 113
WNF. *See* Working Neighbourhood Fund (WNF)
Wolf, A., 66, 81, 84, 103
Wolman, H., 50
Wong, S. S., 52, 53
Woolcock, G., 17, 51
Work and Pensions Committee (WPC), viii, 20, 21, 68, 169, 218, 221, 226

work experience. *See also* Work Programme
  deficit of, 17, 273
  insecure, viii
  placements, vii, 17, 103, 108, 111, 113, 226, 227, 229, 273, 276
  unpaid, vii, 5, 79, 97, 101, 115
  workfare schemes, 97, 101
Work Programme, 7, 20, 79, 82, 83, 95, 96, 99, 101, 106n16, 168, 180, 218, 226, 230, 290
working classes. *See also* capitalism
  class bias in civil service, 134
  and collectivism, 77, 134
  injustice, 174
  low paid women, 96, 107
  and political agency, 174–5
  and poverty, 70, 71, 73, 83, 135, 136, 140, 164, 169
  protection of, 6, 73
Working Neighbourhood Fund (WNF), 100, 111, 112, 149, 168, 180, 183n12, 223, 226, 229, 234, 243, 271, 274, 288, 317
Working Neighbourhood Pilot (WNP). *See also* network cases in Great Yarmouth; network cases in Tower Hamlets
  background, 111, 162, 186, 237
  in case selection, 303, 305
  and evaluation of, 111, 187, 237, 267, 273
  unworkable directives, 186, 187, 190, 196, 202, 237, 238, 244, 250, 264, 266, 270, 275
World Bank, 86, 87, 93, 94
  and neo-liberal policy, 87
World Trade Organisation, 5
WPC. *See* Work and Pensions Committee (WPC)

Y

Yanow, D., 31, 53
Yin, R. K., 302
Young, A. R., 90
Young, M., 163n2, 174, 175

Z

zero hours contracts, 65, 68, 178, 224
Zølner, M., 6

The manufacturer's authorised representative in the EU is Springer Nature Customer Service Centre GmbH, Europaplatz 3, 69115 Heidelberg, Germany. If you have any concerns regarding our products, please contact ProductSafety@springernature.com

Printed and bound by CPI Group (UK) Ltd, Croydon, CR0 4YY

23/03/2026

02076449-0004